THE RISE OF
Artificial Intelligence

Real-world Applications for Revenue and Margin Growth

Zbigniew Michalewicz
Leonardo Arantes
Matt Michalewicz

Published by Hybrid Publishers
Melbourne Victoria Australia
© Zbigniew Michalewicz, Leonardo Arantes, and Matt Michalewicz 2021

Hybrid Publishers,
PO Box 52, Ormond VIC Australia 3204.
www.hybridpublishers.com.au

A catalogue record for this
book is available from the
National Library of Australia

First published 2021
Text design & layout by Midland Typesetters, Australia
Cover design by Marchese Design
ISBN 9781925736625 (p)
Printed by Tingleman Pty Ltd

*Dedicated to the entrepreneurs, scientists, and business leaders
that have paved the way for Artificial Intelligence over the decades past,
and are paving the way for its future in the decades to come.*

PREFACE

What This Book is About and How to Read It

"We're at the beginning of a golden age of AI. Recent advancements have already led to inventions that previously lived in the realm of science fiction—and we've only scratched the surface of what's possible."

Jeff Bezos, *Amazon CEO*

Few terms have captured our imagination in recent times like "Artificial Intelligence." And not just through sensationalized media articles about how AI will soon displace all jobs and rule the world, but also through movies, books, and television shows. It now seems that everyone "knows" about AI; that everyone has an opinion. And yet, in our experience, few people actually understand what Artificial Intelligence is and isn't, where the field came from and where it's heading, and how the technology can be harnessed to generate commercial outcomes.

Given the immense amount of disinformation and misunderstanding, we have written this book to demystify the subject of AI and explain it in simple language. Most importantly, we have written this book with the business manager in mind, someone interested in the topic from a real-world, commercial perspective—a perspective of how the technology can create value and increase competitiveness *today*, rather than what might happen in 25 years' time or how a superior intelligence might overcome the human race in the distant future. Such philosophical treatises are thought-provoking (to say the least) and the subject of many books published each year, but this isn't one of them. Instead, *The Rise of Artificial Intelligence* provides a commercial exploration of AI, with particular emphasis on how AI-based systems can improve decision making in organizations of all shapes and sizes.

As such, this book presents Artificial Intelligence through the lens of decision making for two reasons: First, because the world has reached a level of such unprecedented speed, complexity, and noise, that no one can assess and evaluate all the available data when making decisions; and secondly, because the decisions we make affect the outcomes we achieve. In other words, better business decisions lead to better business outcomes. Although Artificial

Intelligence can be applied to many areas besides decision making—such as automation and robotics, or image and speech recognition—these subjects don't feature heavily in the pages ahead except for Chapter 1, where we provide an overview of the research areas of AI. Ultimately, revenue and margin growth comes down to the decisions an organization makes (or doesn't make), and hence the application of AI to decision making is our primary focus.

To best present the concepts in this book, we've used a *problem-to-decision pyramid* to represent the continuum that exists in terms of an organization's ability to improve its decision making:

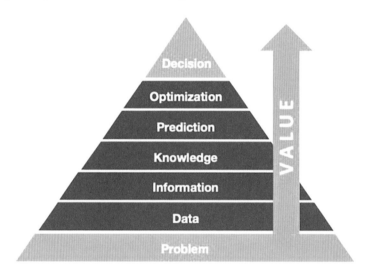

Each layer of this pyramid represents a step in the journey for improved decision making: the higher we go, the better our decisions (and the more value we can create). The structure of *The Rise of Artificial Intelligence* reflects the structure of this pyramid, with the first two parts of the book investigating each layer of the pyramid, and the last two parts illustrating the application of Artificial Intelligence to real-world problems for the purpose of generating revenue and margin growth.

Chapter 1 begins with a high-level overview of Artificial Intelligence—its history, areas of research, and current progress and challenges—before introducing the *problem-to-decision pyramid* in Chapter 2, which conceptualizes the journey from defining a problem to making a decision through the use of data, information, knowledge, prediction, and optimization. Chapter 3 concludes Part 1with an in-depth examination of a complex business problem set in the fast-moving consumer goods industry, which is used to explain the role of objectives, business rules and constraints, and the application of Artificial Intelligence algorithms for improved decision making.

This complex business problem of promotional planning and pricing is then used as a running example throughout Part II, which explores the inner workings of predictive models, optimization methods, and various learning algorithms. Because data and modeling form the basis of prediction and optimization, this part of the book opens with a chapter on data and modeling, along with a discussion of common issues such as data availability, completeness, and preparation. In Chapters 5 and 6 we review various AI and non-AI methods for predictive modeling and optimization, whereas in Chapter 7 we present adaptability and learning concepts—which together (i.e. prediction, optimization, and self-learning) comprise the backbone of any AI-based software system.

As an important aside, Chapters 4 through 7 represent the most technical material of the entire book, attempting to explain the innermost mechanics of several Artificial Intelligence algorithms such as neural networks and genetic programming. Although non-technical readers can easily progress through Part II to gain a deeper understanding of algorithms and models, readers without an interest in data, problem modeling, or how Artificial Intelligence algorithms work, can jump straight to Part III, which presents real-world applications of Artificial Intelligence.

The application areas in Part III explore the problem-to-decision pyramid in the context of real-world problems and business objectives, covering both the lower layers of the pyramid focusing on data and the analytical landscape of an organization (i.e. information and knowledge), as well as the upper layers of prediction, optimization, and self-learning, and how they're enabled by Artificial Intelligence methods. For ease of reading, we've divided Part III into three chapters, each being dedicated to a specific business function—in particular, *sales, marketing,* and *supply chain.* These case studies are based on an enterprise software platform called Decision Cloud®, which is a modularized, cloud-based platform that empowers staff to make better and faster decisions through the use of Artificial Intelligence.

And finally, Part IV concludes the book with common questions and concerns that organizations have on the application of Artificial Intelligence, such as: *"Would AI work for me?"* and *"Where should I start?"* These two chapters provide practical advice for selecting the right business problem, developing a business case, choosing a technology partner, as well as other topics such as digitalization and change management.

To improve the reader's understanding of the content, we've also created a set of supplementary videos that can be accessed at: www.Complexica. com/book/RiseofAI/. These videos bring to life the concepts presented in each chapter—for example, by providing a visual explanation of ant system algorithms in Chapter 1, the layers of the problem-to-decision pyramid in

Chapter 2, the workflow of promotional planning and pricing in Chapter 3, and so on. In these videos we're able to "show" concepts that can only be "told" within the confines of the printed page.

In terms of how to read this book or watch the videos, the ideal way is to progress sequentially from Chapter 1 to 12. For the less technically-inclined reader, however it's possible to jump around in any sequence that best satisfies curiosity and interest. For example, the reader might begin with an overview of Artificial Intelligence in Chapter 1, then progress to the application areas in Chapters 8, 9, and 10, before returning to Chapters 2 and 3 to better appreciate the problem-to-decision pyramid and the intricacies of solving complex business problems (after all, why are complex business problem so difficult to solve?). Alternatively, a reader might start with the application areas in Chapters 8, 9, and 10, then move back into Part II to better understand how algorithms and models work, before progressing to Part IV for practical advice for initiating an Artificial Intelligence project.

However, regardless of the reader's technical sophistication or their interest in the implementation aspects of AI-based software, it's highly recommended that everyone start with the first two chapters for an introduction into the world of Artificial Intelligence and an overview of basic concepts and terminology. From this perspective, the sequence of reading the remaining chapters is of far lesser importance.

Lastly, we'd like to say that the material presented in this book is the result of 40 years of first-hand Artificial Intelligence research within university settings, and more than twenty years of implementing AI-based enterprise software systems in many (often very large[1]) organizations across three continents. With that in mind, we'd like to thank everyone who made this book possible, with our special appreciation going to many Australian companies we collaborated with over the years in the application of Artificial Intelligence, such as PFD Foods, BHP Billiton, BMA, Pernod Ricard Winemakers, Lion Drinks, Bunzl, DuluxGroup, Rio Tinto, Metcash, Pfizer, Janssen, Haircare Australia, Fortescue Metals Group, CBH Group, Roy Hill, Glencore, Polyaire, Treasury Wine Estates, and Costa Group. Within these companies, we'd like to thank Chris Baddock, John Barakat, Renato Bellon, Simon Bennett, Damian Bourne, Warren Brodie, Michael Brooks, Pierre-Yves Calloc'h, Daryl Chim,

---

1   Our experiences of implementing enterprise-grade software based on the latest Artificial Intelligence algorithms and methods are based on many projects with global giants—such as BHP Billiton, General Motors, Bank of America, Pernod Ricard, Unilever, Air Liquide, Ford Motor Company, Glencore, Beiersdorf, Rio Tinto, and ChevronTexaco, among many others—as well as smaller companies that benefited from the research & development and innovation carried out by these larger organizations.

Richard Cohen, Jevan Dickinson, Andrew Endicott, Eglantine Etiemble, Scott Fellingham, Greg Feutrill, Garth Gauvin, Ward Gauvin, Scott Graham, Chris Green, Kylie Grigg, Richard Hansen, Mark Hayden, Kim Heatherton, Mark Ivory, James Jones, Mike Lomman, Brett McKinnon, Stuart McNab, Doug Misener, Luke Mitchell, Stephen Mooney, Aemel Nordin, Mark Powell, Rod Pritchard, Robin Pyne, Mathew Regan, Darryl Schafferius, Mark Shephard, Jon Simpson, Kerry Smith, Richard Taylor, Soner Teknikeller, Lance Ward, John Warda, and Joel Zamek.

We'd also like thank a few individuals who contributed to the content and ideas in this book, namely, Reza Bonyadi, Łukasz Brocki, Tom Heyworth, Xiang Li, Łukasz Olech, Ali Shemshadi, Larisa Stamova, Chris Zhu, as well as members of Complexica's scientific advisory board who we've worked with over the years: Reza Bonyadi, Łukasz Brocki, Longbing Cao, Raymond Chiong, Vic Ciesielski, Carlos Coello, Ernesto Costa, Kalyanmoy Deb, Kenneth De Jong, A.E. Eiben, Xiaodong Li, Masoud Mohammadian, Pablo Moscato, Frank Neumann, Zbigniew Raś, Markus Wagner, Thomas Weise, Adam Wierzbicki, and Mengjie Zhang.

And finally, it was a great pleasure to write about a topic that's been the central focus of our working lives for so many years, and we hope that readers enjoy this book as much as we enjoyed writing it. We believe that anyone in any organization who makes operational, tactical, or strategic decisions— whether on the factory floor or in the boardroom—will find this book valuable for understanding the science and technology behind better decisions. Enjoy!

Adelaide, Australia   Zbigniew Michalewicz
March 2021   Leonardo Arantes
                                                                Matt Michalewicz

TABLE OF CONTENTS

PART I
Artificial Intelligence as Applied to Decision Making

CHAPTER 1

What is Artificial Intelligence?

"We have to face the fact that Artificial Evolution and Artificial Intelligence are hard problems. There are serious unknowns in how those phenomena were achieved in nature. Trying to achieve them artificially without ever discovering those unknowns was perhaps worth trying. But it should be no surprise that it has failed."

Daniel Deutsch, *The Beginning of Infinity*

The recent "rise" of Artificial Intelligence ("AI") isn't really a rise, but rather, a sudden popularity brought on by the media and, to an even larger extent, our curiosity about a poorly understood subject that has been a mainstay of science fiction films and literature. In fact, almost everyone's first "taste" of Artificial Intelligence has been through either books or movies: Who doesn't remember Asimov's *I, Robot* and the three laws of Robotics? Or *Hal 9000* from *2001: A Space Odyssey* ("Sorry Dave, I can't do that") and *Data* from *Star Trek: The Next Generation* ("We must survive to be more than we are"—pictured below, left)? To say nothing of the *Cyberdyne Systems Model 101 Series 800 Terminator* ("I'll be back"—pictured below, right). Besides providing a hefty dose of entertainment, such examples colored our imagination with possibilities of what might be if machine intelligence ever rivaled our own.

But somehow, in recent times, Artificial Intelligence left the big screen and printed page and entered the real world. The abruptness with which Artificial

Intelligence moved from the fringe into the mainstream, and into our everyday vocabulary, startled even those closest to the field, namely, veteran computer scientists. At odds with the hype and hysteria portrayed by the media that "AI has suddenly arrived" and is here to "take over," they are quick to point out that Artificial Intelligence has been steadily progressing for almost 75 years, with plenty of fits and starts along the way (and just as many disappointments and setbacks). Over that period of time, the field has been taught and researched by universities across the globe and applied by corporations and government agencies alike, allowing various forms of Artificial Intelligence algorithms[1] to find their way into countless devices, machines, and software applications.

So contrary to a sudden arrival or rising, Artificial Intelligence algorithms have been gradually pervading our world and everyday lives since the 1980s (and even before), usually behind the scenes, performing tasks such as detecting fraud, translating languages, interpreting handwritten text, recognizing speech, steadying camcorder images, controlling quality on production lines, scoring credit applications, improving fuel economy and ride comfort of trains, designing engineering components, optimizing supply chain operations, and more. Our ignorance of these advances is excusable, in much the same way that our ignorance of the advances in particle physics or neurology is excusable—after all, we cannot stay abreast of every research field, no matter how widely we read. And so now, suddenly, we hear about "AI" everywhere—in the news, magazine articles, movies—and not having followed the progress of Artificial Intelligence research over the years, we can be forgiven for thinking that the field has suddenly "risen," as if from nowhere.

That said, the growing awareness (and to a large extent, current hype) of Artificial Intelligence is quite remarkable. We remember a time—not that long ago, during the late 1990s—when terms such as *neural networks* or *Machine Learning* caused bewilderment in presentations, even if those presentations were to executives or board members of global corporations. The use of "Artificial Intelligence" in those presentations often evoked crude jokes about Skynet or bizarre questions that were difficult to answer (*"What's the relationship between Artificial Intelligence and aliens?"* we were once asked in a boardroom). Most people, back then, had no idea. And now, Artificial Intelligence seems to be everywhere, on everyone's lips, the educated and ignorant alike: *"Let's throw Machine Learning at that,"* or *"Let's apply AI to this,"* people bandy

---

1   A brief overview of *AI algorithms* is provided in Section 1.3—what they are and how they differ from *non-AI algorithms*—along with a more detailed discussion on how they work and their applicability to various problem domains in Part II.

about, joyfully, as if ordering a drink at the bar. So what's changed? Why now? And what exactly is Artificial Intelligence, anyway?

Starting with the last question first, the easiest way to think about Artificial Intelligence—or, more precisely put, what the research field of Artificial Intelligence is trying to achieve—is by comparing it to the human body. In fact, we can think of Artificial Intelligence as "our attempt to artificially replicate the human body" through the use of technology (rather than through biological means, such as cloning or genetic engineering). With this in mind, we can divide Artificial Intelligence into four primary branches that correspond neatly to the major functions of the body:

- *Robotics*: which tries to replicate the function of mechanical movement.
- *Computer Vision*: which tries to replicate the function of seeing and interpreting imagery, both still (photographs) and moving (videos).
- *Natural Language Processing*[2] *("NLP")*: which tries to replicate the function of speaking and listening, along with the nuisances of communicating via spoken and written language.
- *Cognitive Computing*: which tries to replicate the function of "thinking," and includes processes such as analysis, deduction, reasoning, and decision making (and looking further out, more ambitious functions that aren't well understood today, like consciousness and self-awareness).

Again, thinking back on Data in *Star Trek*, or the T-800 in *The Terminator*, all these functions were present—seeing, speaking, listening, moving, thinking— and together, they brought an "authenticity" to AI on the screen. Each of these branches represents a significant research area in and of itself, leaving scientists and research organizations with numerous ongoing challenges to grapple with. For these reasons and others, it's accurate to say that the *goal* of Artificial Intelligence is to artificially replicate the human body, but that goal remains elusive and distant, and is likely to remain so for the foreseeable future (if not permanently, for reasons we'll discuss in Section 1.2 on Cognitive Computing).

In terms of *"What's changed?"* and *"Why now?"* it's important to note that despite the ongoing challenges, obstacles, and setbacks that have beset the field since its inception, Artificial Intelligence research has benefited from a number of tailwinds in recent times, including:

---

2   For the purposes of simplicity, we will consider the research area of *speech recognition* (conversion of speech into text) as a part of Natural Language Processing. Hence, a program like Siri relies on speech recognition to convert acoustic signals into text, as well as the broader research area of NLP to analyze the text for meaning.

- *Increases in computing power*: The face of computing is unrecognizable from its mechanical origins as "counting machines," and its subsequent progression from vacuum tubes through to silicon processors and beyond—a journey that has contributed to the advancement of all research fields, not just Artificial Intelligence. Consider that before 1949 computers couldn't even store commands, only execute them, and the speed and size of those computers could only be described as "painful." But as computers grew smaller, faster, and more affordable, with built-in memories and then in-memory processing, they allowed scientists to carry out more calculations, computations, and experiments, which in turn accelerated their rate of research and improved the usability of real-world applications of Artificial Intelligence (as some algorithms are particularly "computational hungry," so any advancement in computational cost and speed carries over to algorithmic performance).

- *Algorithmic advancements*: Each algorithmic method (such as *fuzzy systems, genetic algorithms,* or *neural networks*—discussed in greater detail in Part II) represents a separate research direction with its own set of dedicated computer scientists, conferences, and peer-reviewed journals. As improvements and advancements are achieved in these areas, they translate into better (and more "accurate") applications of Artificial Intelligence—think of speech recognition or biometric scanners, or systems that predict the outcome for complex scenarios. These applications improve as the underlying algorithmic technology improves, as can be seen through a comparison of speech recognition applications from the 1990s with any present-day example.

- *Availability of training data*: In the same way we learn from our own experience and from the experience and knowledge of others, Artificial Intelligence algorithms can also learn from experience. Instead of taking years, however, the training of AI algorithms (to recognize faces, predict demand, make meaning recommendations, classify biopsy samples, and so on) can be compressed into hours/days/weeks, depending on the problem we're trying to solve and amount of available training data. Until recently, a lack of training data meant a lack of proper learning/training for AI algorithms, leading to poor results and disappointing outcomes. The explosion in Internet data, publicly available government data, as well as proprietary data that can be purchased from third parties, has significantly boosted Artificial Intelligence research, as scientists are better able to tune their algorithms when there is ample training data available.

- *Skills availability*: For decades past, Artificial Intelligence wasn't a popular area for students to venture into, with few career options available upon graduation other becoming another university lecturer on the subject. But with the explosion of interest in AI during the past few years, all this has changed. Master of Science and Ph.D. graduates in Artificial Intelligence are routinely courted by the likes of Apple, Google, Uber, Amazon, and more, signaling that the private sector is willing to pay top dollar for these skills. This "turning of the tide" has encouraged more students to enter the field and more universities to set up specialized AI programs, resulting in a dramatic enlargement of the skills and knowledge available in the marketplace. Whereas 30 years ago the only place such skills and knowledge could be found was within universities, they're now widespread and far more accessible, providing yet another tailwind for Artificial Intelligence.

- *Digitalization*[3]: The process of converting text, pictures, and workflows into digital formats has significantly benefited the field of Artificial Intelligence, because it's difficult to apply AI algorithms to whiteboards, notecards, or manual pen-and-paper processes. Being a digital technology, Artificial Intelligence algorithms must draw on digital inputs. Hence, the explosion of Internet data, along with the intense popularity of social media platforms—to say nothing of the ongoing quest of organizations both large and small to "digitalize" their operation—has been a significant enabler of not only Artificial Intelligence research, but also its application in the real world.

- *Pressures of capitalism*: And lastly, the business world has become faster, noisier, more interconnected, and vastly more complex during the past few decades, creating a challenging environment for corporations of every shape and size. Given the pressures of capitalism (huge bonuses are available to CEOs and executives that can perform in such environments—consider that the CEO of Walt Disney was paid more than US$66 million in 2019), the focus on technologies that can help executives deliver greater results has intensified (what wouldn't the CEO of a telecommunications company pay to automate away their many call centers and customer services reps while simultaneously increasing customer satisfaction? The financial reward for such an achievement would make Aladdin blush). And so capitalism provided another

---

3   The terms *digitization* and *digitalization* are often used interchangeably, but *digitization* is the process of converting data from a physical format into a digital one, and when this process is leveraged to improve business processes, it is called *digitalization*. In this text, we'll use the term *digitalization* to cover both meanings.

tailwind for the field of Artificial Intelligence, as companies rushed in to make investments and start projects that could enable greater business performance (unsurprisingly perhaps, given that shareholders are less lenient these days, eager to push out the old guard in favor of more progressive executives capable of harnessing new technologies to deliver results that can move the share price).

Ironically, the same tailwinds that have contributed to the development of Artificial Intelligence in the past, now represent the limiting factors when it comes to further research. As an example, the trillion-fold increase in computing power since 1956 (the year "Artificial Intelligence" was officially coined as a term and defined as a research direction) has greatly aided researchers within all branches of Artificial Intelligence. But irrespective of Moore's Law[4] and the progress made in computing power, scientists are still hopelessly short on processing speed (and insufficient computing power isn't just a limitation in Artificial Intelligence research, but within many other "computational expensive" disciplines as well, such as seismology, particle physics, and meteorology). The same holds true for skills availability—where there has been an explosion in university programs and training curricula, but there still aren't enough people to enable every business to implement AI projects—as well as algorithmic advancements, where the advent of deep learning improved outcomes in Computer Vision and Natural Language Processing, enabling a "jump" in performance before research plateaued once again, leaving scientists to continue their search for even better algorithms that will one day run on even faster computers.

Another tailwind for Artificial Intelligence (and, by the same token, an ongoing challenge and limiting factor) is our improved understanding of how the human brain operates. Such knowledge has been used to further research in algorithmic areas such as neural networks and deep learning, which attempt to mimic (at a very simplified level) the neuron/synapse structure of the brain. But despite these recent advances in neuroscience, the corpus of knowledge on how the human brain actually works is still speculative in nature and theory based, and thus represents a major limitation of Artificial Intelligence research (if not "the limitation" confronting the entire field). The reason this lack of knowledge might be the ultimate limitation is because it's difficult—some say "impossible"—to artificially replicate something that isn't properly understood in the first place.

---

4   "Moore's Law" comes from an observation made in 1965 by the CEO and co-founder of Intel, Gordon Moore, that the capability of computers will double every two years due to increases in the number of transistors that a microchip contains.

As an analogy, imagine that in the year 2000 B.C. Egyptians were provided a technology much ahead of their time, say a mobile phone, and were even taught how to use it. Irrespective of their fascination with the technology and the undeniable fact that the phone was right there, "in their hand" so to speak, any attempt to "artificially" re-create the mobile phone by building a copy would have been a futile endeavor without a solid understanding of material science, processor chips, electric circuitry, LED technology, and other fields of knowledge related to the inner workings of the phone—fields of knowledge that humankind wouldn't stumble upon until thousands of years later. There is a direct parallel to this when we talk about replicating the human brain, or any phenomena not fully understood by scientists. We'll discuss this point further in Section 1.2, when we move to the subject of Cognitive Computing.

Before proceeding further into the core of this text on the use of Artificial Intelligence to improve revenue and margins outcomes through improved decision making (Chapter 2 and beyond), let's first take a look at the history of the field, and then explain—in simple language—the major research areas, as well as what "algorithms" are and the difference between *AI algorithms* and *non-AI algorithms*, before discussing what Artificial Intelligence means to the modern enterprise, and why the technology will continue to feature heavily in boardrooms seeking revenue and margin growth.

1.1 Artificial Intelligence at a Glance

The idea of Artificial Intelligence isn't new, and one that even the ancients philosophized over with thoughts of mechanical men, automatons, and artificial beings. It wasn't until the 1940s, however, that mathematicians began to conceive of a day when computers could solve problems and make decisions on par with human beings. One of the luminaries of this period was British mathematician Alan Turing, renowned for his leading role in breaking the Enigma code during World War II. He was perhaps the first person to provide public lectures on machine intelligence, describing how a machine could learn from experience by altering its own instructions. In 1950 he published a paper entitled *Computing Machinery and Intelligence,* which opens with the famous line: *"I propose to consider the question, 'Can machines think?'"* And a year later, he was quoted as saying: *"At some stage … we should expect the machines to take control."*

Alan Turing is considered by some to be the founding father of Artificial Intelligence (with a benchmark AI test named after him—the *Turing Test*—for determining whether or not a computer is capable of thinking like a human being), while others consider John McCarthy to be the founding father, who coined the term *Artificial Intelligence* when he held the first academic conference on the subject (the Dartmouth Conference in 1956). In either case, the birth of the field occurred around this time, and then expanded in the decades ahead as interest grew from large corporations and government organizations, and computers became faster and cheaper (the cost of renting a computer in the 1950s exceeded $100,000 *per month*).

During this time of development (1960s–1980s), the field of Artificial Intelligence began to develop branches of specialized research, like Computer Vision and Natural Language Processing (discussed below), as well as areas of algorithmic specialization, like fuzzy systems and neural networks. Some scientists took the route of specializing in a branch of AI (such as Computer Vision) and began experimenting with a wide variety of algorithms, tools, and technologies to see if they could achieve better outcomes within that singular problem domain; while other scientists took the route of specializing in an algorithmic method (such as neural networks) and began experimenting with a wide variety of different problems (e.g. predicting demand, detecting fraud, recognizing speech, and so on) to see if they could achieve better results with some variant of their algorithmic method. Hence, some scientists specialized "vertically" in a problem domain, while others specialized "horizontally" in an algorithmic method that cut across many problem domains:

Areas of Artificial Intelligence Research

Robotics	Computer Vision	Natural Language Processing	Cognitive Computing	
				Algorithmic Methods

As time passed, computer scientists also realized that the original promise of Artificial Intelligence—to create a thinking machine with intelligence and awareness on par with humans—was a far more difficult undertaking than initially envisioned. Throughout the late 1950s and 1960s, they were confident this goal was only twenty years away[5], but when the 1970s and 1980s

---

5   This early optimism is attributable to the number of computer programs developed during this time that seemed "astonishing," such as speaking English, solving math problems, proving theorems, and playing games (with checkers begin an early example in 1959).

arrived, the promise of Artificial Intelligence wasn't any nearer and still "only" twenty years away. Then the 1990s came, followed by the 2000s, and the goalposts kept moving so that the promise of Artificial Intelligence remained the same, being just around the corner, only twenty years away. And today, in the year 2020, numerous prominent computer scientists still maintain that the promise of AI can be realized within twenty years. The point is that we've always been "twenty years away," and next year, next decade, we're still likely to be twenty years away. For reasons we'll explain below (in Section 1.2 on Cognitive Computing), there is some evidence to suggest that we'll never be able to close this gap, just like Achilles in Zeno's paradox.

Because of this continuous shifting of goalposts, the enthusiasm for Artificial Intelligence gradually waned and turned into disappointment, and then over time, ridicule, so much so that many computer scientists began to distance themselves from the term "Artificial Intelligence" and began to publish papers and hold conferences under alternate headings (such as *Computational Intelligence* or *Soft Computing*). These ongoing disappointments also led to "Artificial Intelligence" being redefined into two new terms: *Narrow AI* and *General AI,* with Narrow AI[6] being a specialized implementation of Artificial Intelligence algorithms for a specific (i.e. "narrow") problem, and General AI[7] being the original promise of Artificial Intelligence.

Examples of Narrow AI include Siri (and other digital assistants such as Alexa and Google Home), facial recognition on iPhones, self-driving cars, implementations of IBM Watson (whether tuned for playing Jeopardy or analyzing medical images), as well as all the case studies and examples presented in Part III. In fact, this entire book is about Narrow AI, which has real "here and now" applications for improving business outcomes, particularly around key metrics like revenue, margin, operating costs, and customer engagement.

By the same token, this book is *not* about General AI—to say nothing of *Super AI* where machine intelligence grows exponentially and makes humans obsolete, or the interfacing of biological "wet ware" and technological "hard ware" so that we can "live" forever by uploading our memories and consciousness into machines. At present, such topics remain firmly planted in the realm of science fiction and are of no value to the modern business manager, executive, or board member (for whom this book is written) other than providing entertainment or philosophical reflections.

---

6   *Narrow AI* also goes by other terms, including *Specialised AI, Applied AI,* and *Weak AI.*
7   *General AI* also goes by *Strong AI* and *Full AI.*

Also, while on the subject of terminology, we've observed considerable confusion between the terms *Artificial Intelligence* and *Machine Learning* (often being used interchangeably in many forums). In Chapter 7 we'll cover Machine Learning ("ML") in more detail, along with a discussion on some of the more popular ML algorithms in Part II, but it's important to differentiate these terms upfront, with Artificial Intelligence being the broad, all-encompassing research field with four major branches (Robotics, Computer Vision, Natural Language Processing, and Cognitive Computing) along with a multitude of algorithmic methods (e.g. neural networks) that aim to solve narrow problems (e.g. Narrow AI) as well as continuing the quest to discover a master algorithm capable of intelligence and awareness on par with humans (e.g. General AI).

Within this sprawling field of Artificial Intelligence sits Machine Learning, as a grouping of algorithms that can learn from data to perform specific tasks and then improve their performance through direct experience (without the need for explicitly programmed instructions). Also, these algorithms are not confined to any one branch of Artificial Intelligence. As an example, one major algorithmic area within Machine Learning is deep learning, which has been applied with great success within Robotics, Computer Vision, Natural Language Processing, *and* Cognitive Computing:

Areas of Artificial Intelligence Research

Robotics	Computer Vision	Natural Language Processing	Cognitive Computing	
				Algorithmic Methods

Machine Learning Algorithms and Techniques

Deep Learning

Today, the four primary branches of Artificial Intelligence along with the algorithmic areas that cut across, remain the focus of significant research efforts both in universities and the private sector. Global centers of excellence include the MIT Computer Science & Artificial Intelligence Lab, which consists of more than 20 research groups in AI and Machine Learning; Carnegie Mellon University, which was the first university to establish an undergraduate degree in AI; Stanford University, where AI has been studied since 1962; along with other institutions such as the University of California at Berkeley, Nanyang Technology University, University of Edinburgh, and Harvard, alongside major (and massive) research groups inside technology giants such as Microsoft, Google, and IBM.

As mentioned before, research progress within these organizations is now limited by the same factors that propelled the field of Artificial Intelligence forward in the first place, namely increases in computing power, algorithmic sophistication, training data, digitalization, and skills availability. Many are hopeful, however, that a breakthrough in computing power (e.g. a new paradigm like Quantum Computing) or algorithmic development (e.g. the creation of a master algorithm) will allow the field to leap forward, perhaps bringing it closer to the original promise of Artificial Intelligence.

1.2 Branches of Artificial Intelligence

As mentioned above, the all-encompassing field of Artificial Intelligence can be divided into four major research areas, each representing a critical function of the human body that scientists and researchers are striving to replicate artificially. We'll briefly cover each branch in this section—its history, research direction, and current challenges—keeping in mind that such an overview is cursory, as each branch can be studied for years at the university, and researched for decades more.

Robotics (for mechanical movement)

American physicist and engineer, Joseph Engleberger, is considered to be the father of Robotics, who along with George Devol founded the world's first robot manufacturing company in 1956, *Unimation*. The company went on to commercialize the first industrial robot, called Unimate #001, a 4,000-pound robotic arm that was in production use by 1961 at a General Motors assembly plant. Hence, the field of Robotics began in earnest around the same time as the inception of Artificial Intelligence (1956), with the aim of replicating mechanical movements, particularly within manufacturing environments for jobs that were hazardous for humans to perform.

Over time, robots grew smaller, smarter, more agile, and more affordable, finding their way into numerous household, industry, and military applications. Some of the notable advancements along the way (among *many* examples) include *Shakey the Robot* (the first autonomous, intelligent robot capable of making its own decisions on how to behave, invented at Stanford in 1966—pictured below, left), *Asimo* (the 4'3" robot created by Honda, incorporating "predicted movement control" that allowed it to walk smoothly and climb stairs—pictured below, right), *Roomba* (the first domestically popular robot), and recently, *Baxter and Sawyer* (which could be taught to perform tasks through movement):

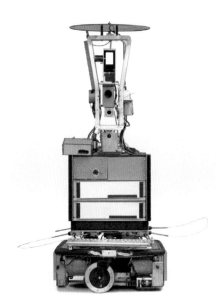 

This multi-decade development has had the greatest impact on manufacturing—particularly assembly plants—introducing a great deal of automation across all sectors (with before and after pictures from the clothing industry shown below):

As the field of Robotics developed in parallel to the burgeoning field of Artificial Intelligence, the two fields began to overlap, with each field being much broader than just this overlap in the middle:

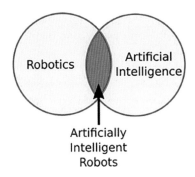

Artificially
Intelligent
Robots

Today, Artificial Intelligence is just one of the challenges facing the field Robotics, alongside the development of an adequate power source, and the fabrication of new materials that can make movement more natural. For example, no battery can yet match our biological metabolism for energy production (in the same way that no computer or algorithm can match our biological brain), and so developing an adequate power source is one of the major challenges for Robotics because the usefulness of a robot is largely dictated by the weight, size, and power of its battery supply. Also, whether researching new battery options or new materials, Robotics engineers are increasingly turning to nature for inspiration (e.g. instead of using mechanical gears and electromagnetic motors for movement, some labs are experimenting with the use of artificial muscles).

Computer Vision (for seeing)

Our ability to see colors, people, have depth perception, differentiate one object from another, and so on—a sense we often take for granted—is quite difficult to reproduce artificially. This is the goal of Computer Vision, which aims to replicate the human visual system through the use of cameras and algorithms to capture, process, analyze, and interpret imagery.

Computer Vision was born during the 1960s within universities that were already pioneers in Artificial Intelligence, and has progressed significantly since that time. In fact, it wasn't that long ago that facial recognition was a clunky, expensive, and often inaccurate technology limited to government use, and now, largely thanks to advances in algorithmic methods (such as deep learning), it has made its way into various consumer devices.

In addition to heavy use within Robotics research—as a visual system is needed to provide robots with sensory information about their environment—other major applications of Computer Vision include:

- *Medical devices*: Faster and more accurate analysis of medical images (e.g. X-ray, MRI, biopsy samples, ultrasound, and so on) can lead to better patient outcomes and reduced clinical costs. As an example, human doctors have an accuracy rate of approximately 87% in detecting melanomas through visual inspection, whereas a 2018 Computer Vision application for skin cancer detection achieved an accuracy rate of 95% (while also making fewer errors than human doctors when assessing benign moles). Such applications reduce clinical costs due to their speed of processing samples, as well as save lives by reducing patient misdiagnosis.

- *Production lines*: Computer Vision has been used for quality control on production lines for decades, visually inspecting products for defects or other quality issues (a task that would have been performed by humans in the past). More ambitious applications of Computer Vision have moved the technology out of factories and into open fields, where algorithms are used to visually search for weeds and pests within agricultural settings, as well as analyze the condition of fruits and vegetables to make better harvesting decisions.

- *Security*: Airports, stadiums, subways, casinos, and military facilities are usually monitored through CCTV cameras. The difficulty of monitoring these environments grows as the number of cameras grows, especially that subjects move from one camera to another and then back. As an example, consider a person who walks around an airport without boarding a flight and eventually leaves their briefcase on a bench before exiting the terminal—being able to automatically identify such suspicious activities from live video footage is an ongoing challenge for Computer Vision research.

- *Consumer applications*: From the iPhone to Google Photos to self-driving cars, Computer Vision is steadily expanding into our everyday lives. As a taste of things to come, facial recognition is already available in China for accepting payments from consumers, so we only need to show our face to pay for items in a store.

Like all real-world applications of Artificial Intelligence, Computer Vision algorithms are most effective when they're highly tuned to a very specific and narrow problem (e.g. interpreting images of skin moles or defective productions in a factory). However, even the best algorithms in the world in Computer Vision often make mistakes that no human being would make (not even a child), as the embarrassing and much-publicized case of Google Photos tagging two black people as "Gorillas" demonstrated.[8]

Natural Language Processing (for hearing and speaking)

The goal of Natural Language Processing ("NLP")—along with the related field of speech recognition—is to help computers understand human language. This research area began in 1950 with the publication of Alan Turing's famous paper *Computing Machinery and Intelligence,* where he proposed a test for determining machine intelligence (which is now called the *Turing Test*). The test evaluates the ability of a computer program to impersonate a human during a real-time written conversation, such that the person on the other end is unable to tell whether they are talking to another person or a computer program.

By 1966, a professor at the MIT Artificial Intelligence lab developed the world's first NLP program, called ELIZA (pictured below, left). The program wasn't able to talk like Siri or learn from conversations, but it paved the way for later efforts to tackle the communications barrier between humans and machines. Natural Language Processing research progressed significantly during the 1980s, which is when the concept of chatbots was invented, and then boomed in the 1990s as the Internet drove the need for advanced algorithms capable of interpreting and summarizing the world's (exponentially growing) depository of textual web pages. Today, NLP research continues to grow as the market for NLP software products expands from US$10 billion in 2019 to US$25 billion by 2024, with many popular consumer devices incorporating the technology (pictured below, right):

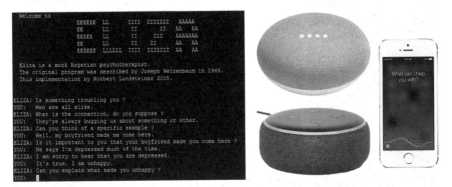

In the Robotics context, advances in Natural Language Processing would allow robots to interact with their environment through listening and speaking, in the

---

8   *Google Photos Tags Two African-Americans As Gorillas Through Facial Recognition Software,* Forbes Magazine, July, 2015.

same way that humans do. Besides consumer devices and robots, other applications of NLP include chatbots, spam filters, sentiment analysis, and recruitment. But like the other branches of Artificial Intelligence research, Natural Language Processing still has a long way to go. As a recent article within *Scientific American* pointed out (*Am I Human?* March 2017), even simple sentences such as "The large ball crashed right through the table because it was made of Styrofoam" illustrate the difficulties in Natural Language Processing, because "it" can refer to either the ball or the table. Common sense will tell us that the table was made of Styrofoam (the "it" in the sentence), but for a machine to reach a similar conclusion would require knowledge of material sciences along with language comprehension, something that is still far out of reach.

Some of the challenges that exist within NLP research include finding the correct meaning of a word or phrase, understanding modifiers to nouns, inferring knowledge, as well as correctly identifying the pragmatic interpretation or intent (as irony and sarcasm may convey an intent that is opposite to the literal meaning). These are not easy problems to overcome, as any regular user of Siri can attest. Despite Apple being a trillion-dollar company by market capitalization and employing some of the best minds in Natural Language Processing, the results are primitive when compared to real speech, as the humorous transcripts below illustrate:

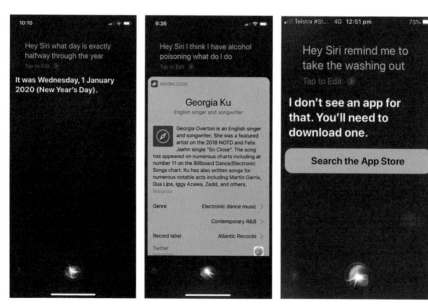

This difference—between the state-of-the-art in Natural Language Processing technology and its biological equivalent—illustrates just how difficult it is to artificially replicate just one element of the human experience (spoken language).

Cognitive Computing (for thinking)

Cognitive Computing attempts to replicate the brain's function of "thinking," and includes such processes as analysis, deduction, reasoning, and decision making. In many ways, the brain "brings everything together" as the command center of the body, interpreting what we see, understanding what we hear, formulating thoughts and speech, and directing our limbs to move. Without the brain, the rest is irrelevant. For this reason, Cognitive Computing research strikes at the heart of the original goal of Artificial Intelligence—of creating a thinking machine with intelligence and awareness on par with humans—and many aspirational computer scientists believe that their research and development efforts will eventually lead to the replication of even higher brain functions, like consciousness and self-awareness.

But with 100 billion neurons and more than 100 trillion connections (with each "connection" called a *synapse*), the brain's structural complexity cannot be overstated. Theoretical physicist Michio Kaku famously said that the human brain is "the most complicated object in the known universe," and we agree. Notwithstanding this complexity, some scientists believe that it's only a matter of time until we create a conscious and self-aware replica—specifically, just a matter of time until computers achieve the necessary speed and affordability to allow for a complete mapping of our neuron/synapse structure, as well as *a complete re-creation of the brain*. Others, however, believe that such a mapping and re-creation—if ever completed—will do little to help us replicate the brain artificially. Their arguments are worth noting, because if correct, they mean the original goal and promise of Artificial Intelligence might never be realized.

First, let's consider that the only organism for which we've fully mapped the neuron/synapse structure is the roundworm, with 302 neurons and 7,000 connections (versus the human brain's 100 *billion* neurons and well over 100 *trillion* connections). More importantly, however, is that after having this detailed map in our possession for more than 25 years, the scientific community eventually concluded that our understanding of the roundworm wasn't materially enhanced because of this neuron/synapse mapping (all that work to build a map, then decades of research trying to understand it—on the simplest of organisms—only to say it didn't help much). So if mapping the 302 neurons and 7,000 connections of a worm was difficult to come by and then proved to be of little value, then where does that leave us with the 100 billion neurons and 100+ trillion connections of the human brain?

The second reason why Artificial Intelligence might be a futile dream is because the neuron/synapse structure represents the first "layer" of the brain. As an analogy, consider that the word "atom" originates from the Ancient Greek adjective *atomos*, which means "indivisible," and was proposed as the smallest building block of matter in 450 B.C. For more than 2000 years, nothing changed, until John Dalton brought the "indivisible" atom into the scientific mainstream in 1800 when he introduced *Atomic Theory*. Textbooks were rewritten, and things remained the same for another 100 years, with the atom featuring as the smallest building block of matter during that time. But then in the late 1880s, the proton and electron were discovered, and lo and behold, the "indivisible" atom turned out to be divisible after all, into smaller pieces. Textbooks were rewritten again, with electrons, protons, and neutrons taking the mantle as the smallest building blocks of matter. But then in 1964, another layer was proposed, namely, that protons and neutrons were made up of even smaller sub-atomic particles called *quarks,* and although we had to rewrite textbooks again, it was all good, because we were done, there was nothing more. But now, again, we suspect there's something more, as the inability of theoretical physicists to reconcile general relativity with quantum mechanics has forced them into a search for a layer beneath quarks, theorizing a layer of "strings" or "membranes," or perhaps "waves of "potentiality"—who knows.

The point is that whenever we master one layer—or think we've mastered it—we suddenly realize that another layer lurks beneath. Such is the argument against Artificial Intelligence, suggesting that we are decades away from mastering even the first layer of the brain (neurons and synapses), at which time we'll encounter the next (such as the role of trillions of microtubules and microfilaments in our brain that might need to be understood, mapped, and modeled), thereby complexifying the problem by orders of magnitude. This will push out the goalposts for AI research yet again, perhaps resetting the timeline so that the realization of Artificial Intelligence is once again "just" twenty years away. And perhaps this cycle will continue to repeat itself, thereby ensuring that we always stay twenty years away from unlocking the secrets of the brain, almost as if Nature is baiting us, refusing to reveal herself.

A related argument to the above is that the world's leading physicists—including those working at the Hadron Super Collider at CERN in Geneva—have been unable to get to the bottom of physical matter and are beginning to doubt they ever will. As a case in point, we don't even know what a kitchen table is made of, because the further down we look through the layers, past the atoms, protons, and quarks, the more empty space we find and the more we question what physical matter is really made of. And if that's the case with a kitchen table, then there might be no hope for computer scientists as they recognize that the brain is ultimately made from physical matter—whether it be quarks or something smaller—and without understanding these smaller components, their efforts might be futile.

There are plenty of other thought-provoking arguments against the realization of Artificial Intelligence (such as consciousness being a quantum phenomenon, or that the "mind" is separate from the "brain"), but they're beyond the scope of this book. And whichever side of this argument you might favor is irrelevant for the remaining chapters, because we're only concerned with practical applications of Artificial Intelligence (in particular, as they relate to decision making), which, thankfully, don't require the replication of the human brain, nor even an understanding of it.

Moreover, no matter how limited current capabilities of Cognitive Computing might seem in comparison to the brain, they're still cutting-edge by historical standards (as Watson winning *Jeopardy!* against two world champions has proven—pictured below, left) with applications being injected into robots, cameras, self-driving cars (pictured below, right), production lines, software—you name it—to provide perhaps not "intelligence," but a level of smarts that can make organizations more productive and profitable, and our everyday lives easier and safer.

1.3 Artificial Intelligence "Algorithms"

The word *algorithm* has a long history, and the word can be traced back to the ninth century. During this time the Persian scientist, astronomer and mathematician *Abdullah Muhammad bin Musa al-Khwarizmi* (who is often cited as "The father of Algebra") was indirectly responsible for creating the term "algorithm," which is best defined as a set of instructions for taking an input and turning it into an output. Cooking recipes are often used as an example of algorithms, because recipes take inputs (i.e. ingredients) and turn them into outputs (e.g. a cake, salad, meat pie, etc.) through step-by-step instructions (e.g. mix 250g of flour with 50mL of water, add 2 eggs, and so on). In the same vein, below is a simple algorithm for adding two numbers together:

Step 1: Start
Step 2: Request first number
Step 3: Request second number
Step 4: Add both numbers

Step 5: Display sum
Step 6: Stop

The inputs are the two numbers, and the step-by-step instructions call for taking these two numbers, adding them together, and then displaying the sum (which is the output of the algorithm). Originally emerging as a part of mathematics, the word "algorithm" is now strongly associated with computer science and Artificial Intelligence in particular. Such algorithms are typically used to carry out exact instructions to solve problems, with *AI algorithms* often differing from *non-AI algorithms* in one or more interesting aspects.

First, AI algorithms are often inspired by nature—consequently, the output "emerges" from the algorithm rather than being calculated through hard-coded rules and mathematical equations. Second, many AI algorithms include a component of randomness. What this means is that for the same input, a conventional, *non-AI algorithm* will produce the same output, whereas for the same input, an *AI algorithm* might produce a different output (in the same way that our brain might arrive at one conclusion to a problem in the morning and then a different conclusion in the evening, even though the problem and the inputs have remained the same—this is because "biological intelligence" is not "hard" and "precise" like classical equations or calculations). Third, many AI algorithms are generic in the sense they can be applied to a variety of problems from different domains (e.g. genetic algorithms have been widely used for various engineering design problems, including automotive design, finance and investment strategies, marketing and merchandising, computer-aided molecular design, and encryption and code breaking, to name a few), whereas non-AI algorithms are usually designed for specific problems that are very "crisp" and well-defined.

As an example of this difference in algorithms, let's consider the famous *traveling salesman problem* (which is discussed further in Chapter 2). Conceptually, the problem is very simple: traveling the shortest possible distance, the salesman must visit every city in his territory (exactly once) and then return home. The diagram below represents a seven-city version of this problem:

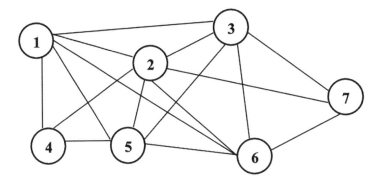

With seven cities, the problem has 360 possible solutions. But with ten cities, the problem has 181,440 possible solutions, and with 20 cities, the problem has about 10,000,000,000,000,000 possible solutions. Although these are difficult problems to solve (due to the large number of possible solutions—something we discuss in more detail in Chapter 2), we could apply a non-AI algorithm to this problem, such as the well-known *Lin-Kernighan algorithm*. Each time we "run" the algorithm, it will produce a sequence of cities the salesman should visit to minimize travel distance (which is the "output," and this output will always be the same, as long the input—in this case, a starting tour of the cities and the initial tour—also remains the same). The algorithm starts with an initial tour and, iteration by iteration, tries to improve it by following a sequence of predefined steps.

Alternatively, we could apply an AI algorithm to this problem, such as *ant systems*. Unlike the Lin-Kernighan algorithm, ant systems are inspired by nature because this algorithmic method attempts to mimic the real-world behavior of ants, in particular, their ability to find the shortest path between a food source and their nest (pictured below, left). Without going into the physiological details of how real ants find the shortest path in nature, what's important is that the mechanism and process by which ants do this is known and understood (by laying down *pheromone trails*), and so computer scientists can artificially replicate these mechanisms and processes when creating an ant algorithm (pictured below, right).[9]

---

9   As a further addition to our discussion on Cognitive Computing in the previous section, this is the difference between replicating something we understand, like ant behavior, versus something we don't, like how the brain works. Like the analogy of a mobile phone provided to Egyptians in the year 2000 B.C., we cannot replicate something we don't understand.

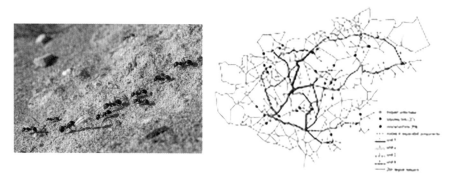

Each time we run this ant algorithm, it might produce a slightly different result, in the same way that the behavior of real ants will be slightly different in nature. Furthermore, ant algorithms can be applied to a variety of other problems (e.g. industrial scheduling, multiple knapsack, bin packing, vehicle routing, to name a few), whereas the Lin-Kernighan algorithm can only be applied to one category of the traveling salesman problem (the so-called symmetrical traveling salesman problem, where the distance from A to B is always the same as the distance from B to A—which is clearly not the case in real-world problems where we must consider differing speed limits, one-way roads, various detours, and so on). Algorithms such as these provide instructions for almost any AI system, which is why the study of various algorithmic methods cuts horizontally across the four branches of Artificial Intelligence:

Areas of Artificial Intelligence Research

Robotics	Computer Vision	Natural Language Processing	Cognitive Computing	
				Algorithmic Methods

To see ant algorithms in "action," and understand a bit more about they mimic nature, we encourage you to watch the supplementary video for this chapter at: www.Complexica.com/RiseofAI/Chapter1. Ant algorithms are also covered in more detail in Chapter 6.9.

1.4 Why Now? Why Important?

Business managers are waking up to the realization that the world is faster, noisier, and more interconnected and complex than ever before. Thinking back twenty years, we can remember a time when things were a bit slower (less disruption), with less noise (no social media), less connectivity (mobile phones weren't yet the devices they are today), and less complexity (fewer moving pieces to consider). By the same token, if we look twenty years ahead, we can be sure the world will be even faster, noisier, and more complex—it's a one-way street and there's no going back. As a consequence, decision-making has become increasingly difficult for managers and executives because the increased speed, noise, and complexity makes it impossible for anyone to process all the available data and information when making decisions. The result is substandard decisions, and consequently, substandard business performance in key metrics such as:

- revenue growth
- margin
- service levels
- quality
- safety
- customer engagement
- share of wallet and market share
- production costs, and
- working capital/inventory levels

Although there are many Artificial Intelligence applications of Robotics, Computer Vision, and Natural Language Processing that can create significant business value through automation and productivity improvements, in this book we'll examine Artificial Intelligence from the perspective of *decision-making*. Why? Because the decisions we make as individuals and organizations define the quality of the future we create for ourselves and our organizations. An implementation of Robotics or Computer Vision might create a one-time performance improvement through automation, but such an improvement pales in comparison to the ongoing gain in revenue growth, margin, and competitiveness that an organization can realize by improving the quality of decisions made by staff, managers, executives, and board members.

Hence, organizations capable of consistently making data-driven, optimized decisions year in, year out, are the ones most likely to positively influence the abovementioned key metrics over the long haul and create the most value for shareholders; whereas organizations that stumble along with gut feel, intuition, repeating what they did last week/month/year, are those most likely to struggle.

To appreciate this point, consider the following question: *Given the complexity of today's business world, what are the odds that everyone in your organization is making optimal decisions day in, day out, 261 working days a year?* We can tell you that the odds are pretty small. And for every decision that is made each day, across each staff member, across the entire organization, a significant amount of value is either lost or left on the table. We can visualize this lost value as a simplified formula ("simplified," because all decisions aren't equal in value or import):

Annual Lost Value = (Value of Optimal Decision − Value of Actual Decision) × 261 working days per year × number of staff

In the same way that calculators have improved our ability to make better decisions—followed by spreadsheets, reporting tools, and countless other software applications—we can think of Artificial Intelligence as the latest "calculator" to assist us, and one that happens to be particularly well-suited for the speed, noise, and complexity of the modern world. With this in mind, let's begin our discussion of how to improve revenue and margin by first understanding the decision-making process within most organizations, along with where Artificial Intelligence can create the most value.

For more information on the material covered in this chapter, including a visual example of ant algorithms and how they mimic nature, please watch the supplementary video at: www.Complexica.com/RiseofAI/Chapter1.

CHAPTER 2

Complex Business Problems

*"The whole universe sat there, open to the man
who could make the right decisions."*
Frank Herbert, *Dune*

Recent years have seen terms like *data science, algorithms, machine learning,* and *big data* solidify their position in our everyday vocabulary, with articles on Artificial Intelligence becoming commonplace in business and mainstream publications. With the growing popularity of websites that make recommendations and smartphones that take voice commands, there is a growing appreciation for how AI-enabled functionality adds value to our day-to-day lives. Parallel to that, in the enterprise space, there is likewise a growing trend of embedding AI functionality into Customer Relationship Management (CRM), Enterprise Resource Planning (ERP), and other corporate systems, so they can handle more sophisticated workflows and deliver more value. In this context, it should come as no surprise that organizations of all shapes and sizes are increasingly asking: *What is Artificial Intelligence truly capable of? What is it best suited for? What could that mean for my organization?*

A good place to start in answering such questions lies in our day-to-day usage of Artificial Intelligence, most likely through mobile apps that recommend books, movies, the best route through traffic, even food we might come to love, despite having never tried it before. What these applications have in common is the use of smart algorithms that analyze data and provide us with recommendations, typically for decisions we make on a regular basis, because when we make decisions over and over, the algorithms can learn our preferences and improve the quality of future recommendations (something that isn't quite possible with one-off decisions like: *Which university should I attend?*)

If we extrapolate the consumer-based use of Artificial Intelligence to large organizations, which usually compete in dynamic environments and must deal with the impact of unforeseen events and a multitude of external and internal forces, we can achieve a similar result: namely, improved decision making through intelligent recommendations. In the same way that AI-based apps can improve our decisions for trivial problems (e.g. where to eat, what movie to watch), Artificial Intelligence can also be applied to complex business problems that are difficult to solve through manual methods, and where the consequence of making the wrong decision is much higher than a bad meal or boring movie.

2.1 Decision Making for Complex Business Problems

To understand why and how Artificial Intelligence can improve business decisions, we first need to look at the decision-making process itself. Although different organizations follow different processes for making decisions, they're usually based on the same fundamental steps:

1. Identify the problem (Problem)
2. Gather data on the problem (Data)
3. Organize and interpret data (Information)
4. Understand the "why" (Knowledge)
5. Consider possible solutions, their pros & cons (Evaluation)
6. Implement a solution (Decision)

The above represents a *problem-to-decision workflow*, which is essentially an analytical workflow at its core. This high-level representation of the decision-making process works well for conceptual explanations, but might be the cause of some fundamental misconceptions. An important one, worthy of attention, is the role of knowledge in the process. Most people are familiar with the popular saying, "Knowledge is power!" but what most businesses have come to realize over time is that knowledge by itself won't guarantee the best, or even right decision, even if a business has more knowledge than anyone else. A business may "know" a lot about its customers, but management may still be unsure of what decision to make!

This is because the vast majority of business problems are inherently complex, and thus, difficult to solve. Hence, the decision-making process often breaks down somewhere between the Knowledge and Decision steps, because knowledge in and of itself, isn't quite enough. A closer look at any real-world business problem, whether in distribution, customer retention, or fraud, will bear witness to this obvious truth. Most complex business problems share the following characteristics, which represent the reasons why they're so challenging to solve:

*   The number of possible solutions is so large that it precludes a complete search for the best answer
*   Real-world business problems are set in dynamic environments
*   There are many (possibly conflicting) objectives
*   The problem is heavily constrained

Of course, the above list can be extended to include many other characteristics, such as incomplete information (e.g. the necessary data wasn't recorded), noisy data (e.g. the data contain estimates and rounded figures) and uncertainty (e.g. the data aren't reliable). However, these four primary characteristics are sufficient for our purposes, so let's discuss each in turn.

The number of possible solutions is so large that it precludes a complete search for the best answer

Let us assume we want to find the best solution to a problem with 100 decision variables. To keep this example simple, let's also assume that each of these decision variables is binary (i.e. each decision variable can only take one of two possible values, such as "yes" or "no"). Each possible combination of these 100 variables produces some result that can be evaluated and labeled with a *quality measure score,* which is a numerical score that tells us how "good" or "bad" each solution is (similar to a KPI[1] measure). Assume, for example, that a sequence:

"yes" & *"yes"* & *"no"* & *"no"* & *"no"* & *"yes"* & *"no"* & ... & *"yes"*

produces a quality measure score of 79.8, whereas the sequence:

"yes" & *"no"* & *"no"* & *"yes"* & *"no"* & *"yes"* & *"no"* & ... & *"no"*

produces a quality measure score of 91.5. The higher the quality measure score, the better the solution, hence the latter solution is better than the former. Our task is to find the combination of values for the 100 variables that produces the highest possible quality measure score. In other words, we would like to find a solution that cannot be improved.

Without any additional problem-specific knowledge, our approach might be to evaluate all possible combinations. However, the number of possible combinations is enormous. Although each variable can only take one of two values ("yes" or "no"), the number of possible solutions grows at an exponential rate: there are four combinations (2×2) for two variables, eight combinations ($2 \times 2 \times 2$) for three variables, and so on. With 100 variables, there are $2 \times 2 \times ... \times 2$ (100 times) combinations—a number that corresponds to 10^{30}. Evaluating all of these combinations is impossible. Even if we had a computer capable of evaluating 1,000 combinations per second, and we began using this computer one billion years ago, we would have evaluated less than 1% of the possible solutions by today!

---

1   KPI stands for Key Performance Indicator; it gives a measurable value that demonstrates how effectively a company is achieving key business objectives.

If we revisit the traveling salesman problem introduced in Chapter 1—where traveling the shortest possible distance, the salesman must visit every city in his territory (exactly once) and then return home[2]—recall that with seven cities, the problem has 360 possible solutions,[3] making it relatively easy to solve:

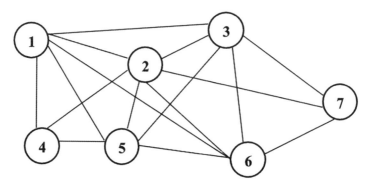

By adding a few more cities, however, the number of possible solutions grows exponentially. To see the maddening growth of these solutions, consider the following:

- A 10-city problem has 181,440 possible solutions
- A 20-city problem has about 10^{16} possible solutions (1 followed by 16 zeros: 10,000,000,000,000,000 possible solutions)
- A 50-city problem has about 10^{62} possible solutions.

By comparison, our planet holds approximately 10^{21} liters of water, so a 50-city problem has more solutions than the number of litres of water on our whole planet! The number of possible solutions to a 100-city problem exceeds by many orders of magnitude the estimated number of atoms in the whole Universe! These numbers are so large they're difficult for us to even conceive of mentally, while most real-world business problems are far more complex than this (in terms of the number of possible solutions). They're defined by a much larger number of variables, and these variables usually take on more values than just "yes" or "no." In such cases, the number of possible solutions is truly astronomical!

---

2   Some closely related problems require slightly different criteria, such as finding a tour of the cities that yields the minimum travel time, minimum fuel cost, or a number of other possibilities, but the underlying principle is the same.

3   For simplicity, we'll assume that the problem is symmetric (i.e. the distance between cities A and B is the same as the distance between B and C). Note also, that solution 1–2–3–4–5–6–7 is the same as solution 3–4–5–6–7–1–2, as both these solutions have a different starting city but represent the same cycle.

So, how can we find optimal solutions to such problems? An exhaustive search that relies on computing power is clearly not the answer, as the number of possible routes, fraud rules, or transportation plans might be so large that examining all possibilities—with even the fastest supercomputers—would take many centuries at best. In the following chapters, we'll explore a real-world business problem where the number of possible solutions is *much* larger than the numbers presented here and show how such problems can be solved using Artificial Intelligence methods.

The problem exists in a dynamic environment

Business managers know that real world problems aren't static, and yet they take static snapshots of the problems they're trying to solve. Such snapshots represent a good starting point for analyzing and understanding a problem, but on their own, they paint a false picture. Because real-world problems are set in dynamic, time-changing environments, we must address the time factor explicitly. To illustrate this point, let's consider a real-world version of the traveling salesman problem with delivery trucks. If the problem is carefully analyzed and a set of delivery routes found, the quality of these routes will be affected by many factors, such as rush-hour and weekend traffic, weather and road conditions, and so forth, as well as random events, such as labor strikes or accidents. Because the problem is influenced by so many external factors, any solution to a static snapshot of this problem might prove inadequate.

We can take another example from sales operations, where significant effort is taken each year to optimize a sales rep's territory and determine the optimum number of visits that each customer should receive within each call cycle. This static approach to the problem is bound to result in the under- and over-servicing of customers during the course of a year, because some customers will change in volume and importance, but the static solution won't consider these changes. Hence, as time passes, the static solution will deteriorate in quality (becoming more sub-optimal) to the point where sales resources could have been put to far better use by pursuing new opportunities or proactively managing customer churn.

This *dynamic environment factor* becomes even more prevalent in manufacturing and distributed supply chains, where static plans and schedules may face steep deterioration in performance when exposed to dynamic variables such as process variability, equipment failure, weather events, and demand spikes, versus more "robust" and "forgiving" plans, which are more tolerant of unexpected events, changes, and modifications.

There are some additional issues related to dynamic environments that are worth noting. Imagine that we are considering two solutions, A or B:

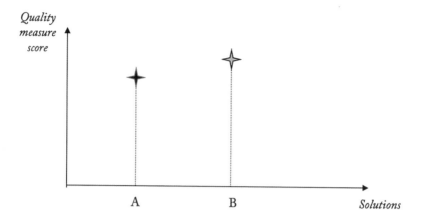

Which of these two solutions should we select? Well, the question seems trivial: Because solution B has a higher quality measure score, solution B is better than solution A. Although this statement is true—solution B *is* better than solution A—the answer might not be that straightforward. It may be the case that solution A "sits" on a relatively flat peak, whereas solution B "sits" on a very narrow peak:

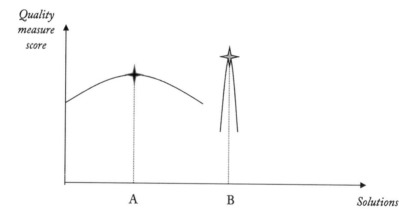

We can interpret the above graph as follows: Although Solution B is better than solution A (there is no doubt about that), it is *peak optimized*—meaning that nothing can go wrong for us to achieve the result (all our assumptions must pan out, no exceptions!). Hence, if we're forced to modify solution B in any way (due to process variability, unexpected maintenance, demand spikes, or some other reason), then the quality of Solution B will deteriorate very quickly. Solution A, on the other hand, has a lower quality measure score to begin with, but achieving this result is far more likely because the solution can tolerate changes and modifications without a sharp drop-off in quality. Given that solution A is less risky than B, should we still select the "better" Solution B?

There are many (possibly conflicting) objectives

It's quite unusual for any real-world business problem to have only one objective. In fact, in recurring cost-cutting environments the go-to position always seems to involve the conflicting request to "do more with less!" Complex problems, especially where there's a lot at stake, often involve a range of objectives that could be working against one another. Such problems are called *multi-objective problems*, as there is more than one objective to satisfy, and an increase in the quality measure score for one objective might come at the cost of another.

A simple example of this phenomenon exists in manufacturing, where companies try to carry just enough inventory to satisfy future customer demand, without carrying too much. By keeping inventory levels high, a manufacturer can be sure to satisfy future customer demand along with any unexpected spike in orders, but this approach can have significant working capital implications, as well as potential obsolescence costs (especially in sectors such as food and electronics, where inventory is either perishable or quickly becomes obsolete). By keeping inventory levels low, on the other hand, the manufacturer can improve its operational metrics (such as stock turns) and realize substantial savings in working capital and obsolescence costs, but is likely to experience occasional stock-outs and lost sales. Hence, there's a trade-off between the competing objectives of minimizing inventory costs and maximizing customer service levels.

In multi-objective problems, maximizing the performance of one objective (such as cost) is likely to come at the expense of other objectives (such as safety, time, or service levels), thereby rendering the concept of a single "best solution" no longer relevant. Instead of a single optimal solution, such problems have many optimal solutions, with each solution performing better or worse against the selected objectives, thereby leaving the decision-maker with the complex task of evaluating these trade-offs.

The problem is heavily constrained

All real-world business problems have constraints of some sort, and for a particular solution to be suitable for consideration, it should satisfy many restrictions imposed by business rules, capacities, contractual obligations, regulations, laws, and/or preferences.

For example, let's consider the problem faced by Australian pharmaceutical wholesalers, which distribute medicines that carry *Community Service Obligation* requirements. Part of these service obligations require the supply of a full set of medicines to pharmacies across Australia usually within 24 hours, regardless of location and cost of supply! Now consider the number of constraints involved in coming up with a delivery plan:

- The number of delivery vehicles and their location (e.g. more than one-quarter of all Australian pharmacies are located more than 100 km from the nearest capital city)
- The desired delivery time window of each pharmacy
- Orders needing to be delivered in less than 24 hours
- Certain medicines are temperature sensitive and require special storage or specialized vehicles

It's also important to note that some of these constraints are mandatory (referred to as *hard constraints*, such as the number of delivery vehicles), while others may be flexible (referred to as *soft constraints*, such as delivery time windows).

Pharmaceutical wholesalers employ teams of people to solve such problems in a way that creates the best outcome for all parties—but what does "best" mean? Well, in this case, it might mean a plan that satisfies all constraints and has the lowest overall cost of implementation (i.e. a plan that is within the total funding provided by the government and is able to meet the contractual service levels). The challenge, however, is that sometimes finding even one plan that satisfies *all* constraints can be quite difficult.

Decision-making process for complex business problems

Consistent, high-quality decisions in any industry can be traced back to the effectiveness of the problem-to-decision workflow discussed earlier:

Problem > Data > Information > Knowledge > Evaluation > Decision

This workflow has traditionally been implemented in a manual way in most organizations, through the use of human experts and analysts. For this reason, the extent to which this workflow effectively bridges the gap between "knowledge" and "good decisions" depends on the nature and complexity of the problem, as well as the amount and quality of resources applied to the "evaluation" step.

For simple problems with few possible solutions, no conflicting objectives, and minimal constraints, the evaluation step can be managed through manual methods. However, as the number of possible solutions grows, as the influence of dynamic variables increases, as multiple (competing) objectives are introduced, and as more and more constraints and business rules are applied, the problem grows exponentially in complexity and the evaluation step becomes more difficult—perhaps even impossible—to undertake through manual efforts alone. As an example, if a complex problem has millions of possible solutions, with many trade-offs among objectives, the time it would take to find and evaluate all these solutions would be prohibitive (i.e. centuries).

This means that the more complex the problem (i.e. the greater the number of possible solutions, dynamic variables, conflicting objectives, and constraints), the more difficult the evaluation step, and throwing more resources at the problem is unlikely to improve the decision for the simple reason that it's difficult just to identify all possible solutions, to say nothing of evaluating them in detail. This puts a ceiling on the complexity of problems that an organization can effectively address through manual efforts, and raises the question of whether we can automate the problem-to-decision workflow for recurring decisions? And if so, how?

To answer these questions, let's look at the various levels of sophistication (and related approaches) that are available to any organization when it comes to decision making, and then discuss the role that Artificial Intelligence can play.

2.2 The Problem-to-Decision Pyramid

The diagram below represents the continuum that exists in terms of an organization's ability to improve its decision making. The best way to understand this diagram is through an analogy of climbing a pyramid, where the higher we climb, the further we can see. Using this analogy, each layer represents a step in the quest for improved decision making, so the higher we go, the better our decisions.

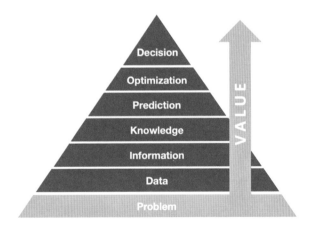

Problem

The *problem* layer is the foundation of the pyramid and represents the specific business problem we're trying to solve. No matter how big or small the problem, this represents the first step in the decision-making workflow, and in many ways, the most important, because we can't climb the pyramid without first identifying and understanding the business problem. For example, we can't collect the necessary data (the next layer of the pyramid) without first defining the problem.

Data

The second step in our pyramid involves the collection and storage of data pertaining to the problem we're solving:

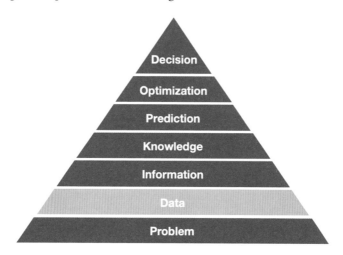

The word *data* means "known facts." As a general concept, it refers to the situation where some existing facts (whether qualitative or quantitative) are represented or coded in a form suitable for usage or processing. Data are collected in the form of bits, numbers, symbols, and objects, and a typical piece of data consists of a pair (attribute, value), such as "color, red." Data can be pre-processed, cleaned, arranged into structures, stripped of redundancy, and organized or aggregated to provide *information*, which is the next layer of the pyramid.

To have a better appreciation of what data looks like, let's have a look at the different *attributes* and *values* in the receipt below, where the "$" sign stands for the attribute "sales price," and "L" stands for the attribute "container volume (in liters)." In the same receipt, we can see a few examples of values, such as "$6.50" for the price, "2L" for the container volume and "1" for the number of units being purchased:

```
WELCOME TO XYZ CONVENIENCE

     THE BEST VALUE IN TOWN
          MELBOURNE, VIC
        PH: +61 3 9863 6115

     --------------------
                                       $
                                     --------
C/C ICE CREAM CONT.  2L.
   1 @ $6.50                           $6.50

       BALANCE DUE                     $6.50

       Cash                           $10.00
       CHANGE                          $3.50

     --------------------
         TRADING HOURS
   MON – WED 9:00AM – 6:00PM
   THU – FRI 8:00AM – 9:00PM
   SAT – SUN 8:00AM – 6:00PM
```

Recent years have seen a growing obsession with data: collecting it, mastering it, reporting on it, and in some cases, even valuing it like a financial asset. That is understandable: after all, data is the first step in the problem-to-decision workflow. However, many organizations—usually well-funded and well-resourced enterprises—have been collecting and storing large volumes of data for years on the premise that "if we collect good data, then good decisions will follow," only to discover that the connection between data and decision making isn't automatic because of the other steps in the problem-to-decision workflow. Hence, we must make a distinction between "good data" (second layer of the pyramid) and "good decisions" (top of the pyramid), and a further distinction between "good" and "bad" data—after all, what is *good* data anyway? There is some temptation to answer this question in terms of the state of the data (i.e. quantity, quality, timeliness, structure, etc.), but in the context of improving business decisions, "good" refers to any data that assists us in diagnosing, explaining, and assessing the problem we're trying to solve.

From this perspective, we should look beyond the boundaries of the organization and consider external data as well, such as demographics, weather, point-of-sale transactions, competitor pricing, government approvals and licenses, and so on. As an example, census data can be used to understand the demographical characteristics of customer groups across various geographic areas, which in turn can be used to better understand why certain product promotions are more effective in some areas than others.

Information

The next layer of the pyramid, *information*, includes facts and relationships that have been perceived, discovered, or learned from the data.

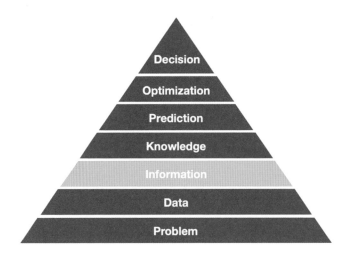

The *information* layer of the pyramid leverages reporting and data visualization techniques to graphically represent data, with the output being reports, charts, graphs, statistical tables, and more. As an example, the supermarket receipt above contains data on ice-cream sales, which the manufacturer could aggregate into an informational report to better understand sales performance in a specific region (in this case, Victoria):

DAY_OF_QUARTER	PROD_CAT	SKU_CODE	SKU_SIZE	REGION	TOT_SALES_DAY (Units)
1	IC_CTN	5290CXA89	2L Container	VIC	1,550
2	IC_CTN	5290CXA89	2L Container	VIC	2,570
3	IC_CTN	5290CXA89	2L Container	VIC	7,080
4	IC_CTN	5290CXA89	2L Container	VIC	1,530
5	IC_CTN	5290CXA89	2L Container	VIC	3,090
6	IC_CTN	5290CXA89	2L Container	VIC	3,520
7	IC_CTN	5290CXA89	2L Container	VIC	6,550
8	IC_CTN	5290CXA89	2L Container	VIC	3,220
9	IC_CTN	5290CXA89	2L Container	VIC	11,050
10	IC_CTN	5290CXA89	2L Container	VIC	3,570
11	IC_CTN	5290CXA89	2L Container	VIC	4,080
12	IC_CTN	5290CXA89	2L Container	VIC	6.020

The same information can also be visualized in a chart:

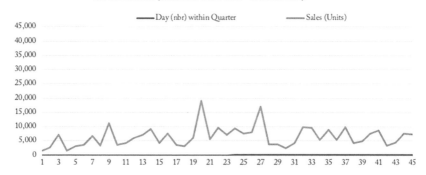

Ice-cream sales (2L Cont. VIC / SKU - 5290CXA89)

Effective outputs of this layer of the pyramid can communicate complex information clearly and efficiently, making it easier for humans to understand trends, outliers, and patterns. While a critical step in the problem-to-decision workflow, this layer of the pyramid represents the most basic level of analytics and is usually referred to as *descriptive analytics*. When done well, it might suffice as a decision support tool for smaller-scale, simpler problems, but would be inadequate for business problems of greater scale and complexity, as discussed above.

Knowledge

If the *information* layer of the pyramid tells us *what* happened, then the *knowledge* layer tells us *why* it happened. This layer of the pyramid builds on the outputs from the previous layers to provide a deeper understanding of both the data and the problem we're trying to solve. Unsurprisingly, many refer to this layer as *diagnostic analytics*:

The goal of the *knowledge* layer is to develop a good understanding of what happened in the past, the factors (i.e. variables) that contributed, relationships between those factors (i.e. correlations), and possibly the extent to which any single variable contributed to the result more than any other (dominant variable). In the ice-cream sales example above, the chart shows us what happened (i.e. information on how many units were sold), but in the *knowledge* layer of the pyramid, we would like to understand *why* it happened by establishing a correlation between consumer demand for ice-cream and other variables, such as changes in price or temperature. This knowledge could then be communicated in a number of ways, such as the chart below, plotting sales units alongside changes in temperature:

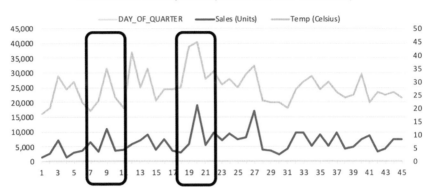

To expand our knowledge, we may want to explore if rainfall has any further effect on consumer demand, and create a complete graph exploring the movement of sales alongside changes in local temperature and rainfall:

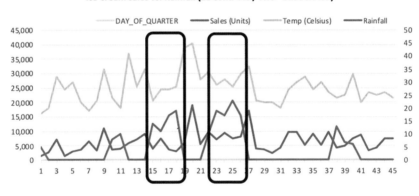

This process can then be repeated, where we search for possible relationships and correlations to other variables, such as competitor pricing or promotional activity, among others.

Prediction

The next layer of the pyramid deals with answering the question: *What will happen next?* So if the *information* layer tells us *what happened* and the knowledge layer tells us *why it happened,* then the prediction layer tells us *what might happen in the future* (with some probability) and is often referred to, unsurprisingly, as *predictive analytics.* Hence, in the problem-to-decision workflow:

Problem > Data > Information > Knowledge > Evaluation > Decision the evaluation step is now expanded to include *prediction* and *optimization,* which are essential for identifying possible solutions, predicting their outcome, and assisting in their evaluation.

A key feature of the *prediction* layer of the pyramid is the ability to predict outcomes for various scenarios that can be interpreted as "what-if" questions. For example, the question might be: *What is the impact on customer services levels, if one of the following happens* (the "what" and the "if"):

- If we buy three additional delivery trucks?
- If we change the overnight location of the delivery trucks?
- If we build another distribution center in Victoria?
- If we have to service 10% more customers?
- If we have to carry 5% more products?

Continuing with the ice-cream sales example from above, let's say the manufacturer is preparing to launch a new product in Australia and needs to identify a territory that meets a specific set of qualification criteria for the launch. For example, the area must have a high volume of existing ice-cream sales, an older and affluent demographic, and a minimum number of retail outlets. To further complicate the search for the right territory, the new product must meet a specific sales target for the month of its launch, and given the size of Australia and the timing, weather is likely to play a role (as well as the advertising budget).

To make a decision for the launch, the manufacturer would need to consider several scenarios (i.e. what-if questions) to find the best territory. Each what-if question, (e.g. *What will sales be if the new product is launched in Victoria during the month of April?*) requires a few core elements, which together constitute a single scenario:

- *Data for each territory:* The number (and characteristics) of retail outlets, seasonal weather patterns, demographics of each catchment area, historical sales volumes for relevant products, field sales staff and territory structures, as well as other pieces of data that might affect the predicted outcome (in this case, sales of the new product).
- *Constraints that define a possible solution:* These could be: (1) the minimum sales target that must be achieved for the new product; (2) that sales of the new product must not cannibalize sales of existing products; (3) the characteristics of the selected territory must be representative of the broader target market (e.g. high volume of ice-cream sales for specific product ranges and an older and more affluent consumer base); and so on.
- *Objective:* The specific metric for which we are predicting the outcome, in this case, finding the territory and month that satisfies the qualification criteria and generates the highest sales for the new product.

While this layer of the pyramid can enable quite sophisticated capabilities (i.e. predicting outcomes), we are still faced with a substantial limitation. Recall that in complex business problems, there is an extremely high number of possible solutions (i.e. scenarios to investigate), and more often than not, we are working with multiple objectives simultaneously. Given the complexity of such an iterative what-if planning process, we'll only have time to create and evaluate a limited number of scenarios. Consequently, the chances of finding the best solution are rather slim. Which raises the question: If we had time to create and evaluate millions of scenarios, could we find better a solution—one that satisfies all problem-specific constraints and has an overall higher level of predicted sales? The answer is yes, which moves us to the next layer of the pyramid, *optimization*.

Optimization

If the *prediction* layer tries to answer the question of *what will happen* in the future for any given scenario, then the *optimization* layer tries to identify the scenario that provides the "best" outcome while satisfying all constraints and business rules—in other words, the best solution to the problem we're trying to solve.

If we go back to the ice-cream sales example, the manufacturer might want to grow market share through the use of promotions. But to make the best possible decision around what promotion to use for what product in what territory (and when), the manufacturer would need to evaluate a huge number of possibilities (scenarios), with each scenario being a promotional plan that can be executed in the marketplace. Given the astronomical number of such scenarios, where each one represents a unique promotional plan that requires evaluation, this workflow is quite intricate and involved, especially that it needs to take into account all key variables, such as:

- *Products*: Do certain products respond better to promotions than others?
- *Promotion type:* What type of promotion (e.g. individual discounts, two-for-one offers, multi-buy discounts) should be applied to each product?
- *When*: What day? week? month?
- *Where*: What territories? retail chains? stores?
- *Duration:* Weekend only? entire week? two weeks?
- *Regularity:* Will the promotion be one-off? or repetitive? If repetitive, what should be the gap, if any, between promotions (e.g. 2 weeks on followed by 2 weeks off)?

Aside from all the possible combinations of different values of the above variables, the manufacturer must also consider various business rules and constraints. As a simple example, some retailers might impose restrictions on the frequency/regularity of promotions for a given product, while others might restrict the entire ice-cream category to only a few types of promotions.

To complicate matters further, the "best" solution might need to consider multiple objectives and their trade-offs. To grow market share, the manufacturer will need to maximize sales volume, but this objective competes against another major objective: margin and profitability—meaning that sales volume can be maximized by increasing the promotional discount, but such discounts drive down the manufacturer's margin (and perhaps overall profitability). On top of this, the retailer's margin needs to be considered, because the proposed promotion will be rejected if it doesn't achieve the retailer's margin and profit objectives, which in turn trade-off against the manufacturer's margin and profit objectives (and so on). These are only a few examples of conflicting objectives, where improvements on one objective come at the expense of another.

Once multiple solutions are created, evaluating the quality of each promotional plan becomes a complicated task, as it requires predicting the outcome of each plan while considering the impact from other variables (e.g. *Does this promotional plan decrease the sales volume of other products? And if so, by how much?*), as well as determining if any constraints or business rules have been violated (which might mean that a particular plan isn't a possible solution after all). Once the predicted sales volume for several promotional plans has been generated, the corresponding plans can be plotted on a graph to visualize their performance and investigate trade-offs:

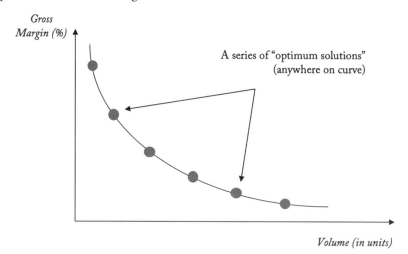

The above curve is termed a *Pareto front*, which takes its name from the famous Italian economist. A solution is *Pareto optimal* if it's impossible to improve any of the objectives without decreasing at least one other objective—therefore, all solutions on the Pareto front are Pareto optimal, and any solution that's not on the curve isn't optimal. In this case, the manufacturer is provided with a set of Pareto optimal promotional plans that show the trade-off between gross margin and unit volume.

Complex business problems like this are not well-suited for manual and spreadsheet-centric approaches for a few reasons: first, given the complexity of evaluating each scenario, manual methods can lead to biased, error-prone, and inaccurate predictions, as they're often based on gut feel and intuition. Second, it's impossible to manually create an extensive set of scenarios that cover a large number of possible combinations of key variable values, as decision-making timeframes don't allow for that. Without a sophisticated tool or system, creating these scenarios, evaluating them, and then analyzing the various trade-offs becomes an impossible task.

To see optimization in "action," and learn more about multi-objective optimization, we encourage you to watch the supplementary video for this chapter at: www.Complexica.com/RiseofAI/Chapter2.

Decision

Climbing past optimization brings us to the *decision* layer, which is the capstone of the problem-to-decision pyramid.

The *decision* layer represents the actual data-driven, optimized decision that is made, and which is enabled through the capabilities at each layer of the pyramid. In the ice-cream sales example, where the manufacturer wanted to grow market share through the use of promotions, the final decision would take the form of an optimized promotional plan. All KPIs, constraints, business rules, and trade-offs would have to be taken into account, and a handful of solutions (i.e. possible promotional plans) would be presented on a trade-off graph, so that the manufacturer can make a well-informed decision. The final plan that's selected and implemented would combine the right mix of products, types of promotions, and timeframes to deliver results against the conflicting objectives of maximizing gross margin and unit volume. Also, once the manufacturer's decision has been implemented and the results are known, the outcome needs to be fed back into the decision-making workflow so that the manufacturer can make even better decisions in the future (in effect "learning" from the outcome of previous decisions).

This problem-to-decision pyramid presents a compelling climb for most organizations, and it's easy to understand why. After all, being able to consistently make data-driven, optimized decisions can unlock significant value in most organizations, as discussed in Chapter 1.4. On the flipside, not making the climb—or attempting the climb with a labor-intensive and spreadsheet-centric approach—is likely to facilitate poor decisions that destroy value and allow competitors to gain an advantage. With that in mind, let's now explore the benefits that Artificial Intelligence methods can bring to the problem-to-decision pyramid.

2.3 AI for Bridging the Gap between Past & Future

Organizations wishing to climb and progress through this problem-to-decision pyramid must realize that a gap exists in the climb, requiring a step-change in capability to bridge and enable further progress. The reason for this gap is because the lower layers in the pyramid deal with the past (reporting on it, explaining why certain things happened), while the top layers deal with the future (predicting it, finding optimized solutions).

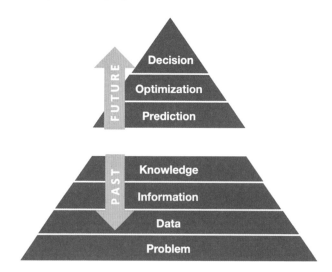

Bridging this gap between the past and future isn't straightforward, because the sophistication required for predicting the future and optimizing decisions is far greater than that of reporting on the past. One example of this difference lies in the fact that past patterns might not continue into the future, which isn't something we need to worry about if we're only reporting on the past, but something we need to deeply consider when trying to predict the future (requiring a greater level of sophistication). For this reason, organizations that have climbed to higher levels of the pyramid usually make better decisions than organizations that haven't.

When it comes to making predictions and recommending an optimized course of action, there are many tools and enterprise software applications available that can help organizations reach the top of the pyramid, and which are usually based on some sort of Artificial Intelligence method. The point is that as an organization climbs up the pyramid, the sophistication of the required tools and technologies also increases, with Artificial Intelligence having the most applicability and delivering the most value within the *prediction* and *optimization* layers of the pyramid, mainly because that's where the greatest complexity resides.

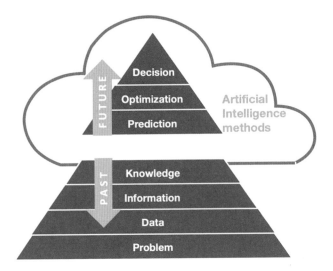

Keeping this in mind, we can use various Artificial Intelligence methods to develop a system capable of recommending optimized decisions, as well as learning from previous actions and decisions. Building such a *Decision Optimization System* involves three fundamental steps:

1. Building a predictive model
2. Building an optimization model
3. Incorporating adaptability (feedback loop)

These steps are briefly discussed below, and then again in Part II of this book, which explores these subjects in far more technical detail.

Building a predictive model

The first step in bridging the gap between past and future is developing a capability to predict what will happen next. To enable such a capability, we first need to identify the relationships and patterns among the various variables in the data, and then use our understanding of these relationships and patterns to build and train a model (or set of models) capable of predicting some outcome. This process is explained in greater technical detail in Chapter 5, but for now, it's only important to understand that the accuracy of the prediction is directly related to the quality and granularity of the underlying model. If the model has too many vague assumptions and approximations, the prediction may be meaningless, or worse.

For these reasons and others, creating such models requires an iterative, flexible, and cyclical approach, involving a set of tasks, usually referred to as a *Data Science Methodology*. One of the most widely used methodologies, the *CRISP-DM's methodology* (Cross-Industry Standard Process for Data Mining) includes the following steps:

- Business understanding
- Data understanding
- Data preparation
- Modeling
- Modeling evaluation
- Modeling deployment

Building an optimization model

The next step is to build an optimization model (covered in more detail in Chapter 6), which requires us to define:

- *Variables and their domains*: For example, a variable in a promotional plan might be the type of promotion (e.g. in-store, catalogue, etc.), and its domain would be the set of possible values: (10% off, 20% off, 25% off, etc.).
- *Constraints and business rules that define a feasible solution*: For example, the min/max frequency (of promotion for a particular product).
- *Objective*: For example, the total volume sold.

Clearly, there are additional details that need to be specified—these include categorization of constraints/business rules into soft and hard (Chapter 3 provides a more detailed discussion on this topic), possible penalties for violation of soft constraints, the relative importance of different KPIs, among others.

It's also important to note that the optimization model will work closely with the predictive model in the following way:

- The *optimization model* will automatically generate many possible (future) scenarios.
- The *predictive model* will evaluate each scenario generated by the optimization model against a single or multiple objectives and constraints, and generate a predicted outcome that is sent back to the optimization model for further action.

This approach allows the two models to "talk" to one another, and find possible solutions that satisfy all problem-specific constraints and business rules. The building, training, and deployment of such models is a significant technical undertaking that involves a great deal of technical expertise from AI scientists that specialize in various algorithmic methods.

And lastly, to fully bridge the gap between the past and future, these two models need to be augmented by the third step: the introduction of *adaptability*. Recall that complex business problems exist in dynamic environments and yet the predictive model is based on a static snapshot of historical data!

Incorporating adaptability (feedback loop)

The third step leverages the arrival of new data and feedback on current performance so that the models can learn from the outcome of past decisions, in order to make more accurate predictions and recommend better decisions in the future. More specifically, the models need a mechanism for "knowing" what actually happened (versus what was predicted to happen), and for updating themselves accordingly by taking these "actuals" into account when making future predictions or recommendations.

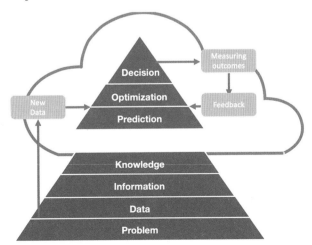

This feedback loop is a critical component of any Decision Optimization System and essential for ongoing optimized decision making in any dynamic environment. If the underlying models remain static, the system would lose accuracy and relevance, and eventually grow obsolete—in some cases, very quickly. By incorporating a mechanism for self-learning, the system can monitor outcome data and calculate the variance between predicted and actual values. When they vary beyond specific thresholds, it will trigger the system's adaptive algorithms to update the underlying models (automatic self-tuning— for example, by creating new (emerging) rules. Chapter 7 explains how this works.

Recommending the best decision

Once the underlying models have been trained and deployed, and the feedback loop enabled, the Decision Optimization System is ready to automate the problem-to-decision workflow and provide optimized recommendations.

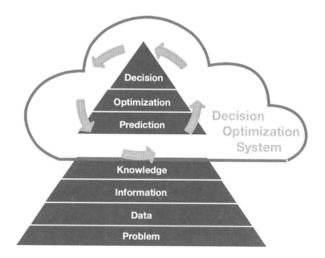

As discussed earlier, business problems grow in complexity as the number of possible solutions grows, as the influence of dynamic variables increases, as multiple (competing) objectives are introduced, and as more and more constraints and business rules are applied. And as these problems grow in complexity and the evaluation step becomes more difficult (and sometimes impossible), they become increasingly difficult to address through manual efforts. Hence, the benefit an organization can gain through a Decision Optimization System based on Artificial Intelligence methods lies in faster, and consistently higher-quality decisions, which will impact key metrics such as margin, sales growth, market share, production costs, and more.

To explain how Artificial Intelligence and Decision Optimization Systems work on real-world problems, we'll delve into a well-known and highly complex business problem in the next chapter, which involves manufacturers and retailers and how they struggle to maximize revenue, margin, and growth. Also, for more information about the problem-to-decision pyramid, along with examples of AI-based systems for prediction and optimization, please visit the supplementary video for Chapter 2 at: www.Complexica.com/RiseofAI/Chapter2.

CHAPTER 3

An Extended Example: Promotional Planning and Pricing

*"Half the money I spend on advertising is wasted;
the trouble is I don't know which half."*

John Wanamaker, *Merchant and Politician*

Having defined the problem-to-decision pyramid—along with each layer and how it contributes to the decision-making process—let's now turn our attention to the challenge of *promotional planning and pricing*[1] in the fast-moving consumer goods (FMCG) industry, which has all the hallmarks of a complex business problem worthy of discussion.

We've all experienced product promotions *(Sale! 50% off! Buy one, get one free!)*, which manufacturers and retailers use to drive foot traffic into stores, increase volume and market share, and build awareness for new products (have you ever bought a product on promotion, liked it, and then switched to that product permanently?). These promotional activities are typically funded by both the retailer and participating manufacturer, and can account for almost 20% of the revenue of FMCG companies. Hence, promotions are a big deal. And not only is the problem of promotional planning inherently complex—where it's difficult to make "good" decisions that generate real improvements in revenue, margin, and overall profitability—but it's also an inherently *high-value problem,* where the difference between good and bad decisions can mean millions of dollars and many percentage points of market share.

Now, suppose we work for a large, global manufacturer, one that produces many household products—ranging from snacks such as ice-cream and peanut butter, through to common toiletries like toothpaste—and most of our sales come through retailers, especially supermarkets, where promotions drive almost 80% of our sales volume. Within this setting, let's say we're in charge of promotional planning for all product categories sold within a single retail

---

1   There are several related terms to the area of *promotional planning*, including *promotional programming, trade promotions, trade promotion optimization (TPO)*, and so on. For the sake of simplicity, however, we'll use "promotional planning" throughout this text even though at times there may be a more fitting term.

chain, called Mary's Market, a fictitious retailer in Australia with 500 stores across the country. We'd like to create the "best" possible plan for Mary's Market (i.e. one that achieves the best result for the objectives we're trying to minimize or maximize), but this isn't easy given the complex nature of the problem and the fact that our objectives (e.g. maximizing profit) can often be in conflict with those of Mary's Market—for example, if they request a deeper discount for an upcoming promotion, it might erode our profit if the volume uplift isn't great enough.

In fact, we can go further on this point by saying that Mary's Market (and retailers in general) are working primarily to build their own sustained profitable growth and are ultimately indifferent to an individual supplier's growth or profit concerns. This creates a critical bottleneck, in that as the manufacturer, we can only "grow through" the retailer (at least for this channel). Hence, in addition to the science of promotional planning and pricing (of finding the "best" possible plan), there's also the "art" of negotiating the final plan with Mary's Market so it can be executed in their stores, rather than being rejected or significantly modified. In this chapter we'll focus on the scientific aspects of the promotional planning problem, while acknowledging that being able to negotiate well with retailers requires a sound understanding of the consumer, shopper, retailer, and the overall category (as discussed further in the Trade Spend Optimization case study in Chapter 9.4).

Basic Terminology

Before exploring this problem in detail, let's first cover some basic terminology in the context of promotional planning:

- *Store*: is an individual retail store that might either be part of a chain or independently owned.
- *Retail chain*: is a network of stores owned by the same company.
- *Banner group*: is a network of independently owned stores that sign up to a collective "banner" (i.e. brand name) to leverage economies of scale in procurement and marketing.
- *Geographic boundaries*: is the geographical region for which a promotional plan is made for a particular retailer, which is typically a state (e.g. VIC or NSW in Australia) but could also be more granular, such as metropolitan areas or rural regions.
- *Planning horizon*: refers to how far out we plan our promotions, which could be yearly, half-yearly, quarterly, or some other time horizon.
- *Promotional period*: is the length of time the promotion runs. For example, a promotional period may be one week for a grocery retail

chain, which means that a plan needs to be created for each week of the year. In other industries, the promotional period might be fortnightly or even longer.

- *Promotional plan*: is the set of all promotional activities during the planning horizon (also called the *promotional calendar*).
- *Variable*: is any value that can be changed when we search for the best plan. For example, variables may include the promotional price, depth of discount, and the promotion type, among many others.
- *Predictive model*: provides a sales forecast for each product within each promotional period. A predictive model for promotional planning may also include external data, such as weather, sporting events, or public holidays.
- *Promotion type*:[2] refers to the type of promotion, such as "% off," set dollar discounts, buy-one-get-one-free (BOGO), or multi-buy (i.e. multiple products offered for a fixed price).
- *Price step*: refers to the minimum amount that a price can be discounted. For example, the discounting of a $19.95 shelf price might occur through 50 cent *price step*s.
- *Objectives*: are measurable business goals we want to minimize or maximize. For promotional planning, typical objectives include maximizing volume and gross profit.
- *Business rules*: are something we might choose to do because there are valid business reasons to do so—for example, we might have business rules for minimum and maximum promotional frequencies for specific products, along with a minimum sell price or margin.
- *Constraints*: on the other hand, are something we're obligated to do, usually because of commercial commitments (such as trading terms), government regulations, capacity limitations, or some other factor. Both business rules and constraints might be "hard" (meaning they cannot be violated), or "soft" (meaning they can be violated in order to achieve a better overall result), and may apply to individual products and/or the overall plan (such as a minimum KPI or business objective that needs to be achieved for the plan to be acceptable).[3]

---

2   Also commonly referred to as the *promotional mechanics*.

3   Even though constraints represent something we're obligated to do (usually because of commercial agreements), it's worth exploring whether the violation of some constraints might produce a significantly better result (and if so, potentially lead us to renegotiate a commercial obligation in order to execute the better plan). For this reason, some constraints are defined as "soft" in order to test if they're detrimental to achieving superior business outcomes.

- *Feasibility and infeasibility*: any promotional plan we create is either *feasible* if it doesn't violate any hard business rule or constraint, or *infeasible* if it does violate a hard business rule or constraint.

3.1 The Problem: Promotional Planning in FMCG

Recall that we work for a global manufacturer that sells household products—including snacks and toiletries—through major retail chains, and we're in charge of promotional planning for one of these retail chains, Mary's Market in Australia. However, it's important to point out that our company has a very limited route to market, meaning that we can only sell our products through a small number of retail chains, with Mary's Market being one of them. Whereas Mary's Market, on the other hand, has an abundant choice of products not only from us, but from all our competitors. As an example, we might offer 32 different ice-cream products and eleven peanut butter products, and Mary's Market is only one of fifteen retail chains through which we can sell these products. Mary's Market, however, has access to hundreds of different ice-cream and peanut butter products from all the various manufacturers that produce those products. This raises the stakes for us on the importance of promotional planning, because there's a limited amount of shelf space available for which many manufacturers are vying, and so if our promotions aren't successful, then Mary's Market might delete our products from the shelf in favor of those offered by our competitors.

Now, let's consider the structural elements of the problem: In addition to the products we offer, Mary's Market also sells many other product categories we don't supply, like fresh fruits and vegetables, clothing, magazines, and pet care products, among others. For the product categories relevant to us, there's a certain length of time for which we must plan our promotions, and let's assume that we plan for the entire year, so our planning horizon is 52 weeks. Let's also assume that the promotional period is one week, which provides us with the plan's granularity and the level of detail we must plan to. We can represent this granularity with a promotional calendar—or *slotting board*—where each column is a week, and each row is a particular product, allowing for promotions to be *slotted* into each column/row combination. For example, the below slotting board is for the snacks product category in NSW:

Retailer:	Mary's Market
State:	NSW
Category:	Snacks

	WK 1	WK 2	WK 3	WK 4	WK 5	WK 6	...	WK 51	WK 52
Product 1									
Product 2									
Product 3									
Product 4									
Product 5									
Product 6									
...									
Product 100									

If there are 100 individual products for us to plan in the snacks category, then the simple decision of whether or not to promote a particular product for any given week requires 5,200 individual "yes" or "no" decisions (52 weeks × 100 products). If we ignore other elements of the problem—such the promotional price, promotion type, ancillary marketing, holidays and seasonality, catalogue placement, and so on—the amount of individual binary yes/no decisions still implies an astronomical number of possible plans (1 followed by 1,565 zeros!), each of which would look something like this on the slotting board:

Retailer:	Mary's Market
State:	NSW
Category:	Snacks

	WK 1	WK 2	WK 3	WK 4	WK 5	WK 6	...	WK 51	WK 52
Product 1	Y	Y		Y	Y				
Product 2	Y		Y		Y	Y			Y
Product 3	Y		Y		Y	Y			Y
Product 4			Y						
Product 5	Y			Y					
Product 6	Y	Y		Y					
...									
Product 100	Y			Y	Y			Y	Y

Our promotional plan must also adhere to business rules for individual products. For example, business rules may prevent the promotion of specific products for less than four weeks or more than twelve weeks during any twelve-month period, known as *minimum and maximum frequency*. Furthermore, these business rules may apply to the "gap" in between promotions for the same product, known as the *minimum and maximum promotional gap*, so that promotions don't happen too often or too infrequently (such as promoting the same product for nine consecutive weeks and then not promoting it for the rest of the year). Such business rules also apply to prices, where the minimum promotional price might be set at no less than 50% of the shelf price (i.e. the non-promoted price) and not more than 90%, and must move by some price step increment (e.g. 50 cents or one dollar, to avoid awkward prices like $8.13). These business rules can be visualized in the following table:

	Shelf Price	Min Freq	Max Freq	Max Promo Length	Min Promo Gap	Max Promo Gap	Min Promo Price	Max Promo Price	Price Step	Allow rounding?
Product 1	$23.50	13	26	2	2	6	$12.00	$20.00	$0.50	Y
Product 2	$14.50	13	26	2	2	6	$10.00	$12.50	$0.50	Y
Product 3	$19.00	13	26	2	2	6	$10.00	$15.00	$1.00	N
Product 4	$22.00	20	39	3	2	4	$11.00	$18.00	$1.00	Y
Product 5	$52.00	20	39	3	2	4	$40.00	$48.00	$1.00	Y
Product 6	$48.00	13	39	3	2	4	$40.00	$45.00	$0.50	Y
...										
Product 100	$65.00	13	13	1	2	6	$55.00	$60.00	$1.00	Y

We need to also bear in mind that different geographic regions may have their own business rules, which adds further complexity to the problem. There may also be specific rules tied to the promotional period itself, such as the minimum and maximum number of products on promotion at any given time (which may differ from week to week as we consider holidays or other events). For example, during most weeks of the year, it may be permissible to have 30% of our products on promotion at the same time, but for certain weeks of the year, such as before major holidays like Easter, Christmas, and New Year's, it may be appropriate to increase this percentage.

These business rules may be quite complex, and yet, we've only considered products in isolation, and not thought about products being constrained by the promotional activity of other products. Hence, we might need to extend our business rules to cover different pack sizes of the same base product (e.g. 45 gram- and 170 gram-bags of the same snack), different varieties or flavors of the same product (e.g. where all go on promotion or none at all), or different products altogether (e.g. where we can promote a subset of our snacks at any given time, but not all of them within the same promotional period). And lastly, the promotional plan as a whole must meet certain KPI thresholds, such as volume or volume growth, revenue or revenue growth, and gross profit (among others), which we must also define as business rules.

Keeping all this in mind, let's turn our attention to the slotting board for Mary's Market in NSW (for the snacks category), which may look like the following:

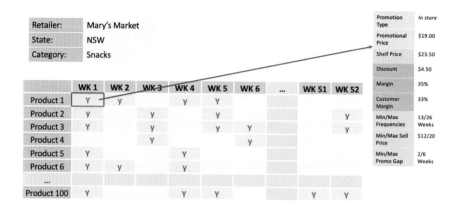

We can see that beyond the simple yes/no decision of whether a product is on promotion during any given promotional period, each cell has a set of values associated with it. These are color coded in the table above on the right, and broken down as follows:

- *Variables (green):* are values we can change and "trial" when searching for the best plan. In the right-hand side table above, these include the promotional price and promotion type, which are often dependant (or linked) because particular promotion types may be associated with particular price points. For example, a "%-off" promotion will always link the promotional price with a corresponding percentage discount level.
- *Reference points (yellow):* are values we don't adjust when creating new promotional plans. In the right-hand side table above, this is the shelf price (i.e. the non-promotional price), which might change only once or twice a year whenever there's an overall price change.
- *Derived values (orange):* are calculated from other values and may sometimes act as business rules and constraints. In the right-hand side table above, these include the discount, margin, and retailer margin, which are all calculated from the shelf price, promotional price, and overall product cost.
- *Business rules* and *constraints (blue):* represent limits imposed on particular values for the purpose of optimization. In the right-hand side table above, these include the minimum and maximum promotional price, the minimum and maximum promotional frequency over any given planning horizon, and the minimum and maximum gap between promotions for the same product.

When we consider all these additional factors, it seems that our job is quite challenging! Just creating a single feasible plan—one that doesn't violate any business rules or constraints—is already a Herculean task, most likely involving a multitude of spreadsheets and endless hours of evaluating new plans against historical promotions to "guess" the likely outcome. Then, on top of all that, the plan won't be executed and produce the desired upside unless we secure buy in from Mary's Market. This means we need to create a "win-win-win" slotting board—one that optimizes marketplace outcomes for ourselves, Mary's Market, as well as for the overall category—a slotting board that represents a needle in an astronomical haystack!

For more information about the myriad complexities of promotional planning and pricing, as well as the use of rules and constraints to create slotting boards, we encourage you to watch the supplementary video for this chapter at: www.Complexica.com/RiseofAI/Chapter3.

3.2 Applying the Problem-to-Decision Pyramid

Considering the difficulty of creating a single feasible plan, what would it take for us to create an optimal plan? And by what measure, or measures, should the plan be evaluated to determine its "optimality"? Before we answer these questions, let's revisit the problem-to-decision pyramid from Chapter 2, and review these layers in the context of promotional planning:

Given that we've already described the problem of promotional planning in some detail, let's move to the next layer of the pyramid: *data*.

Data

Data is the fundamental "raw material" for promotional planning, as any decision we make will be dependent on the underlying accuracy of our predictions, which, in turn, will be dependent on the available data. In the context of promotional planning, what we're most interested in is data that can help

us infer shopper behavior, particularly
consumer demand—after all, promotional
activities are usually undertaken to increase
demand. Hence, we're particularly interested in *point of sale* data, which is the data
closest to the consumer (often called *sell
out* data), as opposed to the retailer's orders
from a manufacturer (often called *sell in*

data), which may be subject to various supply chain related factors such as
inventory policies and minimum order quantities. When such data is overlaid
with promotional activities and pricing history, we can infer long-term trends
at the product or category level, as well as seasonality by examining the baseline
(non-promotional) sales volume over time.

It's common knowledge that "garbage in equals garbage out," thus, data
quality and availability are important considerations. Unfortunately, most
modern organizations struggle with a variety of "data issues" such as:

- *Missing data*: where a critical piece of data has been overwritten or
  wasn't collected in the first place. A classic example is stock on hand
  data, where most inventory management systems only contain the most
  current stock on hand without retaining any historical values. However,
  this historical data is useful for identifying stockouts in the past. When
  overlaid with sales data, it may become evident that some unexplained
  drops in demand were actually due to stockouts rather than actual drops
  in demand. Another example is when a retailer doesn't provide detailed
  point of sale data (by store, by day, by product), which compromises a
  manufacturer's ability to build an accurate prediction model for promotional planning.

- *Dirty data*: which typically stems from master data management issues
  within IT systems (such as ERP or point of sale). Examples of dirty data
  include products assigned to the wrong product category, or the use of
  free text fields to describe product information such as pack size and
  volume. Furthermore, using multiple point of sale systems may result
  in sales data that's in multiple formats, requiring considerable time and
  effort to standardize.

- *Incorrect mappings*: occur when some data mappings change over
  time. A typical example within promotional planning is the mapping
  of individual stores to a particular retail chain. In some industries,
  stores are independently owned and may switch from one retail chain
  banner to another. If these moves aren't recorded over time, then an

incorrect data mapping occurs, which may lead to historical demand not being correctly recorded for the same store. Another example is when new products are launched that replace existing products. For example, a 500g product is discontinued, and 350g and 650g products are launched within the same product line. New product introductions and even simple pack size changes might be incorrectly mapped within IT systems, leading to issues when that data is used to create a predictive model.

Such data issues are typical within most modern organizations and must be dealt with as we climb the problem-to-decision pyramid. Fortunately, there are a variety of methods we can use, which we'll discuss in the following chapter on data and modeling.

Information

Most organizations are overly dependent on informational reports, especially within business functions characterized by complex planning and scheduling activities. Promotional planning is no exception, with spreadsheets and pivot tables often being the norm for pulling data into tables

and charts. Some typical examples of reporting within the context of promotional planning include:

- Sales data in $ (gross revenue)
- Sales data in $ (net revenue)
- Promotions and pricing data
- Volumes sold (units and/or kgs)
- Gross profit
- Trade spend
- Trade spend as a percentage of gross revenue

The above information may be viewed in various ways: *per promotional period, per store, per banner, per state, per product category, per brand, per product,* or any combination thereof. A typical example of data visualization is shown below, where two years of sales data are displayed in units and revenue:

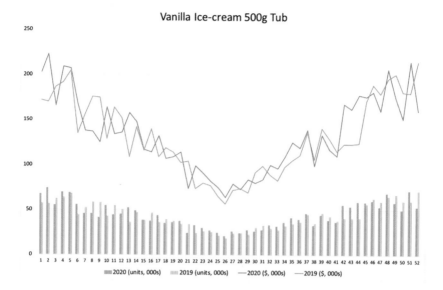

The chart shows a product where no price changes or promotions have occurred over the two-year period, so revenue for any given week is three times the units sold. The steady U-shape movement of sales going down and then up is the *seasonality effect*, with higher sales occurring in summer rather than winter (southern hemisphere data). But beyond this immediately evident seasonality effect, the remaining variability (i.e. peaks and troughs) are not easily explained, requiring further data analysis and investigation.

If we consider another snack product, but one that's promoted at 50% off every four weeks, the chart will look very different:

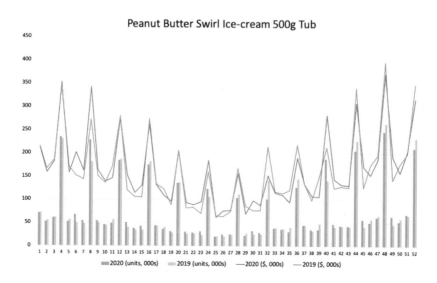

Note the pronounced volume and revenue peaks from the promotional activity every four weeks, along with the seasonality effect from summer into winter and then back again. Such reporting serves as the bare minimum for assessing the effectiveness of promotions and conducting fundamental financial analysis, but is inadequate for optimizing our decisions about upcoming promotions.

Knowledge

Knowledge is typically generated through some deeper analysis of historical data, and attempts to answer the question: *Why did this happen?* Such analysis usually involves the overlaying of several datasets, which can then be visualized in a chart or graph. For example, viewing sales data for a particular product, overlaid with promotional periods and known price changes, may reveal peaks and troughs associated with promotional activity, seasonality, and other factors.

Let's discuss this layer of the pyramid in the context of an everyday product, such as toothpaste. This product is generally used twice a day, every day, in almost the same amount. Not much affects the quantity used by consumers, as the overall demand for toothpaste is relatively fixed, and doesn't fluctuate with time of year, weather, public holidays, or other external factors. However, there are many varieties of toothpaste: regular toothpaste, anti-plaque, anti-calculus, antimicrobial, sensitive, whitening, and children's toothpaste. Most of these sub-categories have basic, mid-range, and premium versions, with some sub-categories even having super-premium versions (such as sensitive and whitening). Furthermore, multiple brands might cover most or all of these niches, and the same product might be available in different sizes.

Using data related to historical toothpaste sales, we can investigate many factors that affect demand, the most important of which is *price elasticity* (or simply *elasticity*). Elasticity has its roots in Economic Theory and is part of the Law of Demand, which states that demand for any given product will go up as its price goes down, and vice versa (with a few exceptions, such as some luxury goods and other limited circumstances). All products have an elasticity curve, such as the one shown below. Some of these curves are steep, where a small change in price causes a large change in demand, and some shallow, where a large change in price causes a small change in demand:

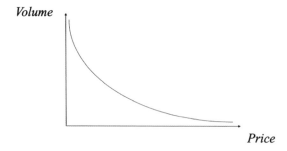

Data related to historical toothpaste prices can help us plot an elasticity curve that shows how permanent changes in price affect demand, while the historical performance of promotions at different price points can help us create a *discount elasticity* curve that shows how temporary (i.e. promotional) changes in price affect demand. Since consumers are much more sensitive to a temporary discount in price than a permanent change, discount elasticity is usually much more pronounced than price elasticity for the same product.

For example, if the toothpaste has a shelf price of $6.50 and has participated in regular "20% off" promotions, as well as less frequent "50% off" promotions, then historical sales data would almost certainly reveal significant volume increases during the "20% off" promotions and even greater volume increases during the "50% off" promotions. With enough historical data of actual promotions, we can plot a discount elasticity curve and extrapolate what the sales volume increase might be for a "30% off" promotion, even though we've never run that exact promotion before.[4]

We can also increase our knowledge by trying to understand *why* a reduction in price caused an increase in sales. For some products, a change in price may have affected a consumer's decision on whether or not to buy the product at all, so the promotion resulted in a real change in consumption. This isn't the case with toothpaste, however, as people tend to brush their teeth twice a day regardless of whether toothpaste is on promotion or not—the only difference being the brand and type used. Therefore, the real consumption of toothpaste doesn't change during promotional periods, but rather, consumers switch between products and this leads to higher sales of one toothpaste over another. However, some products such as ice-cream and champagne may experience a real increase in consumption when placed on promotion, because people not only buy more ice-cream and champagne, but they also consume more as well.

---

4   Such extrapolations are useful until we "overdo it" by promoting too frequently, which may lead to a permanent change of perception for that product's value and when consumers buy it (e.g. they never pay more than $5 for that product, so they'll wait and stock up during the next promotion).

Some products are highly seasonal, like hay fever medicines, and we may find seasonality to be the dominant factor driving sales (with our analysis possibly revealing that any increase in sales wasn't attributable to the promotion). Furthermore, we may discover long-term trends for particular products, brands, or even entire categories (as some categories might be experiencing growth while others are in a state of decline), providing us with even more knowledge of why demand increased or decreased at certain points in time.

And lastly, when toothpaste is on promotion, we can expect consumers to stock up on the product, therefore "bringing forward" future purchases. This is called the *pull-forward effect*, which influences non-perishable products (those that consumers can easily store). The pull-forward effect results in a fall in demand to below baseline levels after the promotion has ended (as shown below), and is something we need to understand and consider when planning future promotions:

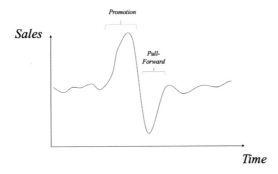

When we're planning our future promotions for Mary's Market, one of our goals is to maximize the "gain" from competitor products (so that consumers switch from a competing toothpaste to our own) and minimize the "loss" from our own product range. If an increase in promotional sales comes at the cost of another one of our products, this is called *cannibalization,* which means that consumers have switched from one of our products they regularly buy, to the one on promotion. Some common types of cannibalization include:

- *Pack-size cannibalization*: which affects sales from other pack sizes. For example, our toothpaste might be available in 100g and 175g variants, with the larger version being more economical on a per gram basis. However, if the 100g version is placed on promotion, and the larger is not, it may result in the 100g version being more economical. We would then expect sales of the 175g version to be significantly reduced during such promotional periods because consumers will "switch" to the 100g version (thereby cannibalizing sales of the 175g version), as shown below:

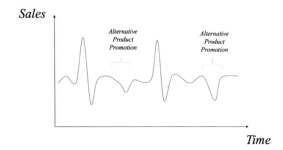

- *Sub-category cannibalization*: which affects similar products within the same product range. If we offer multiple types of whitening toothpaste, at different levels of "premium-ness" and different price points, then a promotion for premium whitening toothpaste may bring the price down on par with the mid-range version, thereby cannibalizing sales of the mid-range version.
- *Cross-category cannibalization*: which affects a different product category altogether. This is unlikely for toothpaste, but may happen for other products. For example, chocolate may cannibalize sales from biscuits, as some consumers are looking for a general snack or dessert, rather than chocolate in particular.
- *Cross-retailer cannibalization*: which occurs when consumers decide to visit a retailer where a specific product is on promotion. This type of cannibalization tends to occur with big-ticket items (e.g. television sets, coffee makers, etc.) and less so with FMCG products, however, there can still be some effect.
- *Delayed cannibalization*: is more prevalent in "treat yourself" categories, and happens when a bad-tasting product is put on special and has a delayed impact on the entire category, which could take multiple promotions to be realized. The theory being that consumers will only buy the next product from that category after finishing the one they bought, and if that product was no good, it will slow down their consumption and delay their next purchase from that category.

Through data analysis, we can create a *cannibalization matrix*, which is a table that outlines the expected cannibalization effect of certain products when placed on promotion. However, the practical challenges of constructing such a table are significant. Without applying any domain knowledge or business rules, it might be necessary to look up each individual product and calculate its cannibalizing effect on every other product in our range. For most manufacturers, which may sell hundreds or even thousands of products, this would result in a table with tens or even *hundreds of thousands of values*. While it's possible to calculate this automatically, there's no easy way to validate these values without going through them line by line.

A better approach is through the application of human knowledge along with various Artificial Intelligence methods; for example, by implementing human rules based on well-founded assumptions, such as cannibalization occurring across individual categories, we could capture 80% of the expected cannibalization effect with just 20% of the modeling effort, and then complement the cannibalization matrix by additional modifications that result from a deeper analysis of the data using AI algorithms.

There's also the opposite effect of cannibalization, where an increase in sales occurs in products that aren't on promotion, but have a complementary relationship with the promoted product (in effect, creating a *cross-sell* between one product and another, even though they aren't bundled or offered together). As an example, toothpaste promotions might cause a small but measurable uplift in sales for toothbrushes, whitening kits, and mouthwash, even if those products aren't part of the promotion. This relationship is even stronger for products that are consumed together, with the classic example being pasta and pasta sauce, where pasta sales data will reveal distinct peaks when pasta sauce is on promotion, as shown below:

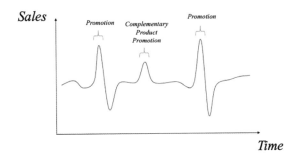

And finally, an increase in sales could be due to external factors; for example, sporting events can increase demand for beer and snacks, while hot weather is positively correlated with higher ice-cream sales. Not only is this type of knowledge important for decision making, but also forms the basis of our predictive modeling efforts.

Prediction

As discussed above, there are many demand drivers we need to understand within the knowledge layer, including seasonality, the correlation between variables, discount elasticity, different types of cannibalization, pull-forward effect, and so on. Furthermore, there may be long-term

trends at play within the product, brand, or entire category, and we can use our knowledge of these demand drivers to develop a prediction model, which brings us to the next layer of the problem-to-decision pyramid.

Prediction models use past data to make forward-looking predictions, in effect answering the question: *Based upon what we know about the past, what's likely to happen in the future?* Prediction models can be as simple as moving average models with one input variable (historical sales), or as complex as Machine Learning models with hundreds of input variables using algorithmic methods such as random forests or neural networks (which we'll discuss in Chapter 5). Irrespective of the algorithmic method used, the goal is to achieve the highest possible accuracy. For example, after testing various algorithmic methods on historical data, we might find that neural networks provide superior accuracy when it comes to predicting the outcome of promotional plans. But after further investigation and experimentation, we might find that we can improve our accuracy even further by combining several algorithmic methods together (through an approach known as *ensemble modeling*, which we'll also explore in Chapter 5.6).

Once we've created a prediction model based on whichever algorithmic method provides us with the highest accuracy, we can then take a promotional plan (let's call it *Plan A*):

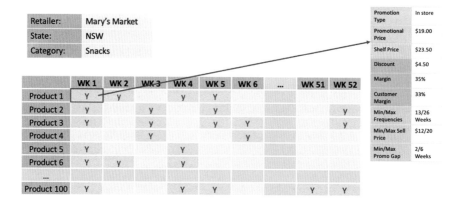

and use our predictive model to "evaluate" this plan by predicting an outcome, such as:

Volume	62,685
Net revenue	$1,034,303
Retailer gross profit	$227,547
Manufacturer gross profit	$268,919

In the table above, a simplified KPI report allows us to compare the effectiveness of *Plan A* against any other plan through the following metrics:

- *Volume*: which is the total unit quantity of products sold in whatever measure is used (e.g. individual units, cases, pallets, liters, etc.)
- *Net revenue*: which is the total revenue generated, based on the promotional sales price and volume
- *Retailer gross profit*: which is the total retailer gross profit, typically calculated as the difference between the promotional price and the retailer's cost of the product (plus any financial contribution the retailer made towards the promotion), multiplied by the total units sold
- *Manufacturer gross profit*: which is our gross profit, typically calculated as the difference between the net revenue and total product cost (including all production and freight costs, as well as any promotional co-funding costs)

If we implement *Plan A* for Mary's Market, then the predicted outcome is 62,685 units sold, $1,034,303 in net revenue, $227,547 in retailer profit, and $268,919 in gross profit. The accuracy of these predictions is of the utmost significance and highly dependent on the quality of our prediction model. Thus, the process of building and training a prediction model is far from trivial, and includes the careful preparation and analysis of data, as well as selecting the best algorithmic method for explaining the variability in question and producing consistent results. We'll return to these topics of data preparation and model building in Chapters 4 and 5.

Optimization

Once we've developed a prediction model for evaluating our promotional plans, we then need to create a number of plans and find the "best" one through a process of optimization. First, however, we need to define what "best" means to us, which in

this case might be "any promotional plan that maximizes overall volume while satisfying our business rules and constraints." Note that some of these business rules and constraints might be for the entire category, while others are just for individual products. In the snacks category, as an example, we may have the following business rules and constraints:

- No less than 30 products and no more than 60 products on promotion in any given promotional period (soft)
- The overall minimum net revenue should be $1,000,000 (hard)
- The overall gross profit growth over last year should be 3% (hard)
- The overall minimum retailer margin growth over the last year should be 2% (hard)

whereas for Product 43, we may have some additional business rules and constraints:

- Minimum promotional price of $4.00 and maximum price of $7.00 (soft)
- Price or discount step: $0.25 (soft)
- Minimum of five and maximum of eight promotional frequencies (hard)
- Maximum of three consecutive promotional periods (hard)
- Maximum of five consecutive non-promotional periods (hard)

Such business rules and constraints typically reside in the minds of human experts within each organization, and it's often a significant undertaking to extract and document them; however, such a process is highly beneficial, because it reduces key man risk, provides visibility of the rules and constraints under which decisions are made, and allows for "testing" of each rule and constraint to ensure ongoing relevance. Recall also that these business rules and constraints can be either "hard" (meaning they cannot be violated under any circumstances) or "soft" (meaning they can be violated, but it's undesirable to do so). We can then define each constraint and/or each business rule (whether hard or soft) within a table, as follows:

	Type
Min number of products on promotion	Soft
Max number of products on promotion	Soft
Min net revenue	Hard
Min retailer margin YOY growth	Hard
Gross profit YOY growth	Hard

and do the same for each product or product category:

	Type
Min freq	Soft
Max freq	Soft
Max promo length	Soft
Min promo gap	Soft
Max promo gap	Soft
Min promo price	Hard
Max promo price	Hard
Price step	Hard
Min promos p/ period	Soft
Max promos p/ period	Soft

In some cases, violating a soft business rule may result in a better overall plan. For this reason, we must apply a "penalty" to such violations, otherwise these soft rules would always be violated and cease to be rules. Also, the penalty for some soft rules can be weighted differently to others, and we'll discuss topic of applying penalties to business rules and constraints in more detail in Chapter 6.8.

There are other important considerations related to optimization. Usually we start the search for the best plan from some starting position, for example, last year's plan or a new promotional plan that we manually create. Sometimes it's desirable to restrict the type of changes made to the original plan (e.g. whether the optimization algorithm can switch between two products, removing one from promotion and adding another) or restrict the number of changes—perhaps due to retailer requirements, or to assist in software adoption and implementation, as people can become disheartened if the slotting board they've worked on for three days comes back with 150 changes! We'll discuss these issues further in Chapter 6.9.

Now that we've defined optimality as volume maximization while satisfying our business rules and constraints, we can create many promotional plans and use our prediction model to evaluate them. At the start of this optimization process, let's say we create just two promotional plans—*Plan A* (presented earlier) and *Plan B*—and then use our prediction model to evaluate each one, with the following results:

Plan A		Plan B	
Volume	62,685	Volume	59,633
Net revenue	$1,034,303	Net revenue	$1,038,027
Retailer gross profit	$227,547	Retailer gross profit	$342,549
Manufacturer gross profit	$268,919	Manufacturer gross profit	$363,309

Assuming both plans satisfy all business rules and constraints, then *Plan A* is clearly better than *Plan B* because it has a higher predicted volume (notwithstanding that *Plan B* generates better results for the other three measures—something we'll discuss in the next section). Whether manually or through a Decision Optimization System, this process could then be repeated in a systematic way until all possible combinations are considered—the so-called *brute force approach*—if it weren't for the astronomical number of possible plans we'd have to consider, deeming such an approach impossible on even the world's fastest supercomputers (as discussed in Chapter 2.1). For this reason, "smart" algorithms are required for complex business problems such as promotional planning, and this is where Artificial Intelligence algorithms can create the most value.

And lastly, as we can see, the optimization process is relatively straight-forward when we only have one objective, such as volume or revenue. To maximize this single objective, a Decision Optimization System would take an existing plan, make changes, evaluate the new plan, and continue this process in a "smart" way until the objective is maximized. This process, however, isn't so straightforward when we introduce multiple objectives.

Decision

The final decision for our promotional planning and pricing problem becomes more complicated if we need to consider multiple objectives simultaneously. This type of optimization problem is *multi-objective,* because there's more than one objective to optimize, and may result in

situations where an increase in one objective results in a decrease in another.

If we return to the predicted output of *Plan A* and *Plan B* from the previous section:

Plan A		Plan B	
Volume	62,685	Volume	59,633
Net revenue	$1,034,303	Net revenue	$1,038,027
Retailer gross profit	$227,547	Retailer gross profit	$342,549
Manufacturer gross profit	$268,919	Manufacturer gross profit	$363,309

we can see that *Plan A* has a higher predicted volume (in aggregate, across all products for the promotional period), but *Plan B* has a higher predicted total net revenue (again, in aggregate, across all products for the promotional period). If we start considering both objectives, then which promotional plan is "better"? Which one should we implement as our final decision?

In such cases, it's no longer possible to choose *the optimal* plan. Instead, we must consider a set of plans, all optimal, with some being better on one objective, while others on another. In other words, there's a trade-off between these plans, all of which are optimal in their own way. Our goal is to *understand* these trade-offs and then make the best decision. For example, a reduction in revenue should result in an increase in volume and vice versa. It's also possible to plot these promotional plans on a Pareto front (which was presented in Chapter 2.2), where all plans on the curve are optimal, meaning that it's impossible to improve any plan on one objective without suffering a decrease on some other objective. We'll also discuss multi-objective optimization problems further in Chapter 6.9.

Another complication in our decision-making process for promotional planning—even if only have a single objective—is evaluating plans where some soft business rules have been violated. As an example, let's say we create two new plans, called *Plan C*:

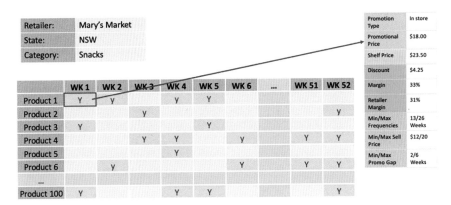

and *Plan D*:

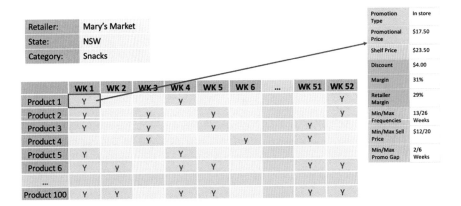

Let's also assume that the only objective we care about is volume, and *Plan C* has a predicted volume of 61,245, while *Plan D* has a predicted volume of 60,985. So, *Plan C* is better out of these two, right? Well, not necessarily.

Both plans might violate some soft business rules, but the better plan, *Plan C*, violates more of these business rules. For example, 63 products are on promotion during Week 17 in *Plan C*, whereas one of the soft rules states that no more than 60 products should be on promotion during any given period. Similar violations occur during Week 23, 38, 39, and 48. On the other hand, there are no violations of this business rule in *Plan D*. Should we still consider *Plan C* as the better plan?

To answer this question, we need to figure out how much each violation of the min/max number of promoted products during a week is "worth," and then apply a "penalty" to the plan. For example, should each violation of the business rule be worth 100 units of volume? More? Less? Also, should we consider the degree of violation? After all, both plans violate the business rule of promoting no more than 60 products during any given period, so what's the difference between 62 products and 63? If this difference is significant, then we should set a penalty that doubles with each additional product added past 60, so exceeding this business rule by a few products would result in a small penalty in comparison to exceeding it by many products. Once we've defined these penalties for our soft business rules and constraints, we can then check the plan for violations and apply the appropriate penalty. Clearly, comparing two plans in the presence of many soft rules and constraints is far from trivial, even when optimizing on a single objective.

Adaptability

Now we can see just how complex promotional planning really is! This process has all the characteristics of a complex business problem, where the number of possible solutions is astronomical, our business rules and constraints differ from retailer to retailer, category to category, and even product to product, and our objectives and KPIs often compete with one another. Underpinning this complexity is also the fundamental challenge of predicting marketplace outcomes, which has to be done as accurately as possible.

Furthermore, all of this inherent complexity exists even if we wanted to create just one promotional plan, just once. But of course, promotional planning isn't done just once, it's an ongoing process with decisions being made in a dynamic environment where the unexpected might happen (as the COVID-19 virus has recently shown). Amidst all these planning decisions, many variables are also in a state of constant flux. New products are launched, and old ones discontinued, retailers open and shut stores, supply chain costs vary, business rules are revised, shelf space and product facings change, brands grow and decline, and consumer preferences and tastes evolve.

That's why the problem-to-decision pyramid introduces the concept of *adaptability*, which leverages a feedback loop and new data to learn from the outcome of past decisions, so that we can make even better decisions in the future:

Any prediction model that's built on historical data and never updated will gradually lose its relevance over time. To maintain accuracy and relevance, the model needs to be updated with new data, and this process usually takes place on a regular but discrete basis—perhaps monthly or weekly, depending on the environment and degree of change from one time period to the next. Furthermore, it may be necessary for the model to "forget" old data that's no longer relevant, forming a moving window of data that's used for generating predictions. We'll revisit the topic of adaptability in Chapter 7, where we'll provide a more in-depth explanation.

To see a visual example of prediction, optimization, and learning within the context of promotional planning, please watch the supplementary video for this chapter at: www.Complexica.com/RiseofAI/Chapter3.

3.3 Competitor Aspects of Promotional Planning

In this chapter, we emphasized the importance of being able to evaluate a promotional plan by accurately predicting its performance in the marketplace. However, without taking into account the strategies and corresponding actions of competitors, our prediction model might produce inaccurate results; which brings us to an entire class of real-world problems where the effectiveness of our strategy is dependent on the (yet unknown) strategies of our competitors.

Interestingly, there are parallels between trying to find the best strategy for a business problem and that of finding the best strategy for a game. Within any game, there are clear rules for what moves can be made, well-defined objectives (e.g. the criterion for winning and losing), and knowledge of who the opponents are and what moves they can make and when. Although real-world situations are far more complex, with unclear rules, many competitors (opponents), and irregular decisions (moves), many similarities exist between these two environments of a game versus the real world. In both situations, we have to devise a strategy for our moves, our opponents' countermoves, as well as decision criteria for choosing one move over another. Furthermore, the process of learning is also based on trial and error in both games and the real world.

Finding the "best" strategy (which means a strategy that is most effective against the strategy of the other player) isn't straightforward, but several algorithmic methods are available for this problem. Before describing one such method, let's first explore some *co-evolutionary* processes that exist in nature to gain a better understanding of the natural system upon which these Artificial Intelligence algorithms are based.

Given that most animals in the wild face the constant challenge of survival, their defensive and offensive survival strategies are genetically hardwired as instinctive behaviors. Some species use coloration to blend into the background, and their strategy is to remain unseen. Other species have developed a strategy based on safety in numbers, while others have learned to seek out high elevations and position themselves in a ring looking outwards, thus providing the earliest possible sighting of a potential predator. These complex strategies emerged over many generations of trial and error and illustrate the process of *co-evolution*—which is not the case of one animal against its environment, but rather an entire species against other species, each competing for resources in an environment that poses its own hostile conditions without caring about which individual animals win or lose in their struggle for existence. Competing species use random variation and selection to seek out superior survival strategies that will give them an edge over their opponents. For example, through evolution and natural selection, antelopes might improve their speed and alertness over time, with slower and less alert antelopes being eaten by lions and their genes being removed from the antelope gene pool. On the other hand, lions might become cleverer over time, so they become better hunters and catch these ever faster and more alert antelopes, with the less clever lions starving to death and their genes being removed from the lion gene pool. Hence, each "innovation" from one side may lead to an innovation from the other, which is similar to an "arms race" of inventions.

We can artificially simulate these natural co-evolutionary processes through the use of *evolutionary algorithms*, which are a type of AI optimization algorithm (covered in more detail in Chapter 6.7). The premise behind this algorithmic method is to run two optimization processes in parallel, where one represents our strategy, and the other represents the strategy of another player. The evolutionary algorithm then simulates our strategy against that of the other player and evaluates the outcome. Strategies that are more effective against the strategy of the other player are then selected, "evolved" further (i.e. optimized further), until an optimized strategy is found for the game we're playing.

In the context of promotional planning, many competing organizations face a similar problem of finding the best promotion strategy to fulfill their objectives, whether maximizing volume, revenue, market share, or some other metric. Of course, the promotional strategies of other companies aren't well known, and because of that, we can't accurately measure the impact of those strategies on our promotions. For instance, if we promote a particular product nationwide during the fifth and sixth week of the year, and one of our competitors runs a promotion during the same timeframe for a similar product, then our sales volume predictions are likely to be inaccurate.

To explore this example further, say we're concerned with the promotional strategy of our major competitor. We know the number of promotions they ran last year, and how those promotions broke down into different products and promotional periods. Using this data, we can construct a "similarity matrix" that groups our products together with those of our competitor on the basis of similarity, so that we can measure the impact on our promotions—in other words, the matrix shows which of our products were negatively impacted by our competitor's promotions of similar products last year:

	WK 1	WK 2	WK 3	WK 4	WK 5	WK 6	...	WK 51	WK 52
Product Category 1	Y		Y		Y			Y	
Product Category 2		Y							
Product Category 3	Y					Y			
Product Category 4			Y					Y	
Product Category 5	Y			Y		Y			
Product Category 6	Y				Y				Y
Product Category 7					Y				
Product Category 8					Y	Y			Y

The interpretation of this "impact table" is straightforward: Some products from Product Category 1 were negatively impacted by competitor promotions of similar products during Week 1. Thus, because of this impact, we need to lower our original sales volume prediction for those products. We can go deeper, of course, improving our predictions even further by modeling promotion types and competitor pricing, but this level of detail is sufficient for now.

Even though this impact table represents our competitor's promotional strategy from last year and is unlikely to be repeated, we can use it as a starting point for the co-evolutionary process—in effect, by starting the process with our promotional plan for next year and our competitor's actual strategy from last year. By grouping all products into eight separate categories, as above, we can then model our competitor's future strategy and the impact it will have on our future promotional plan in the following way: Using evolutionary algorithms, we run two optimization processes in parallel, where the first process attempts to optimize our promotional plan (i.e. maximize volume), while the other optimization process attempts to optimize our competitor's strategy (as represented by the impact table—the better the competitor's strategy, the more impact it will have on our promotional plan). The two optimization processes then compete against each other, mimicking real co-evolutionary processes found in nature, with each process trying to "outdo" the other. In other words, one optimization process is striving to create a promotional plan that maximizes sales volume while taking into account the most damaging impact table from the other optimization process; while the other optimization process is striving to create an impact table that maximizes damage to our best promotional plan.

The "connector" between these two optimization processes is the evaluation of each plan and impact table. For example, to evaluate our promotional plan, we need to know the best impact table from the other optimization process, and then lower our volume prediction accordingly—and vice versa, to evaluate an impact table, our competitor needs to know how much damage their impact table caused to our best promotional plan. Let's illustrate this back and forth process by re-visiting *Plan A* from above:

Retailer:	Mary's Market								
State:	NSW								
Category:	Snacks								

	WK 1	WK 2	WK 3	WK 4	WK 5	WK 6	...	WK 51	WK 52
Product 1	Y	y		y	Y				
Product 2	y		y		y				y
Product 3	Y		y		y	Y			y
Product 4			y			y			
Product 5	Y			y					
Product 6	Y	y		y					
...									
Product 100	Y			y	Y			Y	Y

Promotion Type	In store
Promotional Price	$19.00
Shelf Price	$23.50
Discount	$4.50
Margin	35%
Customer Margin	33%
Min/Max Frequencies	13/26 Weeks
Min/Max Sell Price	$12/20
Min/Max Promo Gap	2/6 Weeks

To evaluate *Plan A*, we can take the best impact table from the second optimization process—as this table represents the best (current) strategy of our competitor—and then use it to measure the impact on our promotional plan and reduce our sales volume predictions accordingly. Because we know our best promotional plan at each moment of the optimization process, we can estimate the quality of our competitor's strategy on the basis of the damage it does to our promotional plan—the larger the damage, the better their strategy. In other words, we can take into account the best strategy of our competitor whenever evaluating our own promotional plan. This means that our promotional plans become better over time, against the superior strategies of our competitor! Thus, the co-evolutionary process may provide us with not only the best promotional plan, but also insights into the likely behavior of our competition.

And finally, co-evolutionary algorithms aren't limited to just one competitor. If we have three primary competitors, we can introduce three additional "players" into the "game," where each player runs a process of optimizing its strategy against those of the other players. To evaluate a promotional plan in this situation, we would take the best strategies from the other three optimization processes, which represent the best current strategies of our competitors. This allows us to improve the effectiveness and robustness of our promotional plan by taking into account the best possible strategy of our competitors, as well as upgrading our prediction model (which was originally based only on our promotional plan) by incorporating data from our competitor's impact table.

The use of co-evolutionary methods is a great example not only of Artificial Intelligence algorithms, but also of how AI scientists go about replicating natural processes in order to solve problems of great complexity (in much the same way that the natural behavior of ants is artificially replicated in order to create ant algorithms, as discussed in Chapter 1.3). In the next part of this book we'll discuss how AI algorithms are used for prediction, optimization, and learning, but because data represents a key part of the whole process, we'll begin with a chapter on data and modeling. For more information on the material covered in this chapter, please watch the supplementary video at: www.Complexica.com/RiseofAI/Chapter3.

PART II
Prediction, Optimization, and Learning

Overview

In Part I we introduced the problem-to-decision pyramid, explaining how the lower layers represent the past, whereas the upper layers represent the future. We also explained why Artificial Intelligence algorithms and systems were ideally suited for bridging these two halves of the pyramid to enable optimized decision making.

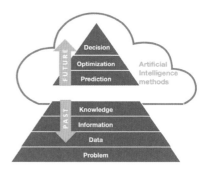

In this part of the book, we'll explore the top half of the pyramid and explain the inner workings of predictive models, optimization methods, and various learning algorithms. However, we'll open with a chapter on data and problem modeling, along with a discussion of common issues such as data availability, completeness, and preparation, because data and modeling form the basis of prediction and optimization. Then in Chapters 5 & 6, we'll review various algorithms and approaches for predictive modeling and optimization, drawing a distinction between traditional methods such as linear programming or hill climbing, and Artificial Intelligence methods such as neural networks, fuzzy systems, and evolutionary algorithms.

And finally, in Chapter 7, we'll discuss adaptability and learning, which is represented by the loop on the right-hand side of the pyramid (feedback). After all, apart from being able to predict and optimize, any Decision Optimization System based on Artificial Intelligence algorithms should also be able to learn and adapt to environmental changes!

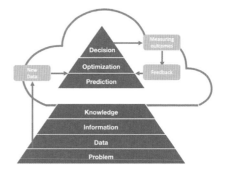

CHAPTER 4

Data

As we discussed in Chapter 2, data is the second layer of the problem-to-decision pyramid and comes directly after the problem has been defined. Many organizations have been collecting and storing large amounts of data over the years, especially as advances in technology have lowered the cost of storing, accessing, and processing  data volumes, making this kind of collection more affordable. The typical goal of such data collection is to replace decision making based on intuition and gut, with decisions that are based on data and science.

Besides using data to create information and knowledge to improve our decision making, we can also use data to build a model of the problem we're trying to solve and then apply prediction and optimization algorithms to generate a solution (which could be a predicted outcome or optimized recommendation). This means that enabling prediction and optimization capabilities in the top part of the pyramid, requires us to use the bottom part of the pyramid to create a model of the problem. In fact, solving *any* real-world problem consists of these same two steps: (1) using data to build a model of the problem, and then (2) using that model to generate a solution:

The direct implication is that whenever we solve a problem, we're only finding a solution to a *model* of the problem, and the quality of that solution is directly related to the quality and granularity of the underlying model (which in turn is often directly related to the quality and quantity of data used). For this reason, we'll first explore the process of model building, before turning our attention to data preparation, and the potential significance of external data sources, as well as dirty, incomplete, or missing data.

4.1 Modeling Considerations

The concept of modeling is quite old. François Viète, a sixteenth century French mathematician, began to use algebraic variables such as x, y, z, etc. to symbolize unknown quantities. This four-hundred-year-old method allowed a new type of reasoning: Let x be the unknown quantity, find an equation which x satisfies, and then solve it in order to find the value of the unknown quantity. In other words, build a model of the problem. We'll use Viète's approach in Chapter 6.1, when discussing an optimization puzzle that involves chairs and tables.

Because of this two-step process—of first building a model for the problem and then using the model to generate a solution—we must realize that our entire problem-solving process is based on accurately representing our problem in a model. If the model accurately reflects the problem we're trying to solve, then the solution will be meaningful. Conversely, if the model is inaccurate, or has too many vague assumptions and rough approximations, then the solution may be useless. For this reason, it's critical to distinguish between models and problems, because a model is *not* the same as the real problem—it's only a *representation* of the problem. In other words, a model "squeezes" the reality into a mathematical, compressed format that is reusable in the same situation over and over.

Furthermore, every model is "incomplete," in the sense that every model leaves something out—it has to, otherwise it would become just as complex as the real world itself. We have to accept that whenever we work with models, we're working with simplifications of the real world and the problems we're trying to solve, not an identical replica. The following story illustrates this point very well:

> *There was an engineer, a biologist, and a mathematician who were asked to improve the productivity of a milk farm. After some months of work, the engineer presented a new feeding system that increased milk production by 10%, the biologist presented a new breed of super-cow that increased milk production by 20%, while the mathematician began his presentation with: "Let's assume the cow is a sphere ... "*

A good model should be precise enough to allow for a meaningful solution, but on the other hand, it shouldn't be so complex that it's impossible to use. In other words, a good model should satisfy two important points:

(1) It shouldn't be too specific, meaning that irrelevant details of the problem are removed

(2) It shouldn't be too general, meaning that it can generate a meaningful solution to the specific problem we're trying to solve.

The first point repeats our earlier observation that if all aspects of the real world are represented in the model, then the model would become just as complicated as the real world. On the other hand (and this is the second point), we have to preserve the essential characteristics of the problem we're trying to solve, otherwise the solution we generate could be useless.

Let's illustrate these two points with a few examples: Imagine we're visiting a friend on the other side of town and need to plan the best driving route. In our planning phase, we would likely use a map, which in this case serves as a two-dimensional model of the real world. This map would allow us to derive a meaningful solution, namely, a route that will take us from point A (our place) to point B (our friend's house). Note that the map, as a model, satisfies the two requirements mentioned above: It's general enough so that irrelevant details of the real word are removed (e.g. the map is a valid model even if a tree is cut down or a new house is built), but specific enough to generate a meaningful solution (e.g. for us to generate a realistic route, the map would need to show one-way streets, highways, etc.).

This simple example illustrates a few interesting points. First of all, whenever we build a model for any problem, we have to decide what data to include in the model and what data to exclude. For example, should the map include speed limits? The exact width of each street? The number of lanes on each highway? Roundabouts? And so on—the list of such questions could be endless. A good rule of thumb is to only include data in the model that's needed for generating an accurate solution to the problem. So, if we want to use a map to find the shortest distance between points A to B, then we don't need speed limit data because we're only interested in measuring the distance, not time. On the other hand, if we want to find the fastest route between points A and B, then we must include speed limit data in the model, otherwise we might generate an inaccurate solution.

That said, some data may seem relevant, but would be impractical to include, as the model would become too complex. As an example, if we want to find the fastest route between points A and B, we may need many additional data, such as speed limits, the duration of green lights, the synchronization of green lights between consecutive intersections, traffic densities for particular times of the day, and so on. These considerations lead us to the second interesting point in selecting the right data for a model: the expected *accuracy* of the solution. A simple model (i.e. a basic map that just contains the street layout) would give us a particular route *r1* between points A and B for which the driving time on a particular day is 43 minutes. On the other hand, a more complex model—which includes speed limits, the duration of green lights on all intersections, etc.—would give us a different route *r2* for which the driving time is 42 minutes on the same day. But is this one-minute difference worth the extra effort and time needed to build and use the more complex model?

To answer this question, we have to consider several other factors:

- How often will the model be used to find the minimum driving time?
- How much time do we have to find a solution?
- What is the "cost" of each additional minute of driving time (i.e. does one minute make any difference)?

For example, suppose we have to drive *immediately* from point A to point B (so the time of finding a good solution is added to our driving time), and it takes several minutes to find the faster route *r2* because of the additional complexities of the model. In that case, it might be better to use a simpler model and find the slower route *r1* in a few seconds, and follow it! For these reasons and others, the process of building a model is still considered to be more "art" than science, because we often choose models based on what's mathematically tractable rather than what's actually a good representation of the real world.

Recall the ice-cream example from Chapter 2.2, where we showed a correlation between local temperature and consumer demand for ice-cream. During the process of building a predictive model for this product, we would need to identify the variables that correlate to changes in consumer demand, including local temperature and rainfall which explain the sales peaks (black rectangles below, showing periods of high temperature) and troughs (red rectangles below, showing periods of high rainfall). If we don't include these variables in the model, then it will be *much harder* to make accurate predictions, if not impossible!

Ice-cream sales vs. Temperature vs. Rainfall (2L Cont. VIC / SKU - 5290CXA89)

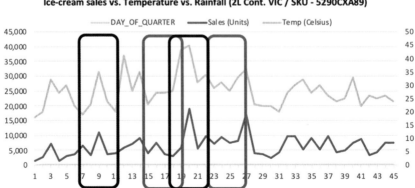

In fact, there's always the danger of missing critical variables. Let's illustrate this point with the following puzzle, which highlights our natural tendency to simplify things:

We're standing next to a door that leads into an empty room. Inside the room are three light bulbs hanging from the ceiling. All three light bulbs are switched off. On the outside wall next to the door are three switches, each of which controls a different light bulb (so there's a one-to-one connection between each switch and each light bulb). All three switches are in the "off" position. Now, our task is to discover which switch connects to which light bulb. We're allowed to play with the switches, but whatever we do with them, we can't see what's happening inside the room. Once we're satisfied, we open the door and enter the room. We can examine the room and, without leaving the room or touching the switches again, we have to determine which switch is connected to which light bulb.

Can we solve this "real-world" problem? Can we determine how to set switches A, B, and C to discover the unique connection between the switches and light bulbs? Let's start by building a model of the problem. We can represent the three light bulbs by x, y, and z, and the three switches on the outside wall of the room by A, B, and C. The task is to determine the connection between the switches and light bulbs by operating the switches in such a way that allows us to determine, for example:

- Switch A controls bulb y
- Switch B controls bulb z
- Switch C controls bulb x

So, what can we do? It seems that there are only four possibilities:

- *Possibility #1*: We can leave all switches in the "off" position. This isn't a good choice because all three bulbs will be off when we enter the room, and there will be no way to determine which bulb is controlled by which switch.
- *Possibility #2*: We can put one switch (e.g. A) in the "on" position, leaving the other two (i.e. B and C) in the "off" position. In this case, we'll know which bulb is controlled by switch A, as this will be the only lit bulb. However, it will be impossible for us to determine which of the remaining bulbs are controlled by switches B or C, as both bulbs will be off.
- *Possibility #3*: We can put two switches (e.g. A and B) in the "on" position, leaving the third switch "off" (i.e. C). In this case, we'll know which bulb is controlled by switch C, as this will be the only dark bulb in the room. However, it will be impossible to determine which of the

remaining bulbs are controlled by switches *A* or *B*, as both bulbs will be lit.

- *Possibility #4*: We can put all switches in the "on" position. This likewise isn't a good choice, because all bulbs will be lit when we enter the room, and there will be no way of determining which bulb is controlled by which switch.

So, what should we do? It seems that the problem is impossible to solve! What's wrong?

What's wrong is *the model*, as there's always a difference between any real-world problem (standing outside a dark room next to three switches) and the model (switches *A*, *B*, and *C*, bulbs *x*, *y*, and *z*, with each switch being in one of two positions, "on" or "off," and each bulb being lit or dark). If we use this model, then we'll never find the solution, because our model is missing an important variable: the temperature of a bulb! By including this variable, the problem becomes easy to solve. We can put two switches, say, *A* and *B*, in the "on" position, wait a few minutes, and put one of these switches (say, switch *B*) back into the "off" position. Upon entering the room, one bulb will be lit (this bulb is controlled by switch *A*) and the remaining two bulbs will be dark—but one of these dark bulbs will be warm (and the warm one, of course, will be controlled by switch *B*, which is the switch we turned off just before entering the room)!

This puzzle is difficult for most people to solve, because the presence of light bulbs suggests a model based only on the "on" and "off" position of switches. As discussed earlier, such a model (suggested by the description of the problem) doesn't lead to a solution, and hence the difficulty in solving this puzzle. Once we re-examine the real-world problem and discover the missing variable of temperature, we can update the model by introducing three additional variables, t_x, t_y, and t_z, which correspond to the temperatures of bulbs *x*, *y*, and *z* at the moment we enter the room. In this puzzle, it's still possible to keep the upgraded model relatively simple and consider only two values for temperature: *warm* and *cold*.

Now, one of the possibilities for this improved model (let's say possibility #7, as the number of possibilities is now much larger) is to put two switches (say *A* and *B)* in the "on" position, leaving the third switch *(C)* in the "off" position. If we wait five minutes, put switch *A* in the "off" position and enter the room. In this scenario, the lit bulb (say, *x*) must correspond to the switch *B*, which is the only switch in the "on" position. Because the two dark bulbs *y* and *z* have different temperature values, say, $t_y = warm$ and $t_z = cold$, switch *A* must control bulb *y* and switch *B* must control bulb *z*.

As this puzzle nicely demonstrates, the quality of our model affects the quality of our solution. If the model is missing key variables or has too many vague assumptions and approximations, then the solution may be useless; but at the same time, a more complex model might prove more difficult to use, and take us more time to find a better solution (e.g. it can take supercomputers months to find solutions to complex cosmological models of the "big bang," or even complex mereological models of long-term weather changes), so there's a trade-off between the quality of the model and the "cost" of finding a solution. The more complex and precise the model, the harder it may be to find a solution. On the other hand, the more realistic the model, the more confidence we'll have in the solution, which might be a key consideration for accepting a prediction result or recommendation (an example of this very point is presented in Chapter 6.1, in the context of optimization). Lastly, we can build several models of varying complexity for any given problem, in the same manner we can build several maps of varying granularity for any given area.

Because models are just *representations* of real problems, let's briefly discuss two important activities related to the evaluation of models: *verification* and *validation*. Verification is concerned with building the *model right*. During the verification phase, we compare the conceptual model to the computer representation and ask: Is the model implemented correctly? Are the input parameters and logical structure of the model correctly represented? Validation, on the other hand, is concerned with building the *right model*. During the validation phase, we want to determine that the model is an accurate representation of the real problem, situation, or scenario. Validation is usually achieved by calibrating the model, which is an iterative process of comparing the model to the real problem and using any discrepancies and insights to improve it. This process is repeated until the model accuracy is acceptable.

As we discussed in Chapter 2.3, creating an accurate model requires an iterative approach that is usually referred to as a *Data Science Methodology*. One of the most widely used methodologies, CRISP-DM (Cross-Industry Standard Process for Data Mining), includes the following steps:

- Business Understanding
- Data Understanding
- Data Preparation
- Modeling
- Modeling Evaluation
- Modeling Deployment

Whether this methodology is used, or another, the available data underpins the process of creating a model, beginning with the first step of data preparation.

For more information on how models are used to represent real-world problems, we encourage you to watch the supplementary video for this chapter at: www.Complexica.com/RiseofAI/Chapter4.

4.2 Data Preparation

As we wrote in Chapter 2.2, raw data are collected in the form of bits, numbers, symbols, and objects. Today's data is collected from an increasingly diverse set of sources and ever-growing variety of forms, where the insights we're looking for might reside within digital images, audio recordings, or real-time sensor data. Most likely, we would also need to combine and analyze more than one data source.

The first step of the model building process is data preparation, where we clean and transform raw data before further processing and analysis. It's an important step and often involves reformatting data, making corrections, as well as combining data sets together. Data preparation is often a lengthy undertaking, but essential as a prerequisite for putting data into the correct context so we can extract insights and eliminate any bias resulting from poor data quality. As every project is likely to be different and involve different data, there are no hard and fast checklists for each step of the process to ensure the data is sufficiently prepped. That said, data preparation usually involves:

- *Data collection*: which is when we pull data from data lakes, warehouses, clouds, and other services to create a large database.
- *Data preparation*: involves the cleaning and organization of data, as well as checking raw data for errors, and removing or correcting any bad data.
- *Data input*: is the loading of clean data into database destinations and transforming it into a usable format.
- *Data preparation processing*: is the processing of data for interpretation.
- *Data interpretation*: is the turning of data into a usable form such as a graph, chart, video, or text.

Rather than following these exact steps, we'll discuss a few aspects of data preparation, beginning with *variables*.

Variables

As discussed in Chapter 2.2, a typical piece of data consists of a pair (attribute, value) such as "color, red." Such attributes (e.g. color) are often called variables. Another term commonly used to describe variables is "feature." Generally speaking, there are only two "types" of variables: *numerical* and *nominal*. The values of numerical variables are numbers (e.g. a price of "$5.49" for a particular product), while nominal variables take their values from a pre-defined set (e.g. "beer," "wine," or "spirit" for the category of alcohol). Because the values of nominal variables are symbols (strings of characters), there is rarely any order between them, and mathematical comparisons and operations do not make much sense (as it's difficult to add "50" to "beer," or to compare which is larger: "beer" or "wine"). Hence, it makes sense to talk about *ordered* nominal variables, where comparisons of the type "greater than," "less than," and "equal to" have meaning.

An additional type of variable is binary (also called a Boolean or "true/false" variable), as it can only be one of two possible values, such as "yes" or "no," or "true" or "false." We may also come across other types of variables, such as those that store free text as a value, or those containing a set of values. Most models require that variables be either binary or numerical (or nominal with numerical codes as values), thus allowing some mathematical order. So what should we do with truly nominal variables, such as promotion type? Well, there are two possibilities: Either the promotion type can be coded as a unique number, or it can be converted into several binary (true/false) variables, with each variable representing a particular promotion type. For example, if the promotion type is "in store," then the variable can take on the value "true" (or "1"); if the promotion type is not "in store," then the value would be "false" (or "0").

To properly prepare the data, it's important to first identify the variable "type" (i.e. to know whether the values of a variable allow arithmetical operations or logical comparisons, whether there is a natural order imposed among them, and whether it's meaningful to define a distance between the values). For example, "very light," "light," "medium," "heavy," and "very heavy" follow a natural order, but the distance between them is not defined. On the other hand, price values have a natural measure of distance: a product priced at $5.95 costs $0.96 less than a product priced at $4.99. Because the goal of any prediction model is to generate an output (i.e. the prediction), it's important to note that the output is also a variable. In the promotional planning example from the previous chapter, the predicted output might be the volume sold or net profit, both of which are numerical variables.

In the data preparation phase, some variables may require *transformation* to be used in the model. For example, it's quite typical for "date of birth" to be recorded as a variable, but many decisions may be based on individual's age. A simple data transformation step would convert the variable "date of birth" into the variable "age" by subtracting a person's date of birth from the current date. In promotional planning, the variable "price" is transformed (i.e. rounded), and the variable "date time" is converted into a promotional period: for example, a transaction on "01/01/2019, 9:07pm" is converted into "W1 2019" (first week of 2019).

Variable selection and data reduction

Although data transformation is an important step during data preparation, *variable selection* and *variable composition* are even more important. In general, the purpose of variable selection and variable composition is to capture hidden business insights and then make the right choice on which variables to choose for the predictive model. Variable composition—which is somewhat similar to data transformation—requires problem-specific knowledge to create new variables. Because these new variables (often called *synthetic variables*) can present the existing data in a better form, they may have a greater impact on the prediction results than the specific prediction model used to produce those results. Simple examples of this include the creation of new variables such as "average unit sales per store per week" or "baseline demand," which are based on a moving average of sales data.

On the other hand, variable selection (also known as *feature selection* or *attribute selection*) is the process of selecting the most relevant variables. This process should be performed carefully, because if we don't select the right variables, then everything else—from data transformation to the final prediction model—will be useless. Conversely, selecting irrelevant variables may reduce the accuracy of a prediction model (and conversely, removing irrelevant variables usually improves the accuracy of a prediction model). This process of selecting the right variables may seem straightforward; after all, there is a finite number of variable subsets,[1] so we can examine them all and select the best one! Unfortunately, it's not quite that simple. First, the number of possible subsets may be too large. For a database with "only" twenty variables, there are over one million possible subsets. Second, to evaluate each subset, we will need to build a prediction model and evaluate it by measuring the prediction error (we'll discuss this in detail in Chapter 5.6, along with some other validation issues). So, what can we do?

Although the best way to select the most relevant variables is still manual (based on problem-specific knowledge), there are some automatic methods for variable selection. Say our prediction problem is one of classification, where we need to accurately classify transactions as either "fraudulent" or "legitimate," and we're trying to evaluate the usefulness of particular variables such

---

1   The concept of subset is simple: it's a set of elements where all those elements are contained in another set. For example, a set of five numbers: {1, 2, 3, 4, 5} has many subsets, e.g. {1, 2}, {1, 3, 4}, {2}, {1, 3, 4, 5}. Actually, the number of all possible subsets (including an empty subset) of these five numbers is 32.

as "time of transaction," or "transaction amount" for predicting the outcome. One of the most popular automatic methods is based on *means* and *variances*, where the means of a variable are compared for the two classes of datasets ("fraudulent" and "legitimate") using a simple statistical test to see whether the difference is likely to be random or not. Small differences in means usually imply that the variable is irrelevant. This method evaluates the variables one by one and should be done before the development of any prediction model.

On the other hand, we could use an automatic method where the variable selection process is an inherent part of building the prediction model. For example, when a decision tree is built (which we'll discuss in the next section), the relevant variables are selected one by one during the tree-building process. We can also use automatic methods to evaluate the entire subset of variables. Many optimization techniques discussed in Chapter 6 would be appropriate for this type of approach, as the variable selection problem is really an optimization problem of finding the optimal subset of variables. There are also a few methodologies for variable selection—such as filter methods, wrapper methods (with forward selection, backward elimination, and/or recursive variable elimination), embedded methods, and so on—but a discussion on these methods is outside the scope of this book.

Many benefits can be derived from proper variable selection, including:

- Faster algorithmic run time
- Simpler predictive model (that's easier to use and interpret)
- Improved accuracy of the model
- Reduced probability of model overfitting[2]

Because the variable selection process removes redundant and/or non-productive variables, we can also consider this step as a form of *data reduction*, the goal of which is to delete nonessential data (as the data may be too large for some prediction models and/or the expected time for building a model might be too long). If we suppose that all data is represented in a table, we can: (1) reduce some variables (columns) in the table, (2) reduce some values present in the table, and/or (3) reduce some records (rows) from the table. We have already discussed the removal of some variables, which is equivalent to the task of variable selection, so let's move on to reducing values.

---

2   A model is "overfit" when it has too many input variables so it "memorizes the noise" instead of finding the true relationship between cause and effect. Such an overfit model would then make predictions based on that noise. It will perform unusually well on its training data and very poorly on new, unseen data. In other words, overfit models don't transition well from historical, training data to new data. Chapter 5.1 provides a more in-depth explanation of this topic.

It's often helpful to discretize a numeric attribute into a smaller number of distinct categories—the main reason being that a variable takes on too many values. For example, the variable "price" (with thousands of possible price values) could be simplified by grouping all individual price values into a few groups of "below $1.00," "between $1.00 and $2.99," "between $3.00 and $4.99," and so on, up to "$50.00 or more." This looks natural, but how can we be sure that such discretization is any good? Moreover, what's a good way to discretize numeric variables into categories? As usual, there are a few approaches to consider. One approach would be to discretize an attribute by rounding: The actual price of the product can be rounded off to the closest 25 cents, thus $1.82 would become $1.75. Another approach would be to create some number of discrete categories (say, 20), and distribute all values to these categories in such a way that the average distance of a value from its category mean is the smallest. For example, the first category may contain prices from $0 to $1.28; the second category may contain prices from $1.29 to $2.71, and so on. Some mathematical methods (such as *k-means clustering*) can deliver near-optimal solutions for such distributions. However, this approach might be a bit risky for time-changing data such as price, as new price data comes regularly and the optimal price distributions might change frequently.

Yet another approach for data reduction is the reduction of data records,[3] as the number of records is often the largest dimension of the data, and it's not unusual to have hundreds of millions of records containing 20 to 30 variables each. This doesn't mean, however, that the process of record reduction is easy. In fact, just the opposite is true: very often, record reduction is the hardest type of data reduction to perform. The general approach for handling record reduction is based on random sampling. Rather than using all the records to build a prediction model, random samples are used instead. Two popular techniques for random sampling include:

- *Incremental sampling:* is where we train the model on increasingly larger random subsets of records, observe the trends, and stop the process when we can't make any further progress.
- *Average sampling:* is where we draw several samples of the same size from the data, create a prediction model for each sample, and then combine the outputs of all the models by voting or averaging (more on this in Chapter 5).

---

3   A record (also called a structure, struct, or compound data) is a basic data structure; records in a database or spreadsheet are usually called "rows". A record is a collection of fields, possibly of different data types, typically in fixed number and sequence.

Other considerations in data preparation

While discussing data preparation, it's also worth mentioning other aspects of this stage, as some problems require *data normalization* (e.g. scaling some values to a specific range, say from 0 to 1). The goal of normalization is to change the values of numeric columns in the dataset to use a common scale, without losing information or distorting any differences in the range of values. Normalization is often applied as part of data preparation for model building, and in most cases, we normalize data if we're going to be using statistical methods that rely on normally distributed data. But not every dataset needs normalization, as it's only required when variables have very different ranges. For example, if we consider a dataset containing two variables, age and income, the range of age is from 0 to 100, whereas income ranges from $0 to $500,000 (and possibly higher). Income values might be about 5,000 to 10,000 times larger than age, so these two variables are in very different ranges. When we do further analysis, like multivariate linear regression, income will influence the result more due to its larger value, but this doesn't necessarily mean that it's the more important variable.

Additional aspects of data preparation are connected with time-dependent data. Because all orders, deliveries, and transactions have some sort of a time-stamp, and most real-world business problems have some time-dependent relationships within their data. Even relatively "stable" data—such as bank customers—change over time. Of course, these changes happen at a much slower rate than changes in the stock market, but they happen nonetheless. Thus, this additional dimension of time—in addition to records and variables—requires the prediction model to be updated at regular intervals. This can be done "live" as new data arrive, or "in batches," by analyzing new data at regular intervals and then updating the prediction model. We'll return to this topic in Chapter 7, where we discuss the process of updating a prediction model.

Time dependencies should be recognized and dealt with during the data preparation phase. Usually, time series models assume that the values for some variables are recorded at fixed intervals. For example, we can record the US Gross Domestic Product at the end of each quarter, the Dow Jones Industrial Average at the end of each business day, the temperature at some location every hour, and so on. However, our time series is far less regular for promotional planning, because products are promoted at different times in different regions with a different subset of other products. So even if we find several exact data points from the past, they may not correspond to regular time intervals. For example, Product 79 might have generated a volume of X during a promotional period in mid-April, a volume of Y during a promotion period in late June, and volume of Z during a promotional period in late August—but we need to make a volume prediction for mid-October or early November. In other words, we're not predicting the value of a variable for the "next" time unit.

Another important issue related to time-dependency is the *time horizon* of the historical data. Simply put, we have to decide how far back to look. It seems natural that we should pay more attention to recent data, as "old" data may have lost their relevance. For example, using pre-COVID 19 virus data to predict air traffic for late 2020 would not yield good results.

Some consider preliminary (exploratory) data analysis to be a part of the data preparation phase, while others consider it a separate stage of the data mining process. In either case, such an analysis is helpful for gaining a better understanding of the data. Preliminary data analysis usually includes graphing the data for visual inspection (e.g. we can graph the prices for particular products based on seasonality and geography), and compute some simple statistics such as averages, minimums, maximums, means, standard deviations, and percentiles for each dataset. We can also use *decomposition analysis* to detect trends, seasonality, and cycles (we'll discuss decomposition in Chapter 5.1), as well as *outliers*, which are values that are abnormally dissimilar from the rest of the data. In a sense, this definition leaves it up to the analyst (or domain expert) to decide what is considered "abnormal." However, the classic definition of an outlier is any data value that lies more than 1.5 standard deviations from the mean. Note that outliers are "innocent until proven guilty," meaning they shouldn't be removed unless there's a good reason to do so.

The main purpose of carrying out this preliminary analysis is to get a "feel" for data. This stage is vital, as understanding the data is fundamental to any further activities related to data preparation. One of the more powerful approaches for getting a better appreciation of the data is visualization, which we'll discuss next.

Data visualization

Data visualization is just a graphical representation of data and information. By the same token that a picture is worth 1,000 words, a visual representation of data might be worth 1,000 numbers. This is because our eyes are drawn to colors and shapes, and we can quickly identify yellow from blue, and rectangles from circles. And data visualization represents another form of visual colors and shapes that grab our attention and interest. When we see a chart, we can quickly see trends and outliers, versus staring at a massive spreadsheet and being lost in the data and details. So, by using visual elements like charts, graphs, maps, scatter plots, and more, data visualization provides an accessible way to see and understand trends, outliers, and patterns in data. In short, meaningful visualizations should make it easier for us to understand the data, and potentially show connections and relationships that are too complex to explain with words.

Probably the simplest way to show the relationship between different features is by using a two-dimensional diagram based on lines, bars, or dots to display the correlation between two variables (one column versus another). For example, in the following graph:

Price vs Volume

there is a negative correlation between price and volume: the higher the price, the lower sales volume. Different colors are used for different products.

Furthermore, it might also be of interest to display the relationship between price, volume, and net revenue for past promotions, as the highest price (resulting in the highest margin) does not correspond to the largest net revenue (very high price may result in volumes close to zero:

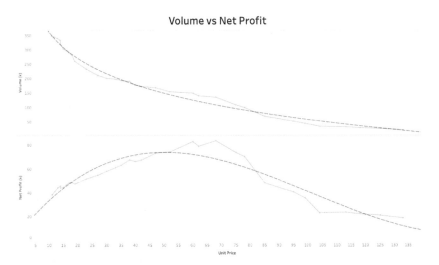

Volume vs Net Profit

As the trend line indicates, the net revenue initially grows with the price, then the highest value of net revenue corresponds to some mid-point of the price, and then it drops again.

Representing data via maps is a rapidly developing area of visualization that has many practical applications. A heat map is one of the most popular map views, which shows how certain activities are distributed across an area. We can use this type of view to display sales volumes, churn rates, customer locations, and many other characteristics. For example, in the following graph:

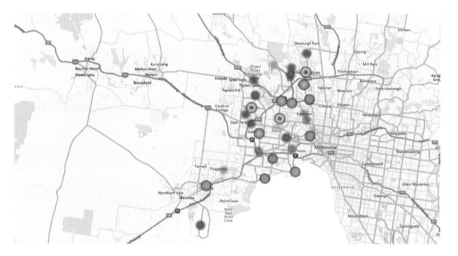

we can see sale volumes in the greater Melbourne area; while the following graph displays customer churn rates (with the color red indicating that churn was higher than the mean rate, in green):

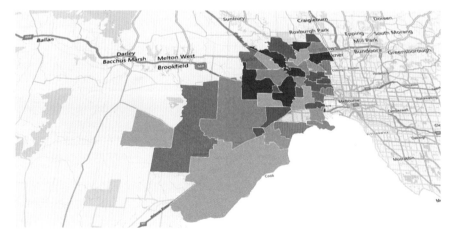

With all colors and shades in place, a data analyst can easily spot areas of interest for further investigation. To show multiple variables at once, we can use an enhanced 3D map view like the graph below, which shows the result of various promotions in Melbourne over some time period:

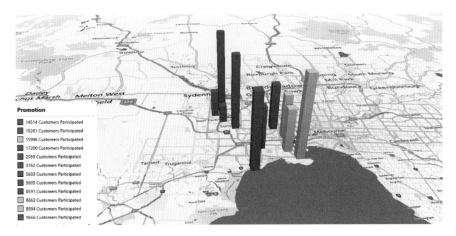

Each bar corresponds to a suburb and the color indicates the type of promotion, while the height reflects the number of customers that participated. To include some other characteristics, we could change the shape or size of the bar.

As these examples show, data visualization can assist in communicating information clearly and efficiently, and also in a way that facilitates further analysis.

Data availability

It would be remiss to talk about data preparation without revisiting the concept of external data, as raised in Chapters 2 and 3. It's quite possible that during our preliminary data analysis or some other activity undertaken during the data preparation stage, we might uncover variability in the data that can't be explained. In such cases, we could consider the hypothesis that external factors are driving this variation. In such circumstances, we might look beyond the boundaries of the organization and consider problem-specific external data (publicly available or purchased) in search of further key variables that should be included in the model.

External data might be essential during the data preparation process, as the following puzzle illustrates: What's the missing letter (marked by "?") in the following sequence:

A ? D F G H J K L

We would analyze this sequence based on characterizing each symbol by its physical features. For example, the letters **A**, **F**, **H**, and **K** each consist of three lines. The letter **L** consists of only two lines, whereas letters **D**, **G**, and **J** include curved lines. Is there any pattern to this? Or is it more important to be more specific and distinguish between longer and shorter lines? For example, the letters **A**, **F**, **H**, and **K** consist of three lines: two longer lines and one shorter one. Is that useful? Or maybe we need a different approach based on numbers, because there's a relationship between letters and numbers in this sequence (as **A** is the first letter, **B** is the second, etc.). So, we can translate the sequence in question into the following sequence of numbers:

1 ? 4 6 7 8 10 11 12

Clearly, this is a growing sequence and the growth is a pattern. If we believe that the pattern is genuine (i.e. it didn't happen by chance), then we can conclude that the second number must be **2** or **3**, thus the missing letter is **B** or **C**. But which of these two?

Actually, none of them … The "obvious" answer is: **S**, as the sequence

A S D F G H J K L

is the middle row of letters on a computer keyboard! This puzzle illustrates the fact that many data analysis activities are based also on some *a priori* knowledge. The puzzle was difficult as the knowledge extracted from the presented sequence wasn't that helpful.

In light of this puzzle, it makes sense that the question of what external data to use is problem specific, and usually addressed through an iterative process (as it might not be immediately obvious). As an example, if we want to predict which cross-sell offers will have the highest likelihood of acceptance by hospitality venues in the food services industry, the process of searching for external data might involve asking ourselves a few key questions: *Why do certain venues (i.e. bars, restaurants, cafes, etc.) sell more or less of any given product at different times? What products are used together? What demographic of the market does each venue serve?* In this context, external factors might include weather, events, seasonality, public holidays, and so on, all of which can be captured from holiday calendars, bureau of meteorology data, and census data. We can go even further and overlay each venue's address with online data about opening hours, number of tables, venue type, social media popularity, and so on, to develop a "profile" for each venue (as both datasets are publicly available). We could then use these profiles to refine our cross-sell offers, thereby enabling a higher degree of personalization and prediction accuracy.

The application of external data brings us to the broader topic of data *completeness,* where we must deal with potentially incomplete data, ranging from missing fields to entire datasets that might be required for the model— for example, to explain non-seasonal variability. Under such circumstances, we should turn to other available data, internal or external, and look for existing fields that could be used as a proxy for the data we don't have. After all, making accurate predictions for promotional planning requires data to answer questions such as: *What products have been sold? At what point of time? How many units? At what price point?* Without point of sale data to answer these questions, we could use *sell-in* data as a proxy, which captures the transactions between the manufacturer and retailer. Once we have sell-in data, we could use our domain knowledge to derive sell-out data via a set of relationships; for example, in the case of perishable products, we might know that X% of products are sold within Y time from being sold in, and so on. We could also use agent-based simulations to augment the available sell-in data, by creating a simulation of shopper behavior and then using the output of this agent-based simulation to "fill in" the data we don't have (we'll discuss this approach further in Section 4.3).

Since there's no silver bullet for dealing with missing data, it's difficult to provide a general solution to this problem. Besides using external data, another possibility would be to ignore all records with missing values (so-called: *listwise deletion*). This might be an option in cases where the missing data is limited to a small number of records, so eliminating this data from the analysis is minimal. However, in some datasets we might lose over 90% of the records by doing this! It's also possible that listwise deletion might introduce some biases in the dataset, as missing values usually aren't distributed randomly (e.g. they depend on some other variables). *Pairwise deletion* (available-record analysis), on the other hand, attempts to minimize the loss that occurs in listwise deletion. Here, for each pair of variables for which data is available, the correlation coefficient will take that data into account. Thus, pairwise deletion maximizes all available data on an analysis-by-analysis basis. The advantage of this method is that it enhances our analysis (so is typically preferred over listwise deletion), however, it also assumes that the missing data are MCAR.[4] Finally, we can consider *dropping variables* rather than records—for example, when the variable is missing from more than 60% of records. This may only work, however, if the variable isn't important.

Because it's better to keep data than discard it, we may consider *imputation methods*, which should be used carefully. In this group of methods, the simplest way of approaching the problem of missing values is to replace them with the variable's mean, median, or mode value. This might be tempting, but it's risky

---

4   Missing data are MCAR (Missing Completely at Random) when the probability of missing data on a variable is unrelated to any other measured variable and is unrelated to the variable with missing values itself.

because the data might become biased (e.g. mean imputation would reduce variance in the dataset). Instead, it's safer to observe a relationship between the variable in question and some other variables, and then replace the missing value with an estimated value. For categorical variables, we can consider mode imputation, or even create a predictive model to estimate the missing values (the *k nearest neighbor* method, discussed in Chapter 5.1, is quite popular for this). And for time series types of problems, we could use linear interpolation (with or without seasonal adjustment). There are also methods such as *last observed carried forward* and *next observed carried backward* for cases when samples have been taken at different points of time, but these methods can introduce bias when data has a visible trend.

We'll also revisit the issue of data availability in Chapter 12, as well as in Part III of this book.

Data cleaning

Thus far, our discussion on data has been underpinned by one silent assumption: that we're dealing with "clean" data. Unfortunately, this often isn't the case. During the data cleaning process, we must:

- Detect (and correct or remove) corrupt or inaccurate records
- Identify incomplete, incorrect, inaccurate, or irrelevant parts of the data
- Replace, modify, or delete the dirty or coarse data[5]

Unsurprisingly, this process (sometimes called *data cleansing*) might be quite demanding, and can sometimes be 90% of the entire data preparation effort! There's no need to argue about the importance of data cleaning, because if we use dirty data to build a model, then we can't expect quality results. But what exactly is "clean" data? In this section, we'll consider a few measures of cleanliness, including:

- *Validity*: meaning that the data must conform to defined business rules and/or constraints. Inaccurate values usually arise from typographical errors. Some of these errors can be "discovered" by analyzing the outliers for each variable, but some of them may be difficult to find. In particular, data values for a particular attribute must be of a particular data type (e.g. integer number, date, Boolean) and values of some attributes must come from a predefined set of discrete values (e.g. male, female). Furthermore, the values should fall within a certain range (e.g. the value of "30 February" is clearly wrong, as is the value of "319" for someone's

---

5   Data is considered to be "coarse" when we don't observe the exact value of the data, but only some set (a subset of the sample) that contains the exact value.

age). Some text fields must be given in a particular format (e.g. date of birth: DD/MM/YYYY).

- *Identifiers, references, and dependencies*: are certain attributes that can't have a null value; for example, the "Product ID" can't be null, otherwise the remaining data fields, like price, are meaningless. When dealing with multiple datasets, we usually need unique identifiers that are a field (attribute), or a combination of fields (attributes) that are unique across the whole dataset. Some reference rules must be met.[6] Finally, there might be some dependencies between various data fields; for example, the price of a case (with 24 bottles) can't be lower than the price of a 6-pack.

- *Accuracy*: means the data values are close to their true values. Validity is the first step towards accuracy, as defining all possible valid values allows invalid values to be easily spotted. However, a valid piece of data might not be accurate. For example, the date of birth "25/10/1991" might be valid, but might be inaccurate because the referenced person was born on 25/10/1992.

- *Completeness*: was discussed towards the end of the previous section, and missing data may occur for a variety of reasons. Incompleteness is almost impossible to fix with any data cleansing method, because we can't infer facts that weren't captured in the first place. However, as discussed above, several methods can be applied to situations of missing data.

- *Consistency*: meaning the data must be consistent within the same dataset or across multiple datasets. It's easier to explain "consistency" by first explaining "inconsistency," which occurs when two values in the dataset contradict each other. The simplest example of inconsistent data is when the same person or product (or some other object) has two different values for the same attribute (such as different addresses). In many situations, fixing inconsistent data might be challenging and require different methods (introducing measures of data reliability and data recency, not to mention manual methods, such as investigating certain products or other objects).

- *Uniformity*: means that the data should be specified using the same unit of measure. For example, weight may be recorded in pounds or kilos and dimensions in centimetres or inches. We can usually deal with issues of data uniformity with ease.

---

6 For example, the so-called foreign-key attributes can't have a value that doesn't exist in the referenced primary key.

In summary, data cleaning is of utmost importance, as incorrect or inconsistent data would lead to false conclusions. Consequently, how well we clean and understand the data has a significant influence on the quality of our results.

The final remark related to data cleaning is that no matter how detailed the cleaning process, it needs to continue when new data comes in (recall the left loop in our problem-to-decision pyramid). In anticipation of such updates that would happen at regular intervals in the future, we should impose some standards on the new data. And with modern data-capture systems, many data cleaning aspects are relatively easy to implement; for example, invalid data arises mainly in legacy contexts where constraints were not implemented properly or where inappropriate data-capture technologies were used, such as spreadsheets, where it's difficult to limit what a user chooses to enter into a cell.

4.3 Less Data, More Complexity

Sometimes we have to make a decision where only a limited amount of data is available, or even worse, no data at all. And by "no data," we don't mean that the data is missing or incomplete (as we explored earlier in this chapter)—we're referring to an entire category of problems for which the data simply doesn't exist. Some examples include:

- Predicting demand for brand-new products where there is no historical sales data to reference, and no equivalent (or similar) product in the marketplace to use as a proxy.
- Predicting the effectiveness of new advertising strategies or product pack sizes, even though that type of strategy or pack size has never been tried before.
- Understanding the performance of new product designs that have never been created before.
- Evaluating government policies for which there's no historical data or precedent; for example, devising evacuation strategies for major cities that have never been evacuated before.

On top of that, these types of problems might include many variables that interact with each other, as well as variables with uncertain values. In such cases, the model might become too complex, with too many "moving parts" and non-linearities that makes analysis too difficult to perform. Only our limited experience might guide us towards a reasonable decision, but we can never be sure whether our decision was the best one under the circumstances. Which raises the question, is it possible to "systemize" our experience in such a way that allows us to test the outcome of many possible decisions? The answer is yes, and to do so, we can turn to *simulation-based methods*.

A simulation is an imitation of something real, whether a process, state of affairs, or something else. In other words, the word *simulation* is defined as the imitation of the functioning of one system or process by means of the functioning of another (e.g. a simulation of an industrial process, military logistics, or traffic patterns, etc.). The act of simulating something real generally requires representing its key characteristics or behaviors, and can be applied to many areas—from physiological systems to war games to safety engineering—to gain insight into the system or process being simulated and also into the effects of alternative conditions and courses of action.

Monte Carlo simulation

A computer simulation is a software program that attempts to imitate a real-world problem and provide predictions on possible outcomes. Every software program (including simulation software) requires a certain number of inputs, and then uses equations to provide a set of outputs (also called the *response variables*). Many computer simulations are *deterministic*, meaning we get the same result no matter how many times we re-run the program. A classic example is simulating compound interest, where we always get the same result for the same investment amount and interest rate. In such cases, we're using a deterministic model of the problem.

Other types of simulation are *probabilistic*, and model the probability of different outcomes in a process that cannot easily be predicted due to the intervention of random variables. Probabilistic simulations are used to understand the impact of risk and uncertainty in prediction and forecasting models, and can tackle a range of problems in virtually every field such as finance, engineering, supply chain, and science.

The best-known method for probabilistic simulation is *Monte Carlo simulation*, which iteratively evaluates a deterministic model using sets of random numbers as inputs. This method is often used when the model is complex, nonlinear, or involves more than just a couple of uncertain inputs. A simulation can typically involve over 10,000 experiments, a task which in the past was only practical using supercomputers. The term *Monte Carlo method* (defined as a technique that involves using random numbers and probabilities to solve problems) was coined by Stanisław Ulam and Nicholas Metropolis in reference to games of chance, which are a popular attraction in Monte Carlo. The concept of Monte Carlo simulation is quite general, and the method has universal applicability to a variety of problems in economics, environmental sciences, nuclear physics, chemistry, and logistics, among others.

Monte Carlo simulations work by sampling the values of a model's variables (from their predefined probability distributions), generating many scenarios, and calculating the outcomes. In other words, Monte Carlo simulation is just one of many methods for analyzing how random variation, lack of knowledge, or error affects the sensitivity, performance, or reliability of the problem being modeled.

The basic steps in Monte Carlo simulation are straightforward:

- Create a model of the problem that includes some number of input variables
- Generate a set of random values for the input variables from a probability distribution that most closely matches the available data or best represent our current knowledge
- Evaluate the model and record the results
- Repeat the last two steps many times
- Analyze the results using histograms, summary statistics, confidence intervals, etc.

To illustrate how Monte Carlo simulation can be applied to a dataless problem, let's imagine that we're playing blackjack in a casino and get an ace and 6. The dealer's up card is a 6. What should we do? Hit, stand, or double? The problem is "dataless" because we don't know what hands were dealt at our table over the last hour or day or month, so we can't use any historical data to build a prediction model. The only data we have are the cards on the table, how the game is played (the rules), and how many decks of cards are being in use by the dealer.

Before discussing the use of Monte Carlo simulation to help us make a decision, let's briefly review the basic rules of blackjack. We are betting that we'll have a better hand than the dealer, with the "better hand" being one where the sum of the card values[7] is closest to 21 without exceeding 21. If our cards total 22 or more, we automatically lose (in casino terms, we "bust"). If our first two cards total, say, 13, and the dealer's up card is a king, we can call for another card ("hit"), hoping for an 8 or less. If the next card is an ace, which counts as 1 or 11, we can hope for a 7 and hit again. However, if the value of our cards is 17 (either with two cards or after hitting), then trying to improve our hand is risky and it's better to "stand." Most casino rules state that

---

7   The value of cards two through ten is their pip value (two through ten). Face cards (Jack, Queen and King) are all worth ten. Aces can be worth one or eleven. A hand's value is the sum of the card values.

the dealer must stop at 17. If our hand is 17, we have to hope the dealer has to stop at 17 or goes bust. If the dealer equals our hand, then it's a draw (in casino terms, a "stand-off"). If the dealer busts, we'll win an amount equal to our bet. In some situations, it's worthwhile to "double," where we can double our bet and get just one extra card. We should usually do this in situations where we initially have 10 or 11 points and hope to get a 10, making the total 20 or 21. We can also consider "splitting" if our two initial cards are of the same value, by doubling our bet and splitting our original hand into two separate hands, and then proceed playing with those two separate hands.

There are many variations of blackjack with additional possibilities (e.g. surrender, insurance), but we'll not go into these; the above information is sufficient. At every stage of the decision-making process, we're limited to one of four decisions: hit, stand, double, or split (the last two decisions can only be made under certain circumstances). With this information, we're ready to address the question and search for the best decision.

By using Monte Carlo simulation, we can generate millions of distributions for some number of decks of cards (say, eight decks, which in casino terms, a "shoe"), implement the dealer's rules (e.g. hit on 16 or below, stay on 17 or higher), implement our own decisions (e.g. hit, stand, or double), and then calculate the number of times we win or lose. After running this simulation, we would find that when holding an ace and 6 against a dealer's 6, we should double to maximize our winnings. We may also discover some other "rules" for optimizing our decisions, such as "always splitting two 8's," "doubling on 11 when the dealer's hand is lower than 10," and so forth.

In short, the Monte Carlo simulation would generate a table of the best possible decisions, with "best" being the decision that gives us the greatest chance of winning. Such a decision table for a multi-deck game is shown on the next page. This table can be interpreted in the following way: The total of our two initial cards is displayed in bold in the left-hand-side column of the table, along with all special cases, which include pairs we can split, and aces, which can be counted as either 1 or 11, while the dealer's single card is displayed in bold in the top row of the table. The intersection of the row (determined by our hand) and the column (determined by the dealer's hand) provides us with the best decision for a specific scenario (in order to maximize our chances of winning). For example, if we hold a king and 5 (total of 15) against a dealer's 6, the best decision is to stand (S). On the other hand, if we hold king and 5 (total of 15) against a dealer's 7, the best decision is to hit (H), and then after getting a new card, we can refer back to the same table to determine the next best decision (unless we bust, of course).

Monte Carlo simulation can also be used to calculate a variety of other

	2	3	4	5	6	7	8	9	10	A
8/less	H	H	H	H	H	H	H	H	H	H
9	H	dbl	dbl	dbl	dbl	H	H	H	H	H
10	dbl	dbl	dbl	dbl	dbl	dbl	dbl	dbl	H	H
11	dbl	dbl	dbl	dbl	dbl	dbl	dbl	dbl	dbl	H
12	H	H	S	S	S	H	H	H	H	H
13	S	S	S	S	S	H	H	H	H	H
14	S	S	S	S	S	H	H	H	H	H
15	S	S	S	S	S	H	H	H	H	H
16	S	S	S	S	S	H	H	H	H	H
17/more	S	S	S	S	S	S	S	S	S	S
A 2	H	H	H	dbl	dbl	H	H	H	H	H
A 3	H	H	H	dbl	dbl	H	H	H	H	H
A 4	H	H	dbl	dbl	dbl	H	H	H	H	H
A 5	H	H	dbl	dbl	dbl	H	H	H	H	H
A 6	H	dbl	dbl	dbl	dbl	H	H	H	H	H
A 7	S	dbl	dbl	dbl	dbl	S	S	H	H	H
A 8	S	S	S	S	S	S	S	S	S	S
A 9	S	S	S	S	S	S	S	S	S	S
2 2	H	H	spl	spl	spl	spl	H	H	H	H
3 3	H	H	spl	spl	spl	spl	H	H	H	H
4 4	H	H	H	H	H	H	H	H	H	H
5 5	dbl	dbl	dbl	dbl	dbl	dbl	dbl	dbl	H	H
6 6	H	spl	spl	spl	spl	H	H	H	H	H
7 7	spl	spl	spl	spl	spl	spl	H	H	H	H
8 8	spl	spl	spl	spl	spl	spl	spl	spl	spl	spl
9 9	spl	spl	spl	spl	spl	S	spl	spl	S	S
10 10	S	S	S	S	S	S	S	S	S	S
A A	spl	spl	spl	spl	spl	spl	spl	spl	spl	spl

blackjack statistics. For example, if we have a hand of 16 points and want to know the probability of busting if we take another card, we can run many Monte Carlo simulations and generate the following table:

Hand value	% bust if hit
21	100%
20	92%
19	85%
18	77%
17	69%
16	62%
15	58%
14	56%
13	39%
12	31%
11 or less	0%

If we bring a printout of this table or the previous decision table to a casino (which is okay in most casinos), sit by a table where multi-deck blackjack is played (usually a combination of 6 or 8 decks of cards), and follow all the recommendations, our odds are just below 50%—the house advantage is quite small and this is why blackjack is often called the "fairest" casino game!

The key point in understanding the above table is that the decision recommendations are based on a simulation with several *complete* decks of cards. However, as pointed out earlier, the probability distribution changes after every hand, as the deck becomes smaller and smaller, and the remaining cards are not uniformly distributed—if 104 cards are left in the shoe, we cannot assume that these contain 8 aces, 8 kings, 8 queens, etc. To optimize our chances of winning at blackjack, it would be ideal to:

1. Start with a full shoe, where the set of available cards would consist of, say, 8 complete decks
2. Create a decision table like the one displayed earlier
3. Play a single hand and make decisions according to the recommendations of the table
4. Remember all previous cards that were dealt
5. Mentally subtract the current set of available cards from those that have already been dealt
6. Update the decision table accordingly
7. Repeat steps 3–6 for as long as possible (e.g. until we are thrown out of the casino!)

By updating the decision table, we can change the odds of the game to our favor and beat the house, especially at the later stages of the game when the set of available cards is relatively small! The only drawback of this "method" is that steps 5 and 6 are quite challenging, to say the least. Step 5 might be doable if we have a great memory: if we succeed, we could maintain the current set of available cards in the shoe. However, the next step (step 6: update the decision table accordingly) is beyond the capability of human beings, as such an update would require *many* simulations (unless you are the Rain Man, of course).

As indicated earlier, we can apply simulations to a wide variety of problems, ranging from economics to logistics, and are useful when the problem is dataless or other types of analysis prove too difficult. As a simple example, we could use Monte Carlo simulations for estimating transportation times within a supply chain operation. To do this, we would begin by estimating the transportation time between two locations as a function of weather (e.g. bad weather usually increases the transportation time). We can then create a lookup table for the transportation time between two distant cities, which might look like the following:

Weather Condition	Time (in hours)
Fair	36
Very hot	40
Windy	38
Rain	44
Snow	48
Ice	56

If we can estimate weather conditions based on historical data (e.g. 10% chance of ice, 30% chance of rain, and 60% chance of fair weather), we can then simulate many scenarios for how this will affect the transportation time. And because different weather patterns occur with higher or lower probabilities (at different seasons), we may generate different scenarios with different weather patterns that follow the same probability distribution as in nature. Then, by including the total transportation time for each of these scenarios, we can use the average in our decision-making process.

Of course, the above example explains only the general idea of applying Monte Carlo simulation for estimating transportation time. In actuality, we would need to be more precise and have answers for many additional questions, such as: What does "very hot" or "windy" mean? Is it possible to have "rain" and "very hot" at the same time? If so, how would that affect the transportation time? Also, since each trip will take a few days, there might be several "fair" days and several "rain" days, and so we would need to consider both categories.

Agent-based simulation

During the past decade, interest has grown in "behavior-based" computer programs called *agent-based modeling* (also known as *agent-based simulation*), which attempt to simulate complex phenomena through virtual "agents." We can look at agent-based modeling as a special prediction method for dataless problems, where the behavior of these agents is determined by programmable rules that reflect the constraints and conditions of a real-world problem. For example, the problem might be to minimize the evacuation time from a convention center by finding the optimal arrangement of tables and chairs for 5,000 people. Clearly, we can't even dream of obtaining such data! We can't load a convention center with 5,000 people, set it on fire, measure the evacuation time and death rate, then change the arrangements of tables and chairs, load another 5,000 people, set it on fire again, and continue this process thousands of times to collect enough data! In situations like these, agent-based modeling is far more appropriate.

The connection between Monte Carlo simulation and agent-based modeling is that both methods generate many possible scenarios according to some probabilistic distributions of variables (e.g. windy weather conditions) or by using agents that follow some probabilistic rules. In the case of finding the optimal arrangement of tables and chairs to minimize evacuation time, we could turn to psychological tests conducted on people to see how they behave during emergency fire situations, which might show 17% of people running in random directions, 38% of people running in a straight line to the closest exit without paying attention to others, 9% of people standing still and screaming, and so on. Additionally, the way people interact with one another during a fire emergency is also important; for example, when two or more people collide, they might fall down and be immobilized for a few seconds. Now, by generating 5,000 agents whose behavior follows these rules and interactions and placing them together in a convention center that's set on fire, we can run many diverse simulations on the placement of people, tables, and chairs, and observe the behavior of the agents and the overall outcome (in terms of evacuation time and injury or death rate).

The general idea behind agent-based modeling is that each simulation is based on the interaction between agents. Furthermore, an agent doesn't necessarily have to be a person—it could be any "component" of the problem we're trying to solve, such as a truck within a supply chain, a threatened species within an ecosystem, or even a machine within a factory—and the entire model includes an environment where the interactions occur. Agents may also have differing capabilities, and their behavior is based on probabilistic rules that determine their actions. The interaction of the agents (with the environment and one another) may result in an emergent behavior that was impossible to foresee due to all the complexity of their interactions.

Let's illustrate the main characteristics of agent-based models using a classic example of two species of birds (*A* and *B*) that fly around while obeying three very simple rules:

Rule 1: If another bird of the same species is close by, then fly towards it
Rule 2: If another bird of the other species is close by, then fly away from it
Rule 3: Keep a minimum distance between any two birds.

Note that this model is extremely simple. There are only two species of birds (with 10 birds in each species) moving at the same speed and their direction is set randomly at the beginning. The figure below illustrates the initial stage, with each "bird" being an individual "agent" within the agent-based simulation:

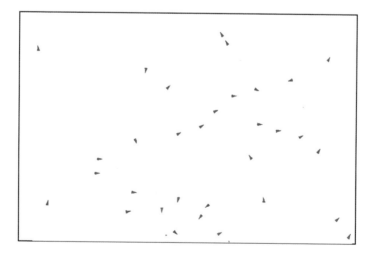

However, the interaction between these simple birds produces an emergent behavior that is complex, organized, and very life-like. After a while, we can observe a flocking phenomenon where birds of the same species are flying together:

To further refine the behavior of these birds, we can add some additional rules to the model. For example, we can modify their speed in some encounters, introduce obstacles in the environment, or create "sub-species" (which have slightly different rules). We can also make the environment more complex and then observe their behavior. One of the properties of this model is *unpredictability* over moderate time periods. For example, although the birds of one species might be flying primarily from left to right at one moment, it's impossible to predict which direction they might be moving at a later time.

In general, an agent can have many behavioral rules and internal states, some of which might be fixed, while others may change. An agent can process many sensory inputs, change its behavior according to these inputs, take into account the interactions with other agents, and make decisions based on the available data. Agents can also operate in an artificial environment, which might be a building, a city, a communication network, or a landscape that changes over time.[8] And yet, as the above example shows, each agent is very simple in comparison with the complex behavior that eventually emerges. For these reasons and others, agent-based modeling is well suited for studying human social phenomena, including the propagation of diseases, trade habits, group formation (as illustrated by the above simple example of birds flocking), evacuation patterns, and migration. Within the context of promotional planning, agent-based models can be used to simulate the effect of new product introductions, shopper behavior for regions (or retailers) where data isn't available, as well as competitor promotions (and their impact), thus enhancing the effectiveness and realism of the prediction model.

---

8   Because of this diversity, there is no general agreement on what an "agent" is, or what key features an agent should have.

For a visual demonstration of agent-based modeling, please watch the supplementary video for this chapter at: www.Complexica.com/RiseofAI/Chapter4.

CHAPTER 5

Prediction

"Anyone," Avallac'h wiped his hands on a rag, "can foretell the future.
And everyone does it, for it is simple. It is no great art to foretell it.
The art is in foretelling it accurately."

Andrzej Sapkowski, *The Witcher series, The Tower of the Swallow*

Given a choice between receiving yesterday's newspaper or tomorrow's newspaper, which one would you choose? Which one would be more valuable? *And why?* The reason why we'd all ask for tomorrow's newspaper is because it would allow us to make better decisions by knowing—with perfect accuracy—what will happen tomorrow.

For this reason, most organizations are interested in trying to understand what is likely to happen in the future, given past trends, events, and patterns. Predictive modeling (also called predictive analytics) seeks to predict future events or outcomes by analyzing patterns that are likely to indicate future results. Which brings us to the prediction layer of the problem-to-decision pyramid for an in-depth discussion of various prediction methods.

To start with, most prediction problems can be categorized as either classification problems, regression problems, or time series problems. When placing a prediction problem into one of these three categories, two major aspects must be considered: the *expected output* and *time*. For some problems, there are only two possible expected outputs: "yes" or "no," "true" or "false," "buy" or "sell," etc. These are classical binary classification problems,[1] because they assign new cases[2] to one of two classes. The best example is the classification of credit card transactions into two classes: "fraudulent" and "legitimate." A classification problem may have, however, more than two outputs; in fact, the number of possible classes can be quite significant—for example, a prediction model that classifies symptoms into a large number of different disease types.

---

1 A prediction model developed for a classification problem is often called a *classifier*.
2 In this Chapter, we'll use the term "case" for a "future record" for which we make a prediction. After some time, these "cases" would become regular records that can be used later for updating the model.

In these classification problems, time doesn't exist; the future is simply when a new (yet unknown) case arrives.

Similar comments are also applicable to regression problems, where the general purpose of (multiple) regression is to discover the relationship between several independent ("predictor") variables and a dependent ("criterion") variable, with the output being a concrete number. For example, we may want to predict salary levels as a function of job role, number of years at that role, size of organization, industry, and so on. A regression model will also tell us which variables are better predictors than others, and we can easily identify outliers.[3] Again, the issue of time is either non-existent or included as a variable.

In contrast to classification and regression problems, "time" is the main variable in a time series problem, where there is a sequence of many values measured over some period of time in the past. In other words, the time dependencies between the data points are so strong that these data points must be kept in a sequential (time) order. In time series problems, the future is referenced explicitly: we would like to predict a variable's value in the future (tomorrow, next month, etc.). A classic example from the field of economics is predicting next year's Gross Domestic Product (GDP). Plenty of historical data are available (released every quarter), and the prediction model may include many additional variables (e.g. unemployment rate, consumer confidence, exchange rates, consumer price index, manufacturing PMI, etc.). In promotional planning and pricing, on the other hand, we would like to predict volume for particular products, in particular retail chains, in particular geographies, for particular promotional periods. To do so, we use historical data to extract the most important variables that influence volume (e.g. identification of trends, seasonality effects, discount elasticity, cannibalization, pull-forward effect, halo effect, etc.), and having this knowledge, we can then apply various prediction methods to develop a prediction model.

As we explore the subject of predictive modeling in this chapter, we'll rely on simple examples and models. The reason for this is twofold: first, so we can focus on the principles behind each prediction method, and secondly, because the construction of a simple model is often the first step before building a more complex model to generate more accurate predictions. Although simple models have many flaws when applied to complex business problems (such as ignoring important variables), they often provide a useful first step in the

---

3   As discussed in Chapter 4.2, an *outlier* is a data point that lies at an abnormal distance from the other data points in a random sample. For example, the annual salary for 1,000 randomly selected people might be in the range of $17,832 to $167,942, with the exception of one (lucky) person, who earns $938,400 per year. Such a salary figure is called an "outlier".

process of building a predictive model by allowing us to better understand the problem and gain insight on the best approach.

In promotional planning and pricing, the prediction model should take into account the relationship between promotional price and promotion type for various products, as well as different types of cannibalization (pack size cannibalization, sub-category cannibalization, cross-category cannibalization) and a variety of other effects such as pull forward effect and halo effect. But these adjustments are usually done in the "second iteration" of building a model, and the entire process usually consists of a few steps:

- *Data preparation*: includes data transformation, normalization, creation of derived attributes, variable selection, elimination of noisy data, supplying missing values, and data cleaning, as discussed in Chapter 4.2. This stage is often augmented by preliminary data analysis to identify the most relevant variables and to determine the complexity of the underlying problem. The data preparation step can be the most laborious, and many people believe that it constitutes 80% of any data science effort.
- *Model building and training*: includes a complete analysis of the data (i.e. the data mining stage), the selection of the most appropriate prediction method, as well as training of the model.
- *Deployment and evaluation*: includes implementing the best-performing prediction model and applying it to new data to generate predictions. However, because new data usually arrives at regular intervals, it's essential to measure the model's performance and tune it accordingly.

Because we covered the first step in the previous chapter, let's examine the remaining two steps in the following sections. Also, in Chapter 7.2, we'll discuss some aspects of updating/tuning a prediction model on the basis of new data and/or feedback.

Note that after the data is prepared, we can begin our search for the most appropriate prediction method. The goal is to build a model that will accurately predict the outcome of new cases. This outcome might be the predicted number of units sold within some time period, the classification of credit card transactions, or the assignment of new customers to the appropriate segment, among countless others. Many prediction methods have been developed over the years that differ from one another in the representation of a solution (e.g. a decision tree versus a set of rules), as well as some other differences, such as whether they're able to explain the predicted result, or the ease with which a solution can be edited. We can group these different prediction methods into two broad groups:

- Classical methods (e.g. statistical, distance, logic), and
- AI-based methods (e.g. random forest, genetic programming, fuzzy systems, neural networks)

The first group is covered in the next section of this chapter, while the second group of AI-based methods will be covered in later sections. Also, some prediction methods for simulation were already explained in the previous chapter—in particular, Monte Carlo methods and agent-based simulations—and the last two sections of this chapter will focus on ensemble models and the evaluation of predictive models.

For additional information on some of the concepts presented in this chapter, we encourage you to watch the supplementary video for this chapter at: www.Complexica.com/RiseofAI/Chapter5.

5.1 Classical Prediction Methods

Classical prediction methods can be divided into three general categories:

- Statistical methods, such as linear regression and time series analysis
- Distance methods, such as instance-based learning and clustering
- Logic methods, such as decision tables, decision trees, and classification rules

In this section we discuss them in turn.

Statistical methods

As mentioned earlier, there are three types of prediction problems: classification, regression, and time series. Classification problems have been the focus of data mining research for the last few decades, and some prediction methods based on distance and logic were explicitly developed for classification problems. For the time being, however, we'll focus on regression and time series problems, with the major difference between these two being that regression problems assume the expected output exhibits an explanatory relationship with some other variables. For example, someone's (predicted) salary might be a function of education, experience, location, and industry. In such cases, an explanatory method would be used to find the relationship between these variables and make a prediction. The goal of time series methods, on the other hand, is purely one of prediction, rather than to discover or explain any relationship between variables.

Probably the most popular explanatory method is *linear regression*. If the predicted outcome is numeric and all variables are numeric, then linear regression is the natural choice.[4] In this method, we build a linear model that uses the values of different variables to produce a predicted value for a *target* variable (i.e. a variable not used in the model). To illustrate this prediction method in more detail, let's consider linear regression for predicting the sales volume (in terms of the number of units sold) for a particular product promoted during a particular week. In this case, the target variable would be the predicted volume. Note, that some variables are *not* numeric, so we have to address this issue first. It's clear that some non-numeric variables, like promotion type, are of key importance, as they influence the outcome of the promotion (which is further influenced by the seasonality, promotional price, etc.). By building a separate regression model for each product for each promotion type, we can eliminate such non-numeric variables.

Another option would be to convert all non-numeric variables into numeric variables. For example, we can make a list of all available promotion types ("non-promotion," "catalogue special," "in-store," "50%-off," "2-for-1," and "3-for-2") and then assign numbers that correspond to the significance of each type. For example, assuming we have six different promotion types, then "non-promotion" might be 6 (the least significant), "3-for-2" might be 3, and "50%-off" might be 1. Similar assignments can be made for other non-numeric variables, while variables such as promotional price and week of the year are already numeric, so there's no need to covert these.

For the linear regression model to answer questions such as "What's the predicted volume for Product 43 during Week 15 for Mary's Market in NSW?" we need to develop a function:

Volume = a + (b × Promotional price) + (c × Promotion type) + (d × Week) + ...

that provides the predicted sales volume for a new case (i.e. Product 43) when supplied with the numeric values of the other variables (promotional price, promotion type, week, etc.). The main challenge here is to find the values (weights) for each parameter *a, b, c, d*, etc. that give the prediction model the best possible performance by minimizing the prediction error. By having several years of historical data, we can extract all records for Product 43 in NSW for Mary's Market and this subset of records (say, we identified 90 such records) would constitute the data set available for training the prediction model. Some of these records would also be used for validation and testing (which we'll discuss in Section 5.7).

---

4   In some situations, we only want to predict one of two values, such as "yes" or "no," "fraudulent" or "legitimate," "buy" or "sell," etc. This type of regression is called logistic regression and a similar methodology is applied (e.g. transformation of variables, building a linear model).

To minimize the prediction error on the training set, there are several standard procedures for determining the appropriate parameter values. Once these parameters are set, the prediction model is ready. For every new case (e.g. a particular promotion for Product 43 for Mary's Market in NSW), we can determine the volume by inserting the appropriate values for promotional price, promotion type, week, and so on, into the linear regression model.

However, the model training process might not be that simple—which is a caveat for any prediction model, not just linear regression. First of all, some values might be missing, such as the promotion type. In such cases, we can:

- Recover the missing value by possibly contacting the appropriate retailer
- Remove the record from the training dataset
- Estimate the missing value on the basis of other variables—for example, based on historical data, we may know the effect of different promotion types for this particular product

Secondly, because the model has to predict more than next week's sales volume (as promotion plans span many weeks and months), the training process might be more complex. The reason for this increased complexity is hidden in the fact that the model's prediction accuracy must be assessed for both shorter and longer time periods. Hence, the process of searching for the best prediction model becomes more challenging if one model provides better short-term predictions while another provides better long-term predictions.

Another reason why the training process might be more involved is because the model might process some "rare" cases from time to time, where very little historical data is available for training purposes. For example, Product 76 was never promoted by Mary's Market in NSW, or Product 13 was never promoted during winter. How can we train a model without any training data? Well, as usual, there are several ways to deal with this problem. One way would be to estimate volume on the basis of (1) volumes of the same product at other states/retail chains, and (2) volumes of similar products at the same retail chain/state. This approach would require some additional, problem-specific knowledge. Another possibility would be to use agent-based simulation (discussed in the previous chapter) as a data mining technique for products without data (i.e. products that were never placed on promotion).

The above example illustrates that the devil is in the details, which is true for any prediction problem, as the development of a prediction model for any real-world problem usually involves the resolution of many issues, including insufficient data. Something else to consider is that linear regression models have one major disadvantage: *they're linear!* Of course, a linear regression model would find the best possible line, but that line may not fit well on real-world data that exhibits non-linear dependencies (recall the non-linear transportation model in Chapter 4.1).

One approach to this problem is to change the line into a curve, which can be done by transforming the variables (by multiplying some of them together, squaring or cubing them, or taking their square root). After completing these transformations, we can determine the new parameters (i.e. *a*, *b*, *c*, *d*, etc.) of the prediction model, although this new model will be more complex, as we're now talking about *non-linear* regression. It's possible to experiment with a wide variety of transformations, and if they don't provide a meaningful contribution to the prediction model, then their parameters will stay close to zero. The difficulty, however, is that the number of possible transformations might be too prohibitive—in other words, the number of possible parameters to explore might be too high, and any training would be infeasible.

Moreover, we should be aware of overfitting, as the use of complex transformations guarantees high precision on the training data which may not carry over to predictions on new cases. As explained in Chapter 4.2, overfitting occurs when a model tunes itself during the training stage to such an extent that all predictions on the training data set are perfect, whereas predictions on new data lack accuracy. In other words, the model is considered "overfit" when it has learned "the noise" instead of "the signal." In the figure below:

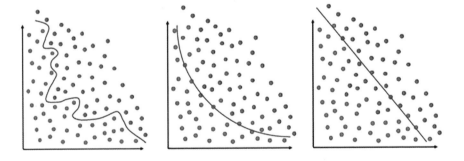

the purple line in the leftmost graph represents an overfitted model and the red line (middle graph) represents a "proper" model. While the purple line ideally follows the training data (displayed as blue and red dots), it's too dependent on that data and is quite likely to have a higher error rate on new unseen data, compared to the red line. In other words, the model represented by the red line should preform better on new data than the overfitted model represented by the purple line. Also, the green line in the rightmost graph represents an *underfitted* model, which performed poorly on the training data and is likely to perform poorly on new data as well; this could be because the model is too simple and is missing key variables.

Let's now turn our attention to time series problems. As mentioned earlier, the main purpose of time series is to forecast a future event by identifying trends, cycles, and seasonal variances; the relationships between the variables themselves are of little interest. Hence, the problem might be expressed as follows:

Given $v[1]$, $v[2]$, …, $v[t]$, predict the values of $v[t+1]$, $v[t+2]$, …, $v[t+k]$

where t is the present timestamp, $t–1$ is the previous timestamp, $t+1$ is the next timestamp, and so on. If we're only predicting the next timestamp ($t+1$), then a time series model is concerned with a function F such that:

$$v[t+1] = F(v[1], v[2], …, v[t])$$

Note that the above function may include some other variables, and not just the values of variable v from earlier timestamps. These are called *composite forecasting models*, which consist of past time series values, past variables, and past errors.

Many statistical time series models have been proposed during the last few decades, including exponential smoothing models, autoregressive (a.k.a. integrated or moving average) models, transfer function models, state-space models, among others. Each model is based on some assumptions and involves a few (at least one) parameters that must be tuned on the basis of historical data. The category of prediction methods collectively known as *exponential smoothing*, for example, generalizes the *moving average method* where the mean of past k records is used as a prediction. All exponential smoothing methods assign weights to past records in such a way that recent records are given more weight than older records (as more recent records usually provide better future direction than less recent records). Hence, it's reasonable to develop a weighting scheme that assigns smaller weights to older records. Such a weighting scheme also requires at least one parameter a. For example, a prediction for the time $t+1$ is calculated as:

Prediction(t+1) = a × Actual(t) + (1–a) × Prediction(t)

which simply means that the prediction for the next (future) case is calculated as a total of two values: the actual last record (*Actual(t)*) with parameter a and the last prediction (*Prediction(t)*) with the weight *1–a*. Note, that parameter a provides the significance of the last record in making the prediction; in particular, if $a = 1$, then the prediction would always report the last actual value as the new prediction. It's easy to generalize this method to include more past records:

$$Prediction(t+1) = a \times Actual(t) + a \times (1-a) \times Actual(t-1) +$$
$$a \times (1-a)^2 \times Actual(t-2) + a \times (1-a)^3 \times Actual(t-3) + ...$$
$$+ a \times (1-a)^{t-1} \times Actual(1) + a \times (1-a)^t \times Prediction(1)$$

so *Prediction(t+1)* represents a weighted moving average of all past obser-vations. Note again, that different values of parameter *a* would result in a different distribution of weights. Also, we assumed the prediction horizon was just one period away (*t+1*), but for longer-term predictions, we can assume a flat function:

$$Prediction(t+1) = Prediction(t+2) = Prediction(t+3) = ...$$

as exponential smoothing works best for datasets without trend or seasonality. However, since some form of trend or seasonality exists in most data sets, *decomposition* methods can identify the separate components of the underlying trend-cycle and seasonal factors. The trend-cycle (which is sometimes separated into trend and cyclical components) represents long-term changes in the time series values, whereas seasonal factors relate to periodic fluctuations of constant length caused by phenomena such as temperature, rainfall, holidays, etc.

Although there are several approaches of decomposing a time series problem into separate components, the basic concept is simple: First, the trend-cycle is removed, then the seasonal components are addressed. Any remaining error is attributed to randomness; thus:

Observed_data = trend-cycle + seasonality factors + residual (error)

In the case of promotional planning, time series analysis allows *decomposition* of the data (e.g. sales over several years) into three components:[5]

- *Trend*: is the long term upward or downward trajectory of sales, often resulting from growth or decline in a category or brand.
- *Seasonality*: is the yearly cycle of peaks and troughs that tend to repeat over the course of year and is representative of typical consumption patterns for certain products. It arises from systematic, calendar related influences such as a sharp rise in retail sales before Christmas, or an increase in certain product sales during summer. Seasonality in a time series can be identified by regularly spaced peaks and troughs, which have a consistent direction and approximate same magnitude every year, relative to the trend.

---

5   The fourth component (called the cyclical component) reflects repeated but non-periodic fluctuations and is often combined with the trend component, hence its name: trend-cycle.

- *Residual (error)*: is the remainder of the deviations that aren't explained by trend and seasonality. A large degree of such "noise" means there are likely additional variables we should consider. It's also important to note that peaks and troughs due to promotional activity will appear as noise when conducting time series decomposition.

The figure below illustrates such decomposition for sales (the original data are displayed in the first part of the figure, labeled: "observed") of a particular product over a period of seven years:

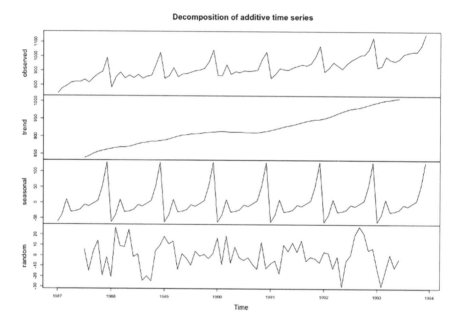

However, the relationship between the data and trend-cycle, seasonality factors, and residual might not be linear; in general, decomposition models search for a function D that would explain a data point at any time t:

Observed_data(t) = D(trend-cycle(t), seasonality factors(t), residual(t))

Decomposition models are typically additive or multiplicative but can also take other forms (e.g. pseudo-additive). In some cases, where the amplitude of both the seasonal and residual variations do not change as the level of the trend rises or falls, an additive model is appropriate. In the additive model, the observed time series (D_t) is considered to be the sum of three independent components: the trend (T_t), seasonal factors (S_t), and the residual (R_t):

$$D_t = T_t + S_t + R_t$$

Each of the three components has the same unit of measure as the original series. In many time series, the amplitude of both the seasonal and irregular variations increase as the level of the trend rises. In this situation, a multiplicative model is usually more appropriate, where the original time series is expressed as the *product* of trend, seasonal and irregular components:

$$D_t = T_t \times S_t \times R_t$$

In this model, the trend has the same unit of measure as the original series, but the seasonal and irregular components are unitless factors, distributed around 1. The multiplicative model cannot be used when the original time series contains very small or zero values. In such cases, a pseudo-additive model is used, which combines the elements of both the additive and multiplicative models.

To choose an appropriate decomposition model, it is necessary to examine a graph of the original series and try a range of models, selecting the one which yields the most stable seasonal component. If the magnitude of the seasonal component is relatively constant regardless of changes in the trend, an additive model is suitable. If the seasonal component varies with changes in the trend, a multiplicative model is usually the better choice. However, if the series contains values close or equal to zero, and the magnitude of seasonal component appears to be dependent upon the trend level, then pseudo-additive model is most appropriate.

Although all these models are capable of characterizing the variables found in actual data, they also assume that the underlying process of data generation is constant. This assumption is often invalid for actual time series, as changing environmental conditions may cause the underlying data generating process to change. In the event that the underlying data generating process does change, the time series data must be re-valuated and the parameter values re-adjusted (in extreme cases, a new model might be required). We'll discuss the topic of adaptability in Chapter 7.

Distance methods

Another method for building prediction models is based on the concept of *distance* between records. Any two records in a dataset can be compared for similarity, and this similarity measure (called *distance*) is assigned some value: the more similar the records, the smaller the distance value. Using a distance measure within a dataset would allow us to compare each new case with the most similar historical record. The outcome of the most similar historical record—for example, the loan was repaid, the transaction was fraudulent— would become the prediction for the new case. Going back to our example of Product 43 for Mary's Market in NSW, we may search our historical data for the most similar record (which has the smallest difference in promotion type and week of year), and then use the volume of this record as our prediction. Ideally, the old record would be relatively recent with many similarities to the new case! Hence, instead of building a function where the variable values are magnified by some weights to determine the outcome, we just keep the historical records.

The essential aspect of this approach is creating a similarity measure between records, because the probability of finding an identical record is very low. Hence, we have to base our decisions on similarities, which is far from trivial. For instance, is Product 43 (chocolate bar) promoted "in store" during Week 15 more similar to Product 47 (also chocolate bar) promoted "2 for 1" during week 17, or Product 81 (a cookie) promoted "in store" during week 15? Or, is the difference in similarity between "in store" and "2-for-1" the same as the difference between "3-for-2" and "catalogue special"? To answer such questions, it's necessary to accurately define the distance between records, where the shorter the distance, the greater the similarity. One of the most popular distance-based prediction methods is *k nearest neighbor*, where *k* nearest neighbors (i.e. *k* most similar records) of a new record (new case) are determined. Clearly, if *k = 1* (i.e. we find only one neighbor), this single neighbor's outcome is the prediction for the new case. If *k > 1*, then a voting mechanism is used (for classification problems) or the average value of the *k* answers is calculated (for regression problems).

However, the most important step of the *k* nearest neighbor method is calculating the distance between records, which is crucial for obtaining high-quality results. There are many ways of defining a distance function, but experimentation is often the best way. Also, during the data preparation stage, it's likely the data will be normalized to equalize the scale for computing distances and/or some weighting will be applied where different variables get different weights. Calculating the distance is then trivial when there is only one numeric variable, (e.g. $5.7 - 3.8 = 1.9$), but with several numerical variables, Euclidean distance[6] can be used, provided that the variables are normalized and of equal importance; otherwise, a weighting must be applied.

The largest problem with respect to distance methods, however, is with nominal variables. Given our earlier question of whether the difference in similarity between "in store" and "2-for-1" for promotion type is the same as between "3-for-2" and "catalogue special," we can assume that different

---

6   *Euclidean distance* is defined as the length of a line segment between two points in an *n*-dimensional space. In particular, the distance *d* between two points (x_1, y_1) and (x_2, y_2) in a *2*-dimensional space is determined by the following function: $d^2 = (x_1 - x_2)^2 + (y_1 - y_2)^2$.

promotion types are different altogether (resulting, say, in a distance of 1), or we can introduce a more sophisticated matrix that would assign a numeric measure for each of them, so that the difference between "2 for 1" and "3 for 2" is smaller than the difference between non-promoted and "in store." These are the two standard approaches for evaluating differences between the values of nominal variables.

In the case of missing data values, a standard approach is to assume that the distance between an existing value and a missing value is as large as possible. Hence, for nominal values, the distance is assigned a normalized value of 1 (all distances are between 0 and 1), and for numeric variables the distance is assigned as the larger distance from 0 and from 1. For example, if an existing value is 0.27 and the other value is missing, then the distance is 0.73; if the existing value is 0.73 and the other value is missing, then the distance is also 0.73.

Yet another issue is the number of stored records. A distance-based method might be too time-consuming for large datasets, because the entire dataset must be searched to evaluate each new case. With larger values of parameter k (for k nearest neighbor method), the computation time increases significantly. For efficiency reasons, it would be beneficial to select a subset of "representative records," so that the process of finding the closest neighbor (or neighbors) becomes more efficient. Also, to make these representative records as "representative" as possible, a new set of representative records can be selected from the current representative records and all misclassified records that produced a prediction error larger than some threshold. In other words, the current representative and misclassified records could constitute an input for some reclassification method (e.g. decision trees), which would be responsible for creating a better set of representative records (Section 5.6 provides more information on this topic).

Some clustering methods can also be used to group records into meaningful categories, so that a new case is assigned to an existing category and the predicted value would be drawn from the records in that category (again, by voting for classification, or averaging for regression). Note that it's unnecessary to store all records per category; again, we can select some representative records instead. A few clustering techniques might be considered for this task, such as k-means algorithm, incremental clustering, or statistical clustering based on a mixture model.

Logic methods

A *decision table* (also known as a *lookup table*) is the simplest logic-based method for prediction. For some cases, like predicting the price of various used cars, there are many such tables published on a regular basis (e.g. *Black Book*, *Kelley Blue Book, Manheim Market Report*). Using these tables, we can locate a specific make/model/year/body style to get a basic price, and then adjust this price for additional variables such as mileage, color, trim, damage level, etc. However, for many other prediction problems (like promotional planning), decision tables aren't feasible.

The most widely used logic method, however, are *decision trees*. Decision tree builds regression or classification models in the form of a tree structure. This method breaks down a dataset into smaller and smaller subsets, while at the same time an associated decision tree is incrementally developed. Decision trees can handle both categorical and numerical data.

Because the structure of a decision tree[7] is relatively easy to follow and understand (especially for smaller trees), its popularity is widespread. To make a prediction for a new case, the root of a tree[8] is examined and a test is performed, and depending on the outcome of the test, the case then moves down the appropriate branch. And by "test," we mean that a node compares a value of a variable with some constant; however, it's possible to include more sophisticated tests, which involve more variables and/or additional functions. This process continues until a terminal node (also known as a *leaf*[9]) is reached, and the value of this terminal node is the predicted outcome.

Although decision trees are used for all types of prediction problems, they're especially popular for classification problems. If the test involves a nominal variable, the number of branches corresponds to the number of possible values that a variable can take, meaning there is one branch for each possible value. If the test involves a numeric variable, there are usually two branches, as the test determines whether the value is "greater than" or "less than" some predefined fixed value[10] (and possibly also "equal to" for integer numbers). For missing values, an additional branch is assigned or some rule is applied, such as selecting the most popular branch or few branches.

---

7   Like in real trees, we have the root, branches, and finally the leaves.
8   The node at the top of a tree is called root. There is only one root per tree and one path from the root node to any other node in a tree.
9   Leaves are the last nodes on a tree; these are nodes without children.
10 It's also possible to have a decision tree with more than two branches for a numeric variable, where a range of values is assigned to each branch.

The following illustrates a decision tree for predicting volumes:

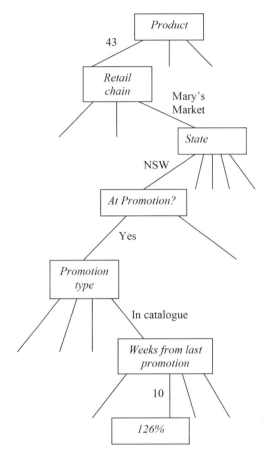

The first three levels of the tree determine product, retail chain, and state. The root has 100 branches for 100 different products, and each node on the next level has 10 branches, assuming 10 retailers. The third level provides branches for states, and further down, the nodes would test whether the product is on promotion; if yes, the promotion type and number of weeks since the last promotion would refer the case to the appropriate branch. The leaf is shown as a percentage uplift in sales volume in comparison to the baseline volume.

Naturally, there are better and more sophisticated ways to use decision trees for numeric predictions. For example, it might not be practical to represent every value or range of values as a separate branch in a decision tree, as the tree's size might become too large. Instead of keeping a single, numeric value at each terminal node (as illustrated above), it might be easier to use a model that predicts a value for all cases that reach this terminal node (e.g. a linear regression model). Such a tree could be used to answer our earlier question of "what's the predicted volume for Product 43 during Week[11] 15 for Mary's Market in NSW?" The variables "product," "retailer" and "state" are used for building the tree (and later for branch determination when processing a new case), whereas the promotional price, promotion type, week, and so on, are used as variables in the linear regression model at each terminal node, as shown illustrated below:

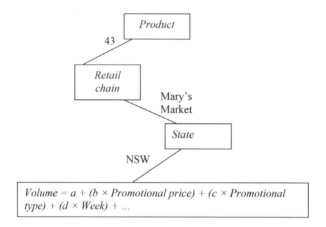

Note that the number of parameters (a, b, c, d, etc.) and their values might differ at each terminal node.

To achieve better prediction accuracy, we could build a linear regression model for each node of the tree (rather than just for the terminal nodes).[12] Note, however, that the root node would now have a function for all variables:[13]

Volume = a + (b × Product) + (c × Retail chain) + (d × State)
+ (e × Promotional price) + (f × Promotion type) + (g × Week) + ...

For second-level nodes, the linear function wouldn't include the variable "product," because the appropriate branch of the decision tree has already been selected. Thus, the linear function would be:

Volume = a + (b × Retail chain) + (c × State) + (d × Promotional price)
+ (e × Promotion type) + (f × Week) + ...

---

11 The variable "Week" stands for the week of the year; its values vary from 1 to 52.

12 Experimental evidence shows that prediction accuracy can be increased by combining several prediction models together, which we'll cover in Chapter 7.

13 This approach usually involves nominal attributes (e.g. product, retail chain, state), as all variables are represented on different levels of the decision tree. Such nominal variables are transformed into binary variables and treated as numeric.

For third-level nodes, where the decision for product and retailer has already been made, the function would be:

Volume = a + (b State) + (c × Promotional price) + (d × Promotion type)
+ (e × Week) + ...

and so on; but note, however, that the parameters *a*, *b*, *c*, etc. are different in all these functions. In our earlier diagram, the terminal node was placed on the fourth level with a linear function of:

Volume = a + (b × Promotional price) + (c × Promotional type)
+ (d × Week) + ...

To compensate for the differences between adjacent linear models at the fourth level, some averaging (also called smoothing) can be applied when processing a new case. Instead of using the predicted value from the terminal node, the predicted value can be filtered back up the tree and averaged at each node by combining the predicted value from a lower level with the predicted value from the current level. This usually improves the prediction accuracy.

Another logic method is based on *decision rules*, which are similar to decision trees—after all, a decision tree can be interpreted as a collection of rules. For example, the single branch of the decision tree displayed earlier can be converted into the following rule:

if *Product = 43* **and** *Retail chain = Mary's Market* **and** *State = NSW* **and** *Promotional price = $19.95* **and** *Promotional type = In store* **and** *Week = 15* **then** *Volume = 115%*

The "if" parts of the rule (e.g. "State" = NSW) are combined logically by the "and" operator, and all tests must be true for the rule to "fire" and be used as the prediction, in this case, *Volume = 115%* that represents a percentage uplift in sales volume in comparison to the baseline volume. There must be several decision rules in the model (as the above rule represents just a single branch) and we can interpret this collection of rules as connected through the "or" operator: if one rule applies to a new case, its conclusion is taken as the predicted outcome. If two (or more) rules fire, then we can combine the conclusions of these rules to determine the final predicted outcome. The other (in some sense, opposite) problem can arise if *no* rules fire for a new case! As usual, some standard approaches exist for this situation, such as the creation of a default rule that will always fire:

if *Product = 77* **then** *Volume = 100%*

which is the overall baseline volume for that product—although we can question the usefulness of this rule ...

These two simple cases, when two or more rules fire or no rules fire, illustrate the point that rules can be difficult to deal with. The reason is that each rule represents a separate piece of knowledge and *all* rules together operate as one model (often called a *rule-based system*); thus, it's essential to understand the consequences of adding or dropping a rule. This also applies to many real-world situations, where human experts add their own rules (from experience) to the data-generated rules. Although dropping and adding rules in a rule-based system is not a trivial task, it's much easier to drop or add a rule than to modify an entire decision tree by cutting or adding new branches. Hence, each method has its advantages and disadvantages.

As mentioned earlier, classification problems have been the focus of data mining research for the past few decades, and the creation of decision rules[14] has been the most popular approach for addressing these problems. Several aspects of generating rules from data have been investigated, including *association rules*, which describe some regularity in the data and can predict any variable, rather than just the class. For example, an associate rule may state that:

if *Product = 76* **and** *Retail chain = Mary's Market* **then** *State* **in** *{NSW, QLD}*

as Product 76 is only promoted in NSW and QLD. It is also possible to extend an associated rule by some exceptions, such as the following example:

if *Product = 76* **and** *Retail chain = Mary's Market* **then** *State* **in** *{NSW, QLD}* **except if** *49 ≤ Week ≤ 52* **then** *State* **in** *{NSW, QLD, TAS}*

which states that Product 76 is also promoted in Tasmania (TAS) around Christmas. Exceptions may also refer to classification rules, such as the following example:

if *Product = 76* **and** *Retail chain = Mary's Market* **and** *State = NSW* **and** *Promotional type = In store* **and** *8 ≤ Weeks from last promotion ≤ 12* **then** *Volume = 130%* **except if** *49 ≤ Week ≤ 52* **then** *Volume = 145%*

as the Christmas season increases volume regardless of any promotional activity.

Rule-based systems, which consist of a collection of rules and an inference engine,[15] are quite popular, because each rule specifies a small piece of knowledge and people are good at handling small pieces of knowledge! Separate rules

---

14 A decision rule for a classification problem is often called a *classification rule*.

15 An *inference system* (*engine*) is responsible for putting the decision rules into the appropriate order and combining the outcomes of the rules that fired.

can be discovered from data mining activities or interviewing experts, and instead of specifying the overall model, only the decision rules and inference system are needed. Note that the rule-based system will try to behave like an expert, performing some reasoning on the knowledge present in the model. We'll also discuss a particular rule-based system in Section 5.4.

5.2 Random Forests

In the previous section we provided an overview of various classical prediction methods, one of which was a decision tree. The concept of a decision tree is quite old, with the first attempts to organize knowledge in a tree-fashion dating back to the year 270.[16] Since that time, decision trees have been commonly used in operations research, specifically in decision analysis, and have enjoyed widespread popularity during the past 70 years.

During the early years of Artificial Intelligence research, decision trees were adopted by the AI community given their intelligibility and simplicity, and over time, became one of more popular Machine Learning techniques. Their application was augmented by smart algorithms for building decision trees that could account for most of the data while minimizing the number of levels (so-called *optimal decision trees*), as well as building decision trees that could generate rules from empirical data. These algorithms furthered the application of decision trees by Machine Learning and Artificial Intelligence scientists, especially given they could be used for both supervised and unsupervised learning (see Chapter 7.1 for a discussion on these types of learning).

In this section we'll examine how individual decisions trees can be combined to build a random forest and discuss the effectiveness of this method. The fundamental concept behind a random forest is simple, but powerful: *the wisdom of a crowd*. This "wisdom," expressed in more formal language, says that a large number (crowd) of relatively uncorrelated models (trees) operating together will outperform any individual model (person). Each tree in the random forest generates either a class or numerical prediction, and the class with the most votes or the average of all predicted values, becomes the overall prediction. Random forests also add additional randomness to the model while growing trees. Instead of searching for the most important variable when splitting a node into new branches, this method searches for the best variable among a random subset of variables, which results in greater diversity and generally a better model.

In a decision tree, each internal node represents a test conducted on a

---

16 Decision trees are widely attributed to the philosopher Porphyry, who wrote an *Introduction to Logic* around 270 AD that used a metaphorical tree to describe and organize knowledge.

variable, each branch represents the outcome of that test, and each terminal node represents the result after processing all variables. However, when discussing decision trees in the previous section, we didn't touch upon the ways to "grow a tree." For example, in the following tree:

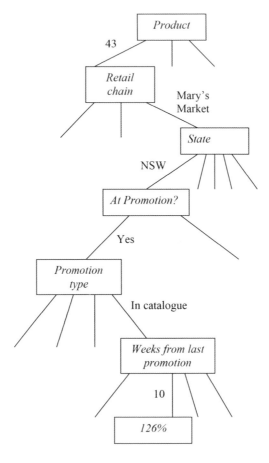

the question is how to select the particular variables for each level—in other words, why should we start with "product" followed by "retail chain" and then "state"? One argument would be that these three variables represent the natural choice, because we'd like to predict each product's sales volume within a retail chain in each state. So, conceptually, we could create 100 × 10 × 5 = 5,000 different decision trees (one for each of 100 products, within each of 10 retail chains for each of the 5 states). But even then, what should be the order of the remaining variables? Should we start by checking whether a given product is on promotion or not, or by checking the season first? Furthermore, should we place the variable "weeks from last promotion" higher in the tree than "promotion type"? Or the other way around?

The logic of constructing a decision tree is always the same: At each node, we ask what variable will allow the splitting of records in such a way that the resulting branches are as different from each other as possible, and the records of each branch are as similar as possible? By following this logic, we would only create one tree, and we obviously need more to grow an entire forest. We can accomplish this through two different approaches, both of which incorporate aspects of randomness.

Because decision trees are very sensitive to training data, a relatively small change to the training data can result in significantly different tree structures. The first approach takes advantage of this observation by allowing each tree to randomly sample the training data. Because the sample is random, the resulting individual trees would be different. This process is known as *bagging*; here it's sufficient to point out that with bagging, we aren't cutting the training data into smaller sets and training each tree on a different set, but rather, we're still training each tree on a dataset of the same size. The difference is that instead of the original training dataset, we take a random sample of the same size *with repetition* (meaning we can select the same record from the dataset multiple times, in effect duplicating those records). For example, if our original training data was {a, b, c, d, e, f, g, h} then we may create a training set for one of the trees as {a, c, c, c, d, d, f, h} and {b, b, c, e, e, e, g, h} for another tree. Notice that both lists are of equal length and that some records (e.g. "c" and "d" in the first set, and "b" and "e" in the second set) are repeated in the randomly selected training data used by the tree (because we sampled the data with repetition). Because of this random sampling, the resulting trees are different.

The second approach relates to the variables rather than the training data. In a standard decision tree, we consider every possible variable[17] when splitting a node and then select one that produces the most separation between records, as discussed earlier. In random forests, however, a tree can only pick from a random subset of variables. This forces even more variation among the trees in the model, providing a further level of diversification. We can make the trees even more random by using additional random thresholds for each variable, rather than searching for the best possible thresholds as we would in a standard decision tree.

So, in random forests we end up with trees that are not only trained on different datasets, but also use different variables to make decisions. This creates uncorrelated trees that buffer and protect each other from their errors (as long

---

17 For practical reasons (to avoid a combinatorial explosion), most decision tree implementations consider binary splits only.

as they all don't constantly err in the same direction). While some trees may be wrong, many other trees will be right, so as a group the trees move in the correct direction. Note that the low correlation between trees in random forest is key, as uncorrelated trees can produce ensemble predictions that are more accurate than any of the individual predictions. In other words, if we input training data with variables and labels into a decision tree, it will formulate a set of rules that will be used to make predictions. On the other hand, a random forest algorithm randomly selects observations and variables to build several decision trees and then averages the results.

Returning to promotional planning, we have many variables to consider:

- *Week of the year*: which is a number from 1 to 52
- *On promotion*: which could be a flag to indicate whether particular product is on promotion during a particular week
- *Variables related to the promotional cycle*: such as "number of weeks since last promotion," "last promotional price," "lowest promotional price during the past 12 months," and "average gap between promotions"
- *Promotion type*: such as "in store," "online," or "in catalogue"
- *Variables related to weather*: such as "lowest temperature," "highest temperature," "rainfall" (for a given week of the year)
- *Retailer*: such as "Mary's Market"
- *Geography*: such as "VIC" or "NSW"
- *Variables related to competitors*: such as "lowest competitor price," "average competitor price," and "highest competitor price"
- *Key sales period*: could be a "yes" or "no" flag to mark an important selling day such Black Friday, Christmas, Boxing Day, or New Year's, as well as school holidays, long weekends, and public holidays
- *Number of stores in the state* (for a given retailer)
- *Average sales per store* (for a given retailer in a state)

For simplicity, let's consider the following six variables from this list for the prediction model (for each product, retailer, and state): key sales period flag, number of stores in the state, average sales per store, average competitor price, weeks without promotion, and promotion type. Also, the decision trees would only consider products on promotion, as some other model would handle non-promoted products (so we don't need to consider the variable "On Promotion").

We then repeat the process of growing individual trees, with each of them being trained on a random sample taken from the overall dataset with repetition, as described earlier. Furthermore, we only take a random subset of variables from the original six variables. Say, the first four decision trees are:

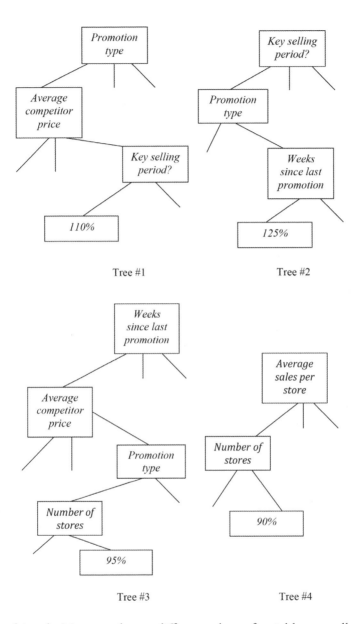

Tree #1

Tree #2

Tree #3

Tree #4

The resulting decision trees have a different subset of variables, as well as a different number of levels (as there were different numbers of selected variables), while the leaf nodes produce the output, which is the predicted sales volume for a particular product, retail chain, state, and week of the year.

In summary, random forests represent a versatile method that can be used for both regression and classification problems, and which produces reasonable results even without much parameter tuning. Another attractive quality of random forests is that we can measure each variable's relative importance on the final prediction and drop the variables that don't contribute much to the final result. This is important, because a general rule in Machine Learning is that the more variables we have, the more likely our model will suffer from overfitting. Thus "deep" decision trees might suffer from overfitting, and random forests prevent this by creating random subsets of the variables and building smaller trees using those subsets.

However, this method can also slow down computation time, depending on the number of trees in the random forest. In other words, these methods are fast to train, but slow to use on new data. Improved prediction accuracy requires more trees, which in turn results in a slower model. In most real-world applications, random forests are usually fast enough for most problems, but there are situations where run-time performance is important and other prediction methods are superior.

5.3 Genetic Programming

Genetic programming is a domain-independent AI method of evolving programs. Specifically, genetic programming iteratively transforms a population of computer programs (usually starting with a population of random programs) into a new generation of programs by applying operations analogous to natural genetic processes. Note that genetic programming is a special type of evolutionary algorithms which we touched upon in Chapter 3.3 (and will discuss further in Chapter 6.7 in the context of optimization), and so many of the same concepts apply. There is a population of individual solutions (i.e. computer programs) that compete for a place in future generations and for the placement of their offspring solutions. The process of evolution is simulated, and the best program emerges after some number of generations.

Genetic programming has been successfully applied as a Machine Learning method for a variety of tasks, including curve fitting, data modeling, symbolic regression, feature selection, and classification. In this section we'll use genetic programming to "evolve" a predictive model where historical data is available containing both input and output values, and our goal is to discover a program that can be used for predicting the output for future cases. In other words, we're searching for a program that would serve as a predictive model, and that we can consider such computer programs as functions that compute output based on a given input.

To begin with, let's say we have historical sales data for a particular product for some period of time, and we'd like to discover a function that describes the relationship between sales volume, price, and marketing expenditure. In other words, we'd like to discover a function F:

Volume = F(Price, Marketing)

Of course, the above example is simplified (e.g. the marketing expenditure has a delayed effect on sales volume), but it will suffice for illustrating how genetic programming works. The typical approach to solving this problem is based on assuming the general structure of function *F* and then directing our efforts to tuning some parameters. For example, we can assume that the relationship between sales volume and the other two variables has the following structure:

Volume = (a × (Marketing / Price)) + (b × (Marketing)) + c

where *a*, *b*, and *c* are unknown parameters. This function states that sales volume increases as marketing expenditure increases and/or product price decreases. The only issue is to find values for parameters *a*, *b*, and *c* that minimize the prediction error for the historical data. And here we have many possibilities: we can use simulated annealing or evolutionary algorithms to find the optimal vector of these three numbers *a*, *b*, and *c* (both methods are discussed in Chapters 6.4 and 6.7, respectively). However, the main weakness of this approach is that we *first* have to make an educated guess about the general form of the function. If our guess is close to the real function, then tuning these parameters will be straightforward. On the other hand, if our guess is incorrect, then we'll spend our time tuning the parameters of the wrong function.

For example, it might be that the function we're searching for is:

Volume = (12.3 × (Marketing² / Price²)) + (3.7 × (Marketing² / Price))
+ (2.9 × (Marketing / Price²)) + (21.8 × (Marketing / Price))
+ (8.7 × (Marketing²)) + (3.3 × Marketing) + 1346

and finding all the parameters (12.3, 3.7, etc.) is easy once we know the general form of this function. However, even in this very simple example, we can consider an almost countless number of possible functions! Through the use of genetic programming, we don't have to make assumptions on what the correct function might look like. This AI-based method allows us to investigate the search space of possible functions for one that fits the problem at hand—in particular, the best function that predicts sales volume based on marketing expenditure and product price. How can we do that? Well, let's explore this method in more detail.

When applying evolutionary algorithms to a specific problem, we have to design the structure of an individual solution, select an evaluation function that measures the quality of each solution, and decide upon the parameters (e.g. population size, probabilities of various crossover and mutation operators). Similarly, when applying genetic programming we must also follow a sequence of steps, including:

1. *Selecting the set of terminals*: which are all variables, parameters, etc. that correspond to the inputs of the function. In our example above, the set of terminals consists of marketing expenditure and product price, as well as a set of real numbers (e.g. 17.4).

2. *Selecting the set of primitive functions*: which are usually standard arithmetic operations like addition or subtraction, standard mathematical functions like log or square root, or domain-specific functions. In our example, a set of standard arithmetic operations extended by the square root function would be sufficient.

3. *Selecting the evaluation function*: is a key decision that ties the genetic program to the problem at hand, as the evaluation function evaluates how well a particular function solves the problem. In promotional planning and pricing, the evaluation function has to estimate how well each function describes the relationship between sales volume, marketing expenditure, and product price (the smaller the error on historical data, the better the fit). Of course, the error is measured on many data points and is often expressed as a total of absolute errors on all data points.

4. *Selecting the parameters of the genetic program*: which would include the population size, number of generations, probabilities of various operators, and possibly some other parameters that would influence the selective pressure of the algorithm (the higher the selective pressure, the smaller the chances that weaker individuals will be selected for reproduction).

Now let's take a closer look at some individual solutions (i.e. functions) that might emerge during a run of the genetic program. Assume that at some generation (say, generation 215), one of the solutions in the population is:

Volume = (Marketing × 21.8 × Marketing) / (Price × Price) – 3.7 × Marketing × Price + 2.9 × (sqrt(Marketing) / Price + 1192

To facilitate a discussion on crossover and mutation operators, we can represent this individual solution as a tree, and any particular solution can be represented by many different trees. For example, we can view the above solution as a sum of two parts:[18]

(Marketing × 21.8 × Marketing) / (Price × Price) – 3.7 × Marketing × Price

---

18 We can also view the above solution as a subtraction of two parts: *(Marketing × 21.8 × Marketing) / (Price × Price))* and *(3.7 × Marketing × Price) + (2.9 × (sqrt(Marketing)) / Price) + 1192*. In this interpretation, the root of the tree (the uppermost node) would represent subtraction (–).

and:

2.9 × sqrt(Marketing) / Price + 1192

Thus, the root of the tree (the uppermost node) represents addition (+), and the first part of the solution is a subtraction between two subparts:

(Marketing × 21.8 × Marketing) / (Price × Price)

and:

3.7 × Marketing × Price

This is represented by the appropriate node (–) in the left sub-tree. This process then continues further, where the first subpart is a division, the enumerator is:

(Marketing × 21.8 × Marketing)

and the denominator is:

(Price × Price)

Hence, the next node down in the tree represents division (/). The correspondence between the original solution and the tree below should now be straightforward:

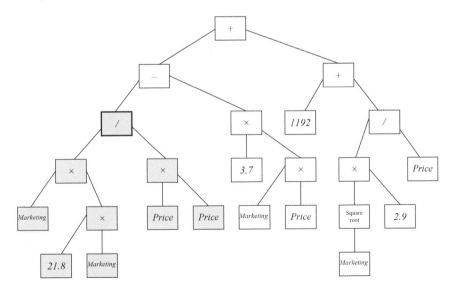

The above solution, which emerged in generation 215, will probably be modified further in subsequent generations. For example, if we assume a mutation operator was applied at the node with a bold outline (this node represents the division operator "/"). In further generations, a randomly generated sub-tree may replace this part of the tree, producing the following result:

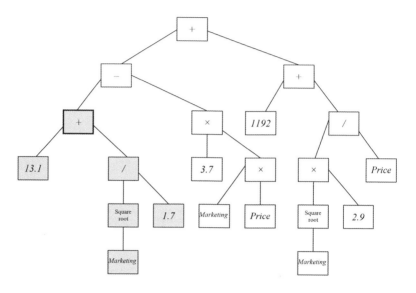

This individual solution corresponds to the following function:

Volume = 13.1 + sqrt(Marketing) / 1.7 – 3.7 × Marketing × Price + 2.9 × sqrt(Marketing) / Price) + 1192

Similarly, individuals may undergo *crossover*, where a sub-tree from one individual solution is swapped with a sub-tree from another solution. In other words, the crossover operator creates two offspring solutions by exchanging two sub-trees from two different parent solutions. For example, if the first parent is:

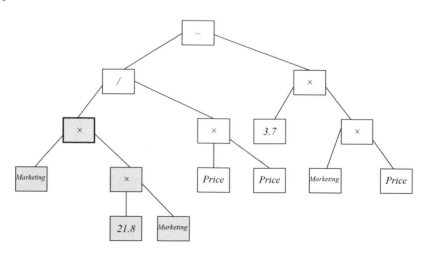

representing the function *Volume = (Marketing × 21.8 × Marketing) / (Price ×
Price) – 3.7 × Marketing × Price*, and the second parent is:

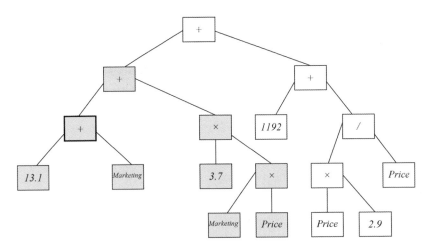

representing the function *Volume = 13.1 + Marketing + 3.7 × Marketing ×
Price + 1192 + (2.9 × Price) / Price*, then the first offspring is:[19]

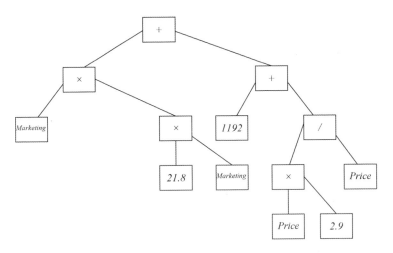

The second offspring is created by the other swap, where the grey sub-tree of
the first parent is replaced by the grey sub-tree of the second parent. After the
crossover, the first offspring represents the following function:

Volume = (21.8 × Marketing × Marketing) + 1192 + (2.9 × Price) / Price

---

19 Note that the node with a bold outline marks the cutting points for the crossover in both
    parents. As for the mutation operator, these cutting points represent where the sub-tree is
    cut and replaced by a sub-tree from the other parent.

There are many issues to deal with when applying genetic programming. First of all, there's a tendency to generate progressively more complex structures (trees), so the functions might become quite complicated and difficult to interpret. The simple function above, which emerged as an offspring solution after the crossover operator was applied, is an exception rather than the rule. Second, there might be parts of the function that are meaningless (like the division *Price / Price* above). Furthermore, most applications of genetic programming require relatively large population sizes, often requiring problem-specific knowledge to be introduced into the algorithm itself to make the evolutionary process more efficient.

Recall that the evaluation function is responsible for measuring how well a particular solution solves the problem. In our case, the evaluation function has to estimate how well the developed function describes the relationship between sales volume, marketing expenditure, and product price. As with other types of evolutionary algorithms (which we'll discuss in Chapter 6.7), the quality measure score is used in the selection process of parents, and the mutation and crossover operators are applied with some probability. All other mechanisms of evolutionary algorithms also apply here. Thus, the genetic program generates subsequent iterations of individual solutions (in our case, a function for sales volume) by creating and evaluating an initial population of individual solutions, selecting parents, applying the mutation and crossover operators, selecting a new population from the existing parent and offspring solutions, evaluating all the solutions in the new population, and continuing this process for some pre-specified number of generations.

There are many possible applications of genetic programming. In promotional planning and pricing, we could use this AI-based method to "evolve" a prediction model with all relevant variables (such as "promotional type," "week of the year," "promotional price," and so on) for each particular retailer and state, and use historic data to assess the accuracy of the models during the evolutionary process. Hence, instead of using this method to evolve a solution for a particular problem, such as a promotional plan for the problem of promotional planning and pricing, we can use genetic programming to evolve a function or model that will be used to solve the overall problem (e.g. evolve a prediction model that will be used to evaluate many possible promotional plans in our search for the best one). For these reasons and others, genetic programming is an important method in our portfolio of AI-based predictive methods.

5.4 Fuzzy Systems

Computers are based on a binary language of zeros ("0"s) and ones ("1"s), with "0" equating to "power off" and "1" equating to "power on." All these "0"s and "1"s inside a transistor chip are used to perform logical calculations according to *Boolean logic,* which treats 0 as "false" and 1 as "true." So, in other words, "0" = false = power off and "1" = true = power on. Now, using this type of logic, let's imagine water running from a tap: Either the water is running ("1") or it's not running ("0"). But unlike computer hardware operations, it's difficult to classify most real-world phenomena with crisp, binary classifications such as "yes" or "no." As an example, what if only a small trickle of water is flowing from the tap? Can we still say that the water is running from this tap? According to Boolean Logic, the water must either be running or not, but in this example, the water is only running to some *degree.* If we had to make a binary decision, we might say the water is still running, and inside the transistor chip this would translate to a "1" = true.

But what about a situation where only a few drops of water fall from the tap every minute? Again, we can ask ourselves if the water is running or not, and try to make another binary decision. However, we can continue this process indefinitely by continuing to reduce the number of drops falling from the tap (e.g. one drop per minute, one drop per hour, and so on). This example exposes the fundamental problem with Boolean logic: Everything must be either true or false, all or nothing. Boolean logic can't deal with something that's true to *some degree.* To deal with this problem, a relatively new type of logic has emerged, called *fuzzy logic.* Because it treats everything as a degree to which something is true, this AI-based method can be used to create effective prediction and classification models.

Overview

Let's begin by taking a look at one of the most common fuzzy logic systems, called the *Mamdani Fuzzy System,* as shown below:

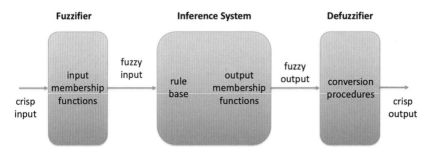

Each component in this flowchart has a separate function:

- The *fuzzyfier* takes crisp input numbers and converts them into a *fuzzy input set* by using input membership functions (explained below) to calculate the degree of something being true.[20]
- The *inference system* takes the fuzzy input set from the fuzzyfier, and applies rules from a rule base and then the output membership functions to create a *fuzzy output set.*
- The *defuzzyfier* takes the fuzzy output set from the inference system and converts it into a crisp output number (i.e. the prediction).

To better understanding how fuzzy logic extends Boolean logic, we'll examine each of these components below.

Fuzzyfier

For fuzzy logic to work, *input membership functions* are used by the fuzzyfier to turn crisp input numbers into a fuzzy input set. There are many types of membership functions, and the three figures below provide examples of the *Gaussian, bell,* and *triangular* membership functions (the horizontal axis covers the range from -10 to 10, but these values were selected arbitrarily, just to illustrate the concept):

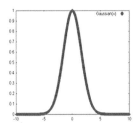 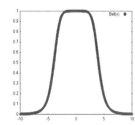 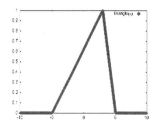

The rightmost figure shows something interesting: Membership functions don't have to be symmetrical and can take on different kinds of shapes. Nevertheless, simple membership functions are often preferred over complex ones because they're easier to understand and explain. We can also see from the above figures that membership functions always return a membership degree between 0 and 1, where 0 equates to "zero degree of membership" and 1 equates to "full degree of membership."

If we imagine a man who is 180 cm, we might say, "He's tall." Boolean logic would attach the value "true" or "false" to this statement, but fuzzy logic

---

20 Note that the *degree* of something being true is *not* the same as the *probability* of something being true.

would try to evaluate to what degree the statement is true with a membership function that defines "tall":

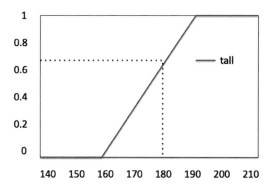

This membership function takes the height of a man and defines to what degree he is tall. If a man is shorter than 160 cm, then his degree of being tall is 0; and if he's taller than 190 cm, then his degree of being tall is 1. If his height falls somewhere between 160 and 190 cm, his degree of being tall is somewhere between 0 to 1. According to the membership function above, a man who is 180 cm is tall to a degree of 0.67.

Now imagine that we also have membership functions that define "short" and "average":

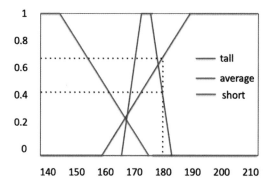

This figure above illustrates an important point: Just because a man is "tall" to a certain degree, doesn't mean that he's not also "short" and "average" to some other degree. In other words, these membership functions can overlap. A man who has a height of 180 cm would be "short" to degree 0, average to a degree of 0.42 and "tall" to a degree of 0.67. These numbers don't have to add up to 1 because "short," "average," and "tall" are completely different linguistic values (some other examples of linguistic values would be "red," "heavy," "fast," "close," "far," "warm," "cold," and so on).

For promotional planning and pricing, we might have the following input membership functions for "margin":

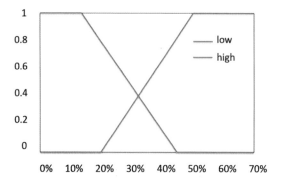

These input membership functions[21] refer to the variable "margin" on the x-axis. To create the fuzzy input set, the fuzzyfier takes the crisp input number in the range of 0% to 70% (let's assume that 70% is the highest margin possible) and then applies the membership functions to calculate the membership degrees. For example, a particular product with "margin" = 25% would be *low margin* to a degree of 0.6 and *high margin* to a degree of 0.2:

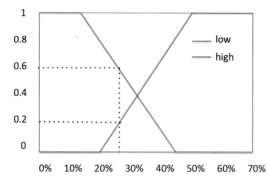

Hence, the fuzzyfier has transformed the crisp input number of 25% margin into a few linguistic values and their degrees of membership:

- "High margin" to a degree of 0.2
- "Low margin" to a degree of 0.6

These membership degrees constitute the fuzzy input set, which is used by the inference system to generate a fuzzy output set.

---

21 Although these input membership functions look simple, this does not have to be the case, and some membership functions might be quite complex. Furthermore, there are usually more than two linguistic values defined for a variable (not just "high" and "low", but also "medium", "very high", etc.).

Inference system

The *inference system* is the heart of a fuzzy logic algorithm and it contains knowledge in the form of rules and output membership functions. If we require exact predictions, then we'll end up with many rules; if we don't require such precision, then our rule base will contain fewer and more general rules. Although more rules usually mean more accurate predictions, the rule base can become too large (and, consequently, too difficult for human experts to comprehend and follow). Hence, there needs to be a balance between generality and precision for any real-world problem.

For promotional planning and pricing, we could build a rule base in one of two ways: The first would be to ask a human expert to define all rules that affect the sales volume for each product. This manually-built rule base might serve as a good starting point for further tuning (as the initial rules might be somewhat inaccurate), but a major drawback of this approach is the amount of time and effort required. For example, consider the countless membership functions that would need to be handcrafted and fine-tuned so that the predicted volume is close to the actual volume!

The second way of building a rule base is by carrying out a data mining exercise on the available data. There might be a few reasons for using this approach rather than querying an expert: For instance, a human expert may not be available for questioning, or the available expert might be unable to define the rules (as many decisions might be based on gut feel and intuition). Another reason could be that we want the rule base to be "unguided" and free of human assumptions and biases that might be flawed. For example, through data analysis we might discover effective rules based on "pack size" and "season"—a combination that human experts might not have considered.

For the sake of simplicity, let's look at a simple rule base for predicting the sales volume of a particular product for a specific retail chain in a specific state:

- Rule 1: *If sale price is medium and season is medium, then volume is average*
- Rule 2: *If sale price is low and season is high, then volume is high*

These two rules refer to the variables "sale price," "season," and "volume," and we assume that linguistic values for "sale price" and "season" were already defined by some triangular input membership functions. We now have to create a simple output membership function for "volume," say, with three linguistic values "high," "average," and "low," where the x-axis indicates the volume for a particular product:

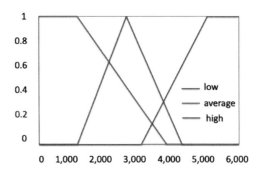

With a fuzzyfier capable of generating fuzzy input sets, and an inference system consisting of a rule base (with two simple rules) and an output membership function, we can now create a fuzzy output set. Let's visualize the fuzzification process for a product with "promotional price" = $49.95 and "week" = 14. As we've already discussed, the crisp input number "sale price" = $49.95 is transformed by the fuzzyfier to calculate a membership degree of 0.43 for the sale price being "medium" (shown below, top-left graph) and a membership degree of 0.6 for the sale price being "low" (shown below, bottom-left graph). Similarly, the crisp input number "week" = 14 is transformed by the fuzzyfier to calculate a membership degree of 0.3 for the season being "medium" (shown below, top-right graph) and a membership degree of 0.53 for the season being "high" (shown below, bottom-right graph). These linguistic and membership values make up the fuzzy input set:

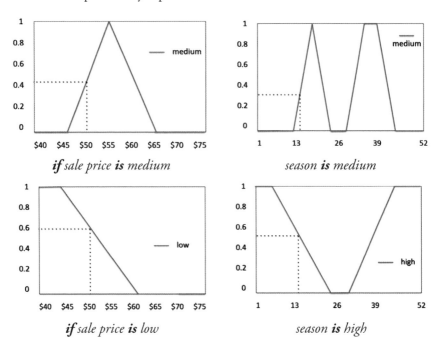

if sale price *is* medium   season *is* medium

if sale price *is* low   season *is high*

The inference system then takes this fuzzy input set and applies a fuzzy *and*,[22] which can be performed in a number of different ways. Two frequently used methods are to take the minimum input membership degree, or to multiply the input membership degrees together. In the above example, the inference system used the minimum input membership degree for fuzzy *and*. If we check the first rule:

> Rule 1: *If sale price is medium and season is medium, then volume* is *average*

the degree of volume being "average" is 0.3 in the output membership function (shown below, in the top-right graph). The inference system then repeats this process for the second rule:

> Rule 2: *If sale price is low and season is high, then volume is good*

but sale price is now "low" and season is "high." Since the input membership degree of season being "high" is lower with 0.53 (while the degree of sale price "low" is 0.6), the inference system sets the degree of volume being "good" to 0.53 in the output membership function (shown below, in the bottom-right graph).

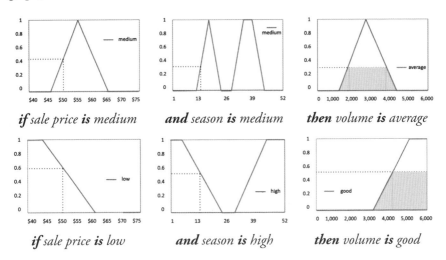

if sale price is medium   *and season is medium*   *then volume is average*

if sale price is low   *and season is high*   *then volume is good*

After all rules are processed sequentially, the next step is to combine the results of the output membership functions into one fuzzy output set for "volume."

---

22 *Fuzzy and* is an extension to the logical *and* in Boolean logic. If we denote *false* by 0 and *true* by 1, then using a *fuzzy minimum* for logical *and* would give the following results: *false* (=0) *and false* (=0) → *false* (=0), *true* (=1) *and false* (=0) → *false* (=0), *false* (=0) *and true* (=1) → *false* (=0), and *true* (=1) *and true* (=1) → *true* (=1). Hence, in the extreme (with "0"s and "1"s), fuzzy logic is reduced to Boolean logic.

This is accomplished by overlapping the gray areas illustrated in the top-right graph and bottom-right graph above. The result of the inference system is a fuzzy output set illustrated by the gray area in the following illustration (this is an enlargement of the top-right and bottom-right graphs above overlaid with each other):

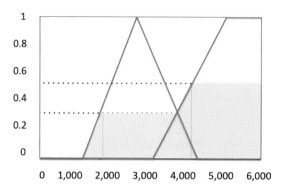

Processing this fuzzy output set (i.e. the grey area) is the responsibility of the defuzzyfier.

Defuzzyfier

The *defuzzyfier* takes the fuzzy output set from the inference system and converts it into a crisp output number. A defuzzyfier can operate in a number of different ways, and one of the most common is the *center of mass defuzzyfier*. It works in the following way: Imagine that the gray area (i.e. fuzzy output set) is a piece of wood that we have to balance on one finger:

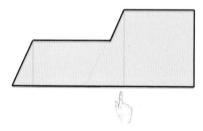

The place where the finger touches the gray area is the *center of mass*. The defuzzy-fier calculates this exact spot, which results in the following defuzzyfication:

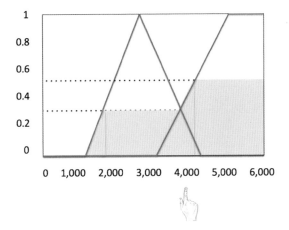

This illustration shows that the center of mass is at 4,150 and so the defuzzyfier would return a predicted volume of 4,150 units (i.e. crisp output number) for this particular product with sale price of $49.95 during Week 14 (or particular retailer and state).

Tuning the membership functions and rule base

Regardless of whether we use human expertise or data mining to construct membership functions, we need to tune them to get the best possible performance; for promotional planning and pricing, this means minimizing the prediction error for the predicted volume. Through the tuning process, we can modify a few components of the fuzzy logic model:

- The output membership functions can be modified, while keeping the input membership functions static. For example, if we use triangular output membership functions, we can adjust the triangles to get more accurate predictions. This would make sense if we knew that the input membership functions were more or less perfect, or if the input membership functions are "given" (e.g. if they correlate to industry standards that we should follow).
- The input membership functions can be modified while keeping the output membership functions static. This would make sense if we knew that the output is relatively static (e.g. when we have to make a binary classification).
- Both the input and output membership functions can be modified. This is the most general way to tune the rule base, and is typically the preferred choice. We would choose this form of tuning for promotional planning and pricing, as both the input (e.g. sale prices) and output (e.g. range of volumes) can change over time.

If we can't reduce the prediction error by tuning the membership functions, then we should consider making larger adjustments. For instance, we can add or delete some linguistic values in the existing rule base, such as adding "*and package size is large*" to the *if* part of the rule "*if sale price is medium and season is medium, then volume is average.*" By adding linguistic values to rules, the rules become more specific; and by removing such values from existing rules (e.g. dropping "*season is medium*"), the rules become more general. If the prediction error is still unacceptable after such adjustments, we should consider adding or deleting entire rules, after which, we can repeat the above steps to fine-tune the entire rule base.

The above steps for tuning the membership functions and rule base are presented in order of increasing "severity," starting with relatively small tweaks of the membership functions all the way to adding/deleting/modifying entire rules. The level of tuning depends on the prediction error, with larger errors typically requiring larger adjustments. Note also that the tuning process should be repeated at regular intervals—the frequency of which is always problem-dependent and can vary from a few hours to a few months—as the fuzzy logic model should *adapt* to changes in the environment (i.e. changes in the economy, product range, competitor activities, etc.)—a topic we'll return to in Chapter 7.

5.5 Artificial Neural Networks

If we spend a significant amount of time trying to solve various business problems by "racking our brain," we might hit upon the idea of automating our thinking process. The idea would be to simulate certain brain functions within a computer program that can solve problems for us. However, our brains don't function in the same way as digital computers. First of all, biological processing is inherently parallel in nature, while traditional computing is sequential, where each step in an algorithm is processed "one at a time" until the termination condition is reached. Secondly, although conceptual similarities exist between the neurons in living brains and logic gates[23] in computers, the firing rates of biological neurons are much slower than computer logic gates: milliseconds for neurons versus nanoseconds for computers. And, third, the response of a biological neuron is somewhat erratic and noisy (with misfiring or no firings at all), while a computer logic gate has very controlled "noise." Because of these fundamental differences, we can conclude that different types

---

23 Logic gates are the basic building blocks of any digital system: they are electronic circuits having one or more inputs and only one output. The relationship between the input and the output is based on a certain logic. Based on this, logic gates are named as AND gate, OR gate, NOT gate, etc.

of devices that transform some input into output, can tackle different problems with different efficiency.

For example, because computers are excellent for quickly calculating arithmetic results, it's better to use a calculator rather than pen and paper for dividing 412.14823 by 519.442. By contrast, computers are not good at generalizing or handling conditions that fall outside the prescribed domain of possibilities. If a friend shaves his beard, we'd probably still recognize him; but a computer program might struggle if it relies on a sequence of "if-then" rules that correspond to the identification of specific features of our friend's face.

Does it have to be this way? Is this a fundamental restriction of computer processing? Or is it possible for computers to function more like biological neural networks? After all, a neural network is an input-output device. Hence, it should be possible to create models of how neural networks perform their input-output behavior and then capture this behavior in a computer program. The resulting *artificial neural network* might yield some of the processing capabilities of living brains, while still providing the computational speed of a computer chip. In this section we'll take a closer look at different artificial neural networks in the context of prediction models.

Overview

The human brain consists of approximately 100 billion *neurons* (i.e. brain cells), each of which is comprised of the following components:

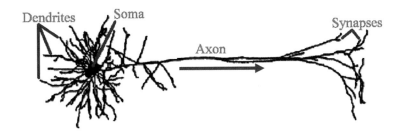

Although the human brain has thousands of different types of neurons, most of them behave in fundamentally the same way: The incoming chemical activity feeds into the *soma* (the body) via the *dendrites*, and if the chemical activity exceeds a certain threshold, then the neuron sends an electrical spike down the *axon*. This spike triggers the "firing" of chemical neuro-transmitters at the *synapses*. Because the neuro-transmitters lie in close proximity to other neurons, the discharge creates a chemical reaction in the next set of neurons. The result of this relatively simple chemical-electrical behavior is responsible for the amazing achievements of the human race.

The above illustration is just one example of a neuron, and below are some others:

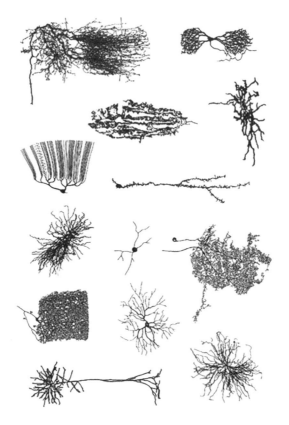

All these neurons are interconnected in a very complex manner, and on average, each neuron in the human brain connects to thousands of other neurons. Hence, the human brain is a complicated network of neurons with more than 100 trillion connections,[24] and science still has great difficulty explaining how the brain learns and performs its magnificent work. As explained in Chapter 1.2, this is one of the major barriers to realizing the original promise of Artificial Intelligence.

Like many other AI-based methods, artificial neural networks are inspired by nature, in that they attempt to mimic the human brain. The high-level idea behind artificial neural networks is quite simple: "create and connect many neurons with one another." However, a closer look reveals significant differences between artificial neural networks and biological neural networks.

---

24 Interestingly enough, recent research has shown that more neurons do not make us smarter. In other words, just because somebody has a bigger brain doesn't necessarily mean that they can process information more effectively.

Not only does the human brain consist of many different types of neurons, but different parts of brain have different architectures that process different types of information. Thus, each part of the brain takes control of a particular task or sense:

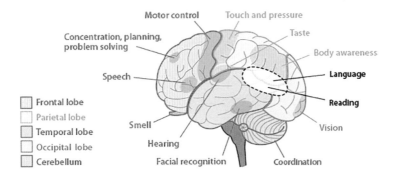

Hence, compared to the complexity of the biological brain, the most advanced artificial neural networks are still primitive toy-things by comparison, with the first artificial neural network (called the perceptron[25]) having only two layers:

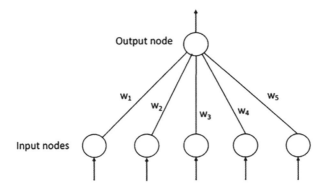

The perceptron was composed of a number of input nodes (the first layer), which directly connected (with some weights[26]) [to an output node (the second layer). Designed for image recognition, the perceptron had an array of 400 photocells that were randomly connected to the neurons. The weights were encoded in potentiometers and then updated by electric motors during the learning process. Back in the 1960's, the expectations for artificial neural networks were very high, and some people believed that the perceptron

---

25 The perceptron algorithm was invented in 1958 at the Cornell Aeronautical Laboratory by Frank Rosenblatt, that marked the beginning of the field of artificial neural networks.

26 A is the parameter within a neural network that transforms input data: as an input enters the node, it gets multiplied by a weight value and the resulting output is either observed or passed to the next layer in the neural network.

represented the embryo of an electronic computer that could walk, talk, see, write, reproduce itself and be self-aware of its existence!

Although the perceptron initially seemed promising, it quickly became evident that perceptrons couldn't be trained to recognize many classes of patterns. This caused the field of neural network research to stagnate for many years, before it was recognised that feed-forward neural networks with more layers (also called a multilayer perceptron) had greater processing power than perceptrons with just one input layer (also called a single layer perceptron).

It's worth noting that the complexity of biological neural networks evolved over millions of years, with a singular and very powerful measure of success: *survival of the human race.* Artificial neural networks on the other hand, only had 60 years to evolve, and that evolution happened within the confines of universities and research labs, and was severely limited by computing power (as the training of artificial neural networks was extremely expensive from a computational point of view). For these reasons and others, artificial neural networks still have a long way to go!

Different types of neural networks

Although many different types of artificial neural network exist, they can all be used to develop prediction models for classification problems. That said, different types of networks are more or less suited for different applications. Each neural network model consists of a number of layers, as opposed to the original perceptron that only had two layers. The different types of artificial neural networks include:

- *Feed-forward neural networks*: are similar to the perceptron, they are composed of an input and an output layer. Furthermore, they include one or more hidden layers in between. The hidden layer enables the network to deal with nonlinearities present in the training examples. In this network, the output of each node moves only to the next layer. Feed-forward neural networks are relatively easy to train. For example, an image labeling system might be based on a feed-forward neural network model.

- *Recurrent neural networks*: are similar to feed-forward neural networks, but the nodes in the hidden layers also have outputs connected to themselves. This enables each node in the hidden layer to feed itself with its own signal produced, thereby enabling them to gain memory-like functionality. In other words, when predicting cases presented in sequential order, recurrent neural networks are able to preserve the weights from predicting the previous case for the current case. Such networks can be used to improve the performance of feed-forward neural networks for predictions required in a sequential order. For example, to recognize activities in a stream of video imagery (e.g. a CCTV feed), the neural network can remember the weights from previous predictions (the previous frames in the stream). Recurrent neural networks are also widely adopted in the area of Natural Language Processing.

- *Long short-term memory (LSTM) neural networks*: represent an upgrade of recurrent neural networks, where memory nodes in the hidden layers can forget the weights of the previous stage. Similar to recurrent neural networks, LSTM neural networks are used when there is a sequential order of cases, but they are easier to adapt to significant changes in parts of the input. They can be used, for example, to segment the pixels in a movie, and weights can then be transferred from one frame to the next.

- *Convolutional neural networks*: represent an architecture where the hidden layers are formed by convolution and pooling layers. A convolution layer provides a tensor (imagine a matrix) to the pooling layer, the pooling layer selects a part of the matrix to be forwarded to the next convolution layer, and this continues until the final result is made. Due to their effectiveness for image processing, these types of networks have attracted substantial interest from researchers.

- *Generative adversarial networks*: are composed of two neural networks: One acts as a teacher and the other as a student. The teacher can be any type of neural network—for example, a feed-forward network that assigns a label for any given image. The student can also be any type of neural network, but it generates images on the basis of a label. This generated image is sent to the teacher for evaluation. For example, if the student (say, a convolutional neural network) is asked to draw the shape of heart, first it draws something random and gets a low score from the teacher. The student then tries another random drawing, gets the score, and so on. The process continues and the scores gradually improve, until the student finally draws a picture that scores close to 100%. This enables the student to generate results that haven't been seen before, and thus, the computer would achieve some level of creativity.[27]

In Chapter 1 we referred to *deep learning algorithms*, which were inspired by the structure and function of the brain and thus are usually associated with artificial neural networks. The term "deep" refers to the number of hidden layers in the neural network. Traditional neural networks only contain two

---

27 An example is the portrait of Belamy, the first AI-generated painting, was sold for $432,000 at Christie's auction in New York in 2018.

or three hidden layers, while deep networks can have hundreds of layers.[28] In other words, the phrase *deep* refers to *large* artificial neural networks with many layers. Deep learning models rely on neural network architectures that learn variables directly from the data without the need for manual variable extraction and are trained using large sets of labeled data. As we construct larger neural networks and train them with more data, their performance continues to increase. Today, some of the best-performing Artificial Intelligence applications are based on deep learning algorithms.

In the remaining parts of this section, we'll take a closer look at artificial neurons and a simple feed-forward neural network to understand some basic principles of their operation.

Node inputs and outputs

As mentioned earlier, biological neurons accumulate chemical input and then create an electrical output. In order to model such living neurons within a computer program, the first step is to calculate a weighted sum of the input activity coming from other nodes:

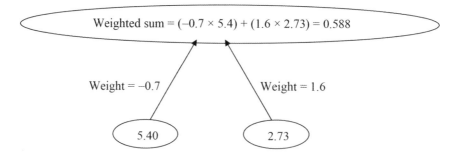

This figure above shows two nodes at the bottom (containing the values 5.40 and 2.73. The node at the top calculates the weighted sum by multiplying the values of the input nodes by their associated *connection weights*, which are parameters that represent the strengths of the connection between nodes; each arrow indicates the direction of the connection. Weights can be positive (exhibitory) or negative (inhibitory), which loosely resembles biological neurons and their connections.

Once the weighted sum is calculated, a node decides whether or not to send an output signal. A *squashing function*[29] is often used to make this determination:

---

28 There is no maximum number of layers in a deep neural network. However, the larger number of layers does not necessarily mean a better neural network model.

29 A squashing function calculates the output of a node by taking the weighted sum as input.

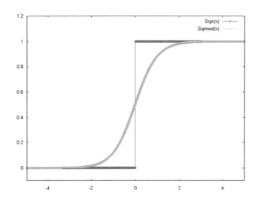

The *sign* and *sigmoid* squashing functions are shown above, with the weighted sum of the input on the horizontal axis and the output on the vertical axis. As shown above, the sign function generates an output when a specific threshold is crossed. In this case, there is no output when the weighted sum is 0 or less, and an output of strength 1 when the weighted sum is greater than 0. On the other hand, the sigmoid function is smoother than the discontinuous sign function, and the output value slowly rises toward the maximum output. In this example, the output values of the sigmoid function are:

- 0 for inputs less than -3
- 0.5 for an input of 0, and
- 1 for inputs of 3 or more

The sigmoid function is typically preferred over the sign function because it can be differentiated,[30] and some neural networks learning methods require differentiability. Nevertheless, the sign and sigmoid functions are just two examples of how nodes generate outputs. The following figure illustrates another type of output generation, where the nodes create a spike instead of a plateau:

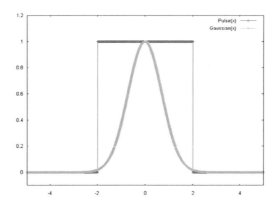

---

30 In layman terms, a "differentiable" function is relatively smooth.

The *pulse* and *Gaussian* squashing functions are shown above, with the weighted sum of the input on the horizontal axis and the output on the vertical axis. The pulse function has a distinct area where it spikes (from -2 to +2 in the above graph) to full strength of 1, and outside of this area, the output is 0. The Gaussian also has a spike, but it's much smoother than the pulse function. When comparing these different squashing functions, the sign and pulse functions are conceptually very simple, returning either 0 (no output) or 1 (full strength), but are discontinuous and non-differentiable (which may exclude some learning methods). The sigmoid and Gaussian functions, on the other hand, are more complex, but they provide continuous output values.

Feed-forward neural networks

Let's consider a simple feed-forward neural network, which doesn't have any recurrent connections between nodes:

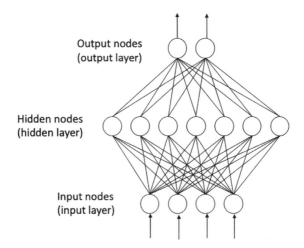

In this figure, there are three layers of nodes: The lowest layer is called the *input layer*, as these nodes accept the input values; the middle layer is called the *hidden layer*, and the top layer is called the *output layer*, as the activity that traversed the network finally arrives here. Many different neural network architectures are possible, and the number of nodes in the input and output layers are usually determined by the problem. However, the number of hidden layers and the connectivity between nodes and neighboring layers are design decisions that may vary from one model to the next. Thus, the number of possible architectures for any given problem is quite large.

The illustration above depicts a "fully connected" neural network, as every node on a lower layer is connected with every node on the next level (but this needn't be the case). Each of these connections also has its own weight.

A feed-forward neural network that's fully connected and has assigned weights for all connections works in the following manner:

1. The input values are fed directly into the nodes at the input layer.
2. The hidden nodes perform their calculations by summing up the weighted input received from the input layer nodes and by applying their squashing function to determine their output values.
3. Each output node determines its final output value by calculating the weighted sum using the values from the hidden nodes.[31]

For promotion planning and pricing, a feed-forward neural network that uses two input variables (promotional "price" and promotion "type") could look like:

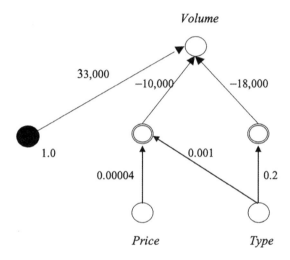

In this example, the variable "type" is numerical; as discussed earlier (Section 5.1) all promotion types can be assigned numbers that correspond to the significance of this particular promotion type. For example, assuming we have 6 different promotion types (non-promotion, catalogue special, in-store, 50% off, 2-for-1, and 3-for-2) then "non-promotion" might be the least significant as "6," "3-for-2" might be middle of the road as "3," and "50%-off" might be the most significant as "1." There's also a bias node (the black circle), which is added to increase the flexibility of the model to fit the data; the bias node in a neural network is always "on." That is, its value is set to 1.0 regardless of the data.

Now, the two hidden nodes (marked with double circles) use the sigmoid squashing function.[32] Rather than using a squashing function, the output node calculates its own weighted sum (using the black leftmost input unit with

---

31 Note that the output from the hidden nodes is the input to the output nodes.
32 The exact function is: Sigmoid $(x) = 1 / (1 + exp(-x))$.

a constant value of 1.0 to calculate the *threshold* for the output node). The output of the above neural network can also be described mathematically as a weighted sum:

Output = (33,000 × (1.0))
+ (-10,000 × Sigmoid (0.00004 × Price + 0.001 × Type))
+ (-18,000 × Sigmoid (0.2 × Type))

For this particular neural network prediction model, the following table shows the output values for different inputs of "price" and "type":

Price	Type	Volume
$19.95	6	14,150
$19.95	3	16,369
$21.95	1	18,098

Not surprisingly, the predicted output (volume) increases together with the significance of the promotion type. Further, determining the various weights (such as 0.00004 between the "price" input node and the hidden node) occurs during the training process, which we'll discuss later in this section. Note also that no connection exists between the "price" input node and the rightmost hidden node. Here we can assume there *is* a connection between these two nodes, but the assigned weight is 0. Although this neural network only has a single hidden layer, more often than not, additional hidden layers are used to improve the precision of the predicted output (in much the same way that more rules in fuzzy system usually mean more precise predictions). The exact number of hidden layers, connections, and hidden nodes—as well as the input-output function of the nodes—is dependent on the problem. This is why creating an effective and accurate neural network model still requires a considerable amount of scientific experience.

When discussing various types of neural networks, we can illustrate the structure by showing how the different layers are connected. For example, feed-forward neural networks would have the following structure:

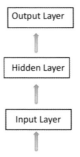

The bottom arrow indicates that input values are fed into the input layer nodes, and the middle arrow indicates that all activities are processed through the hidden layer toward the output layer. Once the nodes in the output layer have calculated their final output values, these values are often used for prediction or classification.

While feed-forward neural networks are fast, precise, and able to generalize well, their results are difficult to explain because the learned features are "hidden" in the weights. For this reason, neural networks are often referred to as "black boxes," because they do a great job of making predictions but it's difficult to understand how those predictions were made. Furthermore, because feed-forward neural networks can only process the entire input, one input at a time, they have no history or memory of earlier inputs, outputs, or processes—a deficiency that's addressed by recurrent neural networks.

Over the years, many different methods have been developed to train, adjust, and update neural network models. One category of such training methods is called *supervised learning*,[33] where historical data is available with both input and output values: for example, a collection of historical promotion plans along with actual results. In such situations, given all the data we possess about the results of different promotions over the years, we can apply supervised learning methods to train a neural network model for predicting volume. Let's say we've trained such a model, and the actual volume was 15,567 (for a particular product for a particular week) while our neural network predicted it would be 12,372. Obviously, the model made a significant error and requires further adjustment/training. This is often done using the quadratic error function *least mean square error* (LMS error), such that:

$$LMS\ error = 0.5 \times (actual\ volume - predicted\ volume)^2$$
$$= 0.5 \times (15,567 - 12,372)^2 = 5,104,012$$

The point of using the LMS error function is to *slightly* update the node weights so that the predicted sales volume would be closer to the actual volume. Although the LMS error function is most commonly used to update the weights,[34] the most well-known method for training a feed-forward neural network is *back-propagation*. This learning method corrects the error at each layer of the neural network by adjusting the node weights, starting at the output layer and moving back towards the input layer.[35] The training process

---

33 We'll discuss different types of learning methods and paradigms in Chapter 7.
34 To apply the LMS error function, we have to calculate the "gradient" by differentiating the squashing function of the nodes.
35 The volume prediction moves from input to output, while the error propagates in the opposite direction.

is typically repeated many times, until the LMS error is sufficiently low and the neural network has *learned* how to predict the volume of all possible promotion plans for all retailers and states.

Let's illustrate this back-propagation method with a simple example. Earlier in this section we described a neural network model for predicting sales volume of a particular product. This feed-forward neural network accepted two variables as inputs (promotional "price" and promotion "type") and six weights were assigned to the different connections between nodes:

- The three weights between the input nodes and hidden nodes were 0.00004, 0.001, and 0.2
- The three weights between the hidden nodes and the output node were 33,000, -10,000, and -18,000

The question is: How did we determine those weights?

Let's assume that at some earlier stage of the training process (i.e. during the back-propagation process), the neural network had the following connections and weights:

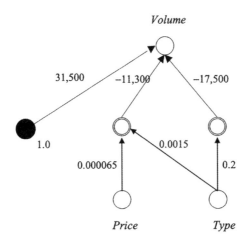

And let's also assume that for an input (say, "price" = $19.95 and "type" = 6) the model predicted a volume of:

12,372 = (31,500 × (1.0))
+ (-11,300 × Sigmoid (0.000065 × Price + 0.0015 × Type))
+ (-17,500 × Sigmoid (0.2 × Type))

while the actual volume turned out to be 15,567. Thus, the LMS error is:

5,104,012 = 0.5 × (actual volume − predicted volume)²
= 0.5 × (15,567 − 12,372)²

We now have to update the weights between the output node and the hidden nodes. The output node doesn't use a squashing function, so the update rule is quite simple:

$$weight_{new} = weight_{old} + \alpha \times error \times input$$

where α is the learning rate (larger values of α would result in larger adjustments of weights; in this example we assume $\alpha = 0.0001$). So, the new weights for the connections between the output node and the hidden nodes are:

31,500 + 0.0001 × error × input = −31,500 + 0.0001 × 5,104,012 × (−1) = **32,010**

−11,300 + 0.0001 × error × input = −11,300 + 0.0001 × 5,104,012 × Sigmoid (0.000065 × price + 0.0015 × type) = **−10,174**

−17,500 + 0.0001 × error × input = −17,500 + 0.0001 × 5,104,012 × Sigmoid (0.2 × type) = **−15,295**

To update the weights of the hidden nodes, we use the same rule:

$$weight_{new} = weight_{old} + \alpha \times error \times input$$

However, the error is calculated differently, as the hidden nodes include the squashing functions. Without going into the function details for error calculation, let's say the new values for these weights are:

0.000065 + 0.0001 × error × input = **0.000089**

0.0015 + 0.0001 × error × input = **0.0008**

0.2 + 0.0001 × error × input = **0.27**

So, after a single iteration (e.g. after adjusting for a single piece of data, where "price" = $19.95 and "type" = 6), the model is updated to:

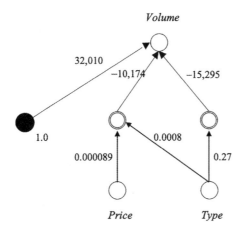

And this model, for "price" = $19.95 and "type" = 6 would produce the output of 14,138 units—a higher value than the earlier prediction of 12,372 and closer to the actual value of 15,567. This process is repeated (as the training process takes samples one by one) and the model becomes more accurate.

Several issues need to be addressed to make this training process work smoothly. First of all, we need to decide on the neural network's structure (also referred to as its architecture). Although many methods and heuristics exist for constructing a neural network's layers and connections, no efficient way of selecting the optimal structure is known. Furthermore, because the training process is usually quite slow (especially when multiple hidden layers are used), complicated learning methods are needed to accelerate this process. Lastly, a very large training dataset is usually required to train a neural network.

Despite these difficulties, neural networks can produce excellent predictions once properly trained, and retraining is straightforward when new data becomes available. Neural networks also interpolate[36] very well, especially if they don't contain too many nodes. Lastly, if enough training data is available, neural networks can efficiently handle noise. Within promotional planning and pricing, for example, it might happen that actual sales volumes were incorrectly recorded, and the resulting prediction error would constitute noise in the data.

Data representation

Neural networks are sometimes referred to as *universal approximators*, because of their efficiency in learning from data with an unknown underlying distribution. However, in order to construct an efficient neural network, some data representation issues first need to be resolved.

To start with, the typical input to a neural network is a sequence of numeric values (such as −5.3425 or +7.935) and the output is another sequence of numeric values. Note that the input doesn't include nominal values[37] like "in store," "yellow," "dog," "cat," etc. Simply assigning a numeric value to each nominal value would be disastrous, because neural networks cluster values that are numerically close together. If color is one of the attributes of a promotional product and we assign random numerical values to nominal values, then we could have the following situation: "white" = 6, "green" = 2, "red" = 3, and so forth. Using only one numeric input, it would be difficult for a neural

---

36 Interpolation is a type of estimation, a method of constructing new data points within the range of known data points.

37 Nominal values are values without ordering or distance between them, as discussed in Chapter 4.

network to separate products with these different colors,[38] and consequently, to make accurate predictions. If we instead translated colors into a Boolean-like vector using three inputs (where 1 is true and 0 is false) then we could express a "white" as "1, 0, 0", "green" as "0, 1, 0," and "red" as "0, 0, 1." Such a Boolean translation would make it significantly easier to train a neural network to predict sale volume, because different hidden nodes could *fire* more easily for products of different colors.

Another data representation issue is *high input dimensionality*, where an extremely large number of inputs is fed into the neural network. If we wanted to train a neural network to recognize faces, we might assume that the task would be easy. After all, a high-quality camera has millions of pixels, and each pixel could be represented with three integers: one describing the intensity of red, another for blue, and the third for green. But millions of inputs require millions of connection weights, and each weight requires training data to correctly distinguish different cases. This results in the need for an extremely large number of training pictures, which is not feasible in practice since the training process would take too long. For this reason, a special form of "compression" is often applied to reduce the input dimensionality—in this example, high-quality pictures are compressed by reducing their resolution, applying retina pre-processing[39] or a Fourier transformation.[40] This compression significantly reduces the number of inputs to the neural network and reduces the number of connection weights, which in turn makes it more feasible to train the neural network model within a reasonable time.

To summarize, we must keep the following in mind about the input for a neural network:

- Inputs consists of numeric values
- Numeric values that are "close" may be considered to "belong together"
- Input dimensionality must be reduced to a manageable size

In addition to these input issues, the output also requires special consideration, especially for classification problems. For facial recognition, the output consists of a numeric value. Using a single node, we could assign specific output values to specific faces, but this would create immense difficulties for a neural network. Numerically close values would be regarded as "similar

---

38 Recall that the hidden nodes use a squashing function that makes it inherently difficult to separate nominal values.

39 *Retina preprocessing* reduces the effective resolution to a neural network by averaging the values of neighboring pixels.

40 *Fourier transformation* creates waveforms from a picture that are used as inputs into a neural network.

faces," making it difficult to correctly identify different faces. We can solve this problem by using a *Boolean output vector translation*, which would be a separate output node for each face. Once all the nodes complete their calculations, we could then select the node (and hence the face) with the largest output value in a "winner-takes-all" fashion. Although this approach works well for a limited number of faces,[41] consideration has to be given to unknown faces.

5.6 Ensemble Models

As different prediction methods have different properties and may perform better or worse when trained on different data, it may be difficult to select "the best" prediction method for any given problem. The safest way to make consistently accurate predictions is based on the concept of *ensemble models,* where instead of using a single prediction method to build a single prediction model, we use a few methods to build a few models, and then use all the models together to reach a consensus. This consensus might be achieved through voting or averaging, with the final prediction being the one with the largest number of votes (for classification problems) or some weighted average (for regression/time series problems). After all, this is commonly done within most organizations, where several people in a team express their opinions on "what may happen" before a final decision is made.

Ensemble models can combine many different prediction methods—both classical and AI-based—each of which might be specialized for a particular purpose. For example, one model may capture the correlations between variables in labeled data, while another model handles uncertain or unlabelled data. In short, the models within an ensemble can complement each other's strengths and weaknesses. *Ensemble learning* is the process of generating and training multiple individual models, before combining them to generate the final prediction. There are many benefits of ensemble learning and the resulting ensemble model, which include:

- *More options for model development*: Through the use of ensemble learning, we can use different model structures for different problem types, thereby providing more flexibility during the implementation process. The models can simply be "stacked" on top of one another, or the output from one model can be used as the input into another model. We'll discuss these ensemble structures later in this section.
- *Comprehensiveness*: Ensemble models can usually cover a broader range of problems than any individual model. In Natural Language

---

41 This approach also works well for the recognition of handwritten characters and numbers.

Processing, for example, one model might perform better on shorter texts while another performs better on longer texts. The combination of these two models can cover a broader range of problems (i.e. both short and longer texts) than either model on its own.

- *Cost effectiveness*: Ensemble models can be used for sub-problems, and then combined to solve the larger, overall problem. As an example from Computer Vision, identifying various activities within video imagery can be divided into two sub-problems: recognition of body movements and classification of facial expressions, each of which are easier to handle separately and more cost-effective.
- *Improved prediction accuracy*: Different models within an ensemble can mitigate each other's weaknesses—for example, merging an overly optimistic prediction model with an overly pessimistic one, can potentially alleviate the deviation from either.

In short, the use of ensemble learning for constructing a more sophisticated model can be highly beneficial. However, as the scale and complexity of the model grows, we need to also consider the following:

- *Complexity*: while ensemble learning is a powerful approach for many problems, the additional complexity of the model can result in additional risks and costs. Hence, it's important to avoid any unnecessary complexity whenever possible; for example, by combining two models when one of them is already performing very well.
- *Dependency*: the performance of the overall ensemble model depends on the performance of its constituent models, which means that to get an accurate prediction from the top-level model, all constituent models must contribute. For example, if the top-level model for predicting sales volume includes a weather prediction model as one of its constituent models, then the top-level prediction model will be impacted if the weather prediction model fails to generate a prediction (or generates an inaccurate prediction).

In practice, the performance of an ensemble model is usually superior to any single constituent model. Furthermore, the use of ensemble models has become more attractive with the advent of affordable multi-processor servers, as constituent models can run in parallel using different processors. This provides all the benefit of an ensemble model (including improved prediction accuracy) at virtually no extra computational time, as all constituent models perform their calculations in parallel.

Different types of ensemble models

When building an ensemble model, there are three primary approaches we can take: *bagging, boosting,* and *stacking.* In this section, we'll provide a high-level description of each approach. Bagging—a name that stands for bootstrap aggregating—is a Machine Learning ensemble method designed to improve the stability and accuracy of Machine Learning algorithms used for classification and regression. It also reduces variance and helps to avoid overfitting.

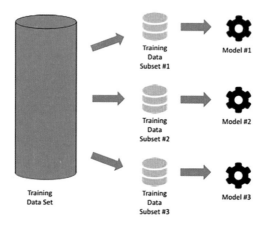

Bagging is based on the *bootstrap* technique (discussed further in the next section) which selects cases from the whole training dataset with repetition. Given a standard training dataset, bagging generates several new training data subsets of the same size by sampling from the original dataset uniformly with repetition (as shown above). Some records may appear more than once within each training data subset, and these differing subsets are then used to develop individual models (as shown above, where three models are built from the three training data subsets). As discussed earlier, the final prediction is generated by averaging or voting.

Ideally, an effective ensemble model consists of several complementary models that cover the vast majority of all possible cases. While bagging assumes that an "appropriate" type of prediction model has been selected, boosting addresses this issue directly by finding models that complement each other. Unlike bagging, where each model is developed independently, boosting takes advantage of pre-existing models. This makes it easier to find a model that addresses cases that earlier models didn't handle well, and is usually accomplished by assigning weights to cases so that misclassifications of higher weight cases would have more serious consequences than misclassifications of lower weight cases. Thus, each new prediction model focuses more on the higher weight cases (meaning those that were the most difficult to predict).

This iterative process of boosting can be best explained as follows: The first prediction model is built on the original dataset, where all records have the same weight; then, for records that were classified correctly (i.e. where the prediction was accurate), the weights are *decreased*, and for records that were misclassified (i.e. where the prediction was inaccurate), the weights are *increased*. The second prediction model is then built on the original dataset, however, the records now have modified weights, so the second model would focus more on the difficult cases (i.e. records with higher weights). Note that the weights are adjusted after each iteration of creating a new model, so that each subsequent iteration produces a prediction model that focuses more on the harder cases from the previous iteration. In other words, every new model *boosts* the previous one in the sequence.

So in *boosting*, models are created through an iterative process that applies higher weights to cases that are more difficult to predict. Repetition of this process creates a sequence of prediction models, where each new model focuses on the cases that were not accurately predicted in the previous model. This iterative process is the main difference between bagging, where all models are developed separately, and boosting, where each new model is influenced by the performance of the previous model. This approach allows us to cover a greater number of difficult cases and develop models that address such cases directly. As with bagging, several prediction models (usually of the same type) are used together, as shown below:

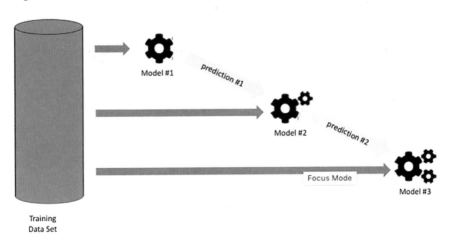

Although there are many similarities between bagging and boosting—as both techniques usually employ prediction models of the same type, and both techniques use voting or averaging—bagging usually creates a democracy of equal voters, while boosting uses weights to influence the model's performance.

We'll also return to bagging and boosting in our discussion on adaptability and learning in Chapter 7.

The third approach for building an ensemble model is *stacking*, where one prediction model uses the outputs (i.e. predictions) of several other prediction models to come up with the final prediction. The best example of stacking is where the outputs from several prediction models become an input into a neural network model, which then generates the final prediction. Note that this final prediction isn't the result of a vote or weighted average (as is the case in bagging and boosting), but rather of a higher-level prediction model that takes the predictions of the lower-level models into account! Therefore, this higher-level prediction model makes the final prediction on the basis of the preliminary, lower-level prediction models.

The training of a *stacked ensemble* model is done in two phases: During the first phase, a subset of training data (training data subset #1 in the figure below) is used to create a few models:

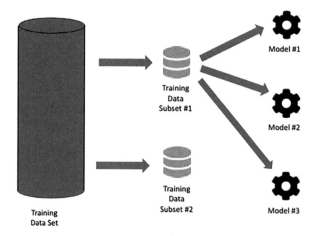

Whereas during the following phase, the second training data subset (training data subset #2 in the figure above) together with "ground truth assignment" data (predicted outcomes versus real outcomes of all trained models) are used to train the meta model (i.e. the higher-level model). The meta model then takes predictions of all the constituent models and combines their result to reduce the final prediction error, as shown below:

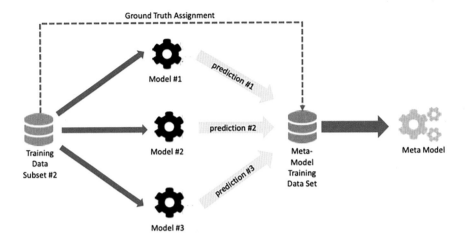

To summarize, ensemble models are developed to address different types of prediction problems. And although the constituent models of two ensemble models might be the same, the result can differ significantly between them.

Once an ensemble model is built, regardless of the approach, it makes predictions as illustrated below:

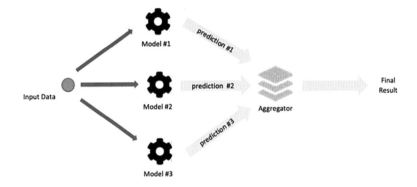

The input data is passed into the individual models to generate individual predictions. Next, the predictions are collected by an *aggregator* that generates the final result (the final prediction). How the aggregator calculates the final result depends on the nature of the problem and the architecture of the ensemble. In classification problems, the aggregator may simply count the number of models that predicted the same class, and the class with the highest number of predictions (votes) becomes the final result. In regression problems, on the other hand, the aggregator might take the average of the individual predictions, or a weighted average, where each weight is associated with past performance of the corresponding model. A model with more accurate predictions (with respect to the other models) may be assigned a higher weight. Another possibility for the aggregator is to build a separate, higher-level model that generates the final result based on the predictions made by the individual models. Such an aggregator is called a meta model, and is used in the stacking ensemble described earlier. Furthermore, the aggregator (whether a meta-model or a set of weights for calculating a weighted average) must be regularly updated with new data or direct feedback so the ensemble can learn and adapt to changes in the environment—a topic we'll return to in Chapter 7.

Let's conclude this section with an example of using an ensemble model for predicting price elasticity.

Using ensemble models to predict sales volume

To predict sales volume for any given product, we have to first understand its price and discount elasticity. As discussed in Chapter 3.2, price elasticity has its roots in Economic Theory and is part of the Law of Demand, which states that demand for any given product will go up as its price goes down, and vice versa. For example, when the price of a product increases, consumers may find an alternative and demand might fall sharply. For other products, the fall in demand might be minimal as consumers continue purchasing the product despite price increases (which is often true for luxury goods and tobacco products). Most products fit into one of three categories of price elasticity:

1. *Elastic*: is where a small change in price causes a disproportionately large change in demand for a product. Highly commoditized products fit into this category, where substitute products are readily available and the cost to switch from one product to another is low.
2. *Unitarian*: is where a change in price causes a proportionate change in demand, so that as price continues to rise, demand continues to fall in a lockstep manner.
3. *Inelastic*: is where a large change in price causes a disproportionately small change in demand for a product. Necessities like milk and bread fit into this category, as do luxury goods like exotic cars or gold watches, where a large increase in price results in a disproportionately small drop in demand.

While marketing departments focus on branding and positioning to reduce elasticity, sales departments often search for price points at which profit is maximized. This requires a good understanding of price elasticity for a given product, product specification, geography, week of the year, etc. The following figure shows a sample price elasticity chart:

The figure shows the predicted sales volume for different price points for a particular product. In this example, the orange line represents predicted sales (the orange *y axis*) while the blue line represents predicted profit (blue *y axis*) at the specified price in the *x-axis*. In the chart above, a $21 price would result in 46 units being sold. Comparably, a $20 price would result in 52 units being sold, but the increased volume doesn't offset the decreased margin and as a result, overall profit falls. On the other hand, if the price is increased to $22, volume remains steady and overall profit increases to $310. This might be due to a lack of competitors at this price point, allowing an increase in price from $21 to $22 without a corresponding fall in demand.

Once we understand price elasticity (how demand is affected by permanent changes in price), we then need to model discount elasticity to predict how demand is affected by temporary changes in price through promotional activities. To make such predictions, we can use any of the beforementioned types of ensemble models, or even merge a few types together. For this particular example we'll use a stacked ensemble model, the training of which takes place in two phases. During the first phase, we split the training data into two subsets: data for the first year and data for the second year (assuming we only have two years of historical data). The first year's data is then used to train two models—a time series model and a decision tree model—as shown below:

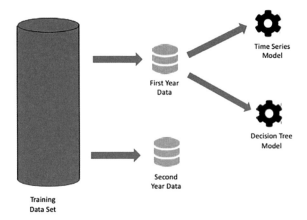

Each of these two models serves a certain purpose. The time series model is used to predict natural demand (i.e. the demand without any promotions) for a given product by looking at the average sales volume over time, which reduces the impact of any short-term promotions. As discussed in Section 5.1, based on the observed data (the graph in white below), the time series model focuses on trend and seasonality (the graph in green below) because the sales volume prediction would follow these trends if there weren't any promotions or changes in price. The decision tree model, on the other hand, focuses on the residuals (the graph in yellow) and is used to explain sudden and short-term spikes in sales as they relate to promotions or changes in price.

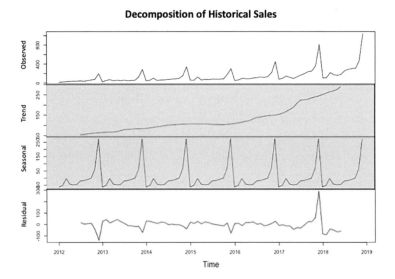

During the second phase of the training, a meta model (responsible for the final sales volume prediction) takes the output of the time series and decision tree models from the second training set, uses them as inputs to make the final prediction, which is then compared with the actual values, as shown below:

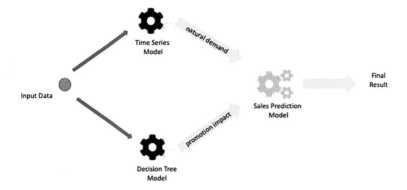

In other words, the second-year data is used to train the meta-model for predicting sales volume by combining the time series and decision tree predictions in such a way that the final prediction error is minimized. The meta model might be a feed-forward neural network, which, in addition to the two predictions made by the time series model and the decision tree, relies on many other inputs, such as the promotion flag, promotional period, promotion type, promotional price, among others. For example, the value of the promotion flag ("Y" for "on promotion" verses "N" for "not on promotion") would have a major impact on the final prediction. Another binary variable, "key sales period," would reduce the influence of the decision tree model during promotional periods that don't include a key selling date, and increase its influence otherwise. Furthermore, for products belonging to the "lower price" category (which are more elastic), the neural network would increase the decision tree's influence on the final prediction, while for products that belong to the luxury or premium categories (and which are more inelastic), the neural network would decrease its influence.

In the above example, both models receive input data such as the product, promotional period, promotion type, price and other required variables, and the time series model generates a prediction for natural demand using the trend and seasonality of the product, while the decision tree model receives more inputs and predicts the impact of promotions or price changes. Finally, the meta-model makes the final sales volume prediction by deciding to what extent each lower-level model should influence the final result. For additional information on some of the concepts presented in this section, we encourage you to watch the supplementary video for this chapter at: www.Complexica. com/RiseofAI/Chapter5.

5.7 Evaluation of Models

Several additional considerations should be taken into account when selecting a prediction method for a Decision Optimization System. Although the prediction error is quite possibly the most important measure, it provides only one dimension of a model's quality. For most real-world problems we need to consider several other factors, including:

- *Response time*: is an essential consideration, as any Decision Optimization System must have a defined response time. Fraud detection systems, for example, process millions of transactions per second, so the frequency of classifying new transactions into "fraudulent" or "legitimate" is very high. Other systems, on the other hand, might be used on a weekly basis—such as long-term forecasting models—and so the response time isn't as critical.
- *Editing*: is another important consideration, as it might be necessary to add human knowledge to the final model. Keeping this in mind, some prediction models are difficult to edit—such as neural networks—while others are easy, such as rule-based systems.
- *Explainability*: is an often-overlooked aspect of evaluating the usefulness of a prediction model. For some real-world problems, such as credit scoring, it's important to explain and justify the prediction; in some cases, it might even be a regulatory requirement (e.g. justification for rejecting a loan application).
- *Model compactness*: refers to the fact that prediction models shouldn't be overly large and complex, because they're likely to be slow when generating predictions and difficult for humans to understand. Hence, if two models provide very similar prediction accuracy, then the more compact of the two models is preferred.
- *Tolerance for noise*: all prediction methods require some approach for handling missing values (as discussed in Chapter 4.2), but some methods do a better job than others. Also, some values might be present, but noisy (i.e. imprecise), like stating that sales volume is "high."

Because of these factors, it may be difficult to select "the best" prediction method for the problem at hand—especially that different prediction methods have different properties, so some of them perform better or worse when trained on different data sets. It might therefore be worthwhile to use a few methods to build a few models, and then use all the models to reach a consensus, as discussed in the previous section.

Although it's possible to use a variety of different prediction methods to build a variety of different prediction models, the key issue is which method should be applied to a particular problem. To answer this question, it's necessary to evaluate and compare different models. Because these comparisons must be unbiased, the evaluation methodology should be fair and just. At first blush, this may seem easy: we just train a few models, test them on historical data, and measure the prediction error. The best model would then be selected.

Unfortunately, it's not that simple.

For starters, the amount of available data might be limited. In promotional planning, for example, if we take into account all the different products sold through all the different retailers in different states, each with a different promotion type and price, our "large" dataset (of say, three million records) might actually represent quite a small number of relevant records. For example, 100 products sold across 10 retailers in 6 states translates into 6,000 different combinations, and our three million records suddenly becomes just a few hundred records for each product/retailer/state combination. Also, because some product/retailer combinations are very common while others aren't, the rare product/retailer combinations will only have a few records per state at irregular intervals (we'll discuss the issue of reasoning from insufficient data later in this section).

Secondly, the performance of any prediction model on training data might be very different from its performance on new data. This might be due to overfitting—which is a common phenomenon—as a model is trained to such an extent that all predictions on the training dataset are perfect! However, the goal is to build a prediction model that performs well on *new* data!

Thirdly, models that provide different outcomes require different techniques for measuring prediction error. For instance, a prediction model may indicate whether a new case belongs to class A or class B, or, if we have a larger number of classes, the probability that a new case belongs to each class. Alternately, a prediction model may predict a number (e.g. sales volume) or a sequence of numbers (e.g. sales volume and net revenue), and in each of these examples, we have to consider what the outcome of the prediction model is and apply the appropriate error measurement technique.

Furthermore, as indicated earlier, the aspect of self-explainability might be important for particular problems. In general, there is an emerging trend called "explainable AI," as users of AI-based systems want to understand why certain predictions or recommendations were made by the underlying model. Apart from the explanation itself—which is often very valuable—explainable AI also helps to build trust with users. There are a few methodologies used for explaining predictions, which are based on either:

- Generating an importance measure for each variable, so we can understand which variables play a major role in generating the final prediction

- Discovering rules that define a subset of the variables that determine particular types of prediction, or
- Considering results from game theory that apply to situations where each variable is a member of a team, and these teams are making predictions

Finally, we have to consider the cost of a potential error. When classifying records into two categories ("yes" or "no," "fraudulent" or "legitimate," etc.), there are two types of errors: a *false-positive* is where the outcome is incorrectly predicted as "yes," and a *false-negative*, where the outcome is incorrectly predicted as "no." Clearly, the cost of these errors is very different. By classifying a legitimate transaction as fraudulent (false-positive), there's a small cost to check the transaction. On the other hand, classifying a fraudulent transaction as legitimate (false-negative) usually carries a much higher cost, especially if the transaction is significant (e.g. approving a fraudulent transaction for $5,000).

Because different models may generate a different number of false-positives and false-negatives on test data, the costs of these two types of errors must be taken into account. Although many techniques exist for error measurement—such as mean-squared error, mean absolute error, relative squared error, or relative absolute error—it's much harder to measure the consequence of an error. For example, the error in a volume prediction of a particular product for a particular retail chain might only be 450 (less than 1% of the total predicted volume), but this error may influence other promotional decisions.

Because we're interested in the *future* performance of a prediction model— in other words, the model's performance on *new* data, rather than on training data—we cannot take a model's performance or error rate on the training data as a foolproof indicator of future performance. The reason for this is simple: The most "reliable" prediction model would be a lookup table where all previous records are stored. Such a model would score exceptionally well on old records, but this score would tell us nothing about the model's performance on new data! When prediction models are overtrained, they often behave like a lookup table. Hence, a model's performance on the training data will always be better than the model's future performance.

To predict a model's performance on new data, we need another set of data (usually called a *test dataset*) that didn't participate in the process of building, training, and/or tuning of the model. This is important, because we need *fresh* data (i.e. data unseen by the model) to evaluate the model's performance. The most popular way of doing this is by randomly dividing the original dataset into a training dataset and test dataset. The prediction method uses the training dataset to select variables, compose additional variables, calculate ratios, parameters, etc., but doesn't have access to the test dataset. Once the prediction model is built on the basis of the training dataset, it can be evaluated for performance on the test dataset.

In many cases, the process of building a prediction model consists of two phases: building a model, and then tuning its parameters. For this reason, it's worthwhile to split the training data set into two further subsets: the primary training dataset and a *validation dataset*—the former for building the model, the latter for tuning its parameters. So, altogether, it's ideal to have three independent datasets, with the third one being the test dataset to evaluate the model's performance. Each of these three datasets should be selected independently, and each of them plays an important and independent role:

- The training dataset is used for building the prediction model
- The validation dataset is used for tuning the parameters of the model (i.e. for optimizing the performance of the model[42])
- The test dataset is used to evaluate the model's performance

If we had plenty of data for training, validation, and evaluation, then the result should be a better model. However, if there's only a limited amount of data, then what can we do? Note again that the general idea is to split the data and use two thirds for training and validation, and one third for testing.

The first issue to consider is whether each data subset is representative of the entire dataset. For example, it may happen that the training dataset has no records with "in store" promotions, while the test dataset contains many. If a particular value is missing in the training dataset, then the prediction model might have serious difficulties in predicting the right outcome when this value is missing (as the learning process is based on data). Moreover, the evaluation of the prediction model would be biased, because all records with "in store" promotion would only appear only in the test dataset!

Clearly, it would be beneficial to "guarantee" that the distribution of records is uniform across all datasets. One way of approaching this problem is through *stratification*, by using an algorithm to split the data into training and testing subsets in such a way that each value is properly represented. The other approach is to repeat the training and testing with different datasets, and then average the performance of the prediction model from all these iterations. A popular statistical technique, called *cross-validation*, is often used in connection with the latter approach. In this technique, we divide the dataset into some number (say k) of disjoined subsets (called *folds*). Then $k - 1$ folds

---

42 If several prediction models are constructed from the same training dataset, then the validation dataset is sometimes used for selecting the best model.

are used for training and one for testing, and we can repeat this process k times, each time with a different group of folds selected for training and a different fold for testing. If $k = 3$ (i.e. the dataset is partitioned into three subsets), then the technique is called *three-fold cross-validation*. It's quite common to use $k = 10$ (*10-fold cross-validation*),[43] as 10 is a reasonable number of folds to estimate the prediction error.[44]

One extreme (and, in many cases, useful) application of the cross-validation technique is when the number of folds equals the number of records in the dataset, which is called the *leave-one-out* approach. In a dataset with three million records, there would be three million folds. Hence, we would repeat the following process three million times: A prediction model is built on a training dataset of 2,999,999 records and the error estimate is made on the remaining single record. Once this process is completed, the average of all errors will give us the error estimate for the prediction model. In this technique, the largest possible amount of data is used for training, and, because the approach is deterministic, there's no need to repeat the process. However, the computational cost might be prohibitive for some datasets.

The final model evaluation technique we'll mention is bootstrapping, which we discussed in the previous section and which has a reputation for being one of the best techniques when data is limited. In bootstrapping, a collection of records is selected as the training dataset with repetition, with the number of records in the training dataset equal to the total number of records in the original dataset. During this process, some records are selected more than once, while some records aren't selected at all. It's relatively easy for a mathematician to calculate the probability of a record *not* being selected for the training dataset by dividing the constant e by 1, which equals to 0.36787944117 ≈ 0.368. This means that approximately 36.8% records won't be selected, and 63.2% of records will be selected at least once.[45] If we apply bootstrapping to a dataset of three million records, then approximately 1,896,362 records would be selected at least once for the training dataset, whereas the remaining 1,103,638 records would become the test dataset. Like cross-validation, the bootstrap procedure is usually repeated several times with different samples.

As mentioned earlier, most real-world problems have some time-dependent relationships within their data, such as transactions, orders, deliveries, sales,

---

43 *10-fold cross-validation* is often used with stratification. Stratified 10-fold cross-validation is generally held as a standard evaluation technique when the amount of data is limited.

44 This estimation, however, needn't be perfect, as different fold selections may give different error estimates. Thus, it's a standard procedure to repeat the cross-validation process 10 times, which results in building and testing a prediction model 100 times altogether.

45 Because 63.2% of the records (on average) will be selected for the training dataset, this technique is also called the *0.632 bootstrap*.

and so on—all of which have a time stamp. These datasets change over time, with some datasets changing quickly (e.g. the closing price of S&P 500 stocks) and others more slowly (e.g. the average income in a particular region). As a matter of fact, some changes are so slow that we consider the dataset to be static, even though small changes are constantly taking place. In any case, it's important to organize records in such a way that training records have an earlier timestamp than test records. This is done so that the predictions go from the past into the future. In other words, we should identify a particular point of time, and take all relevant preceding records for the training dataset and all relevant subsequent records for the test dataset. Note also, that the time dependencies among records might be so strong that we should treat the dataset as a time series, where all records are kept in a sequential time order.

These inevitable changes that occur in data—from which we must create a prediction model—have powerful consequences. If the changes are slight, then the sampling and evaluation techniques discussed in this section would work. However, if the changes are significant, like after a major stock market crash or natural disaster, then it might necessary to build a new model altogether. Also, different prediction methods produce different models of varying complexity. For this reason, it might be safer to select a simpler model that has a higher degree of generality, which allows for better adaptation to small changes that occur in the data. Another approach, which we'll discuss further in Chapter 7, is to use a learning mechanism to automatically adjust the various parameters in the model.

5.8 Closing Remarks

In this chapter, we discussed various prediction methods that are all based on the same premise: patterns from the past will repeat in the future. Thus, the discovery of these past patterns (and training the predictive model to recognize them) is of utmost importance. The point is that once we've identified a pattern, it might be easier to predict the next value and/or event. There's no need to argue about the importance of such activities, as most of our predictions in life are based on patterns or trends that we've observed. However, there are also some dangers here, as illustrated by the following example of flipping a coin many times and getting the following sequence of heads (H) or tails (T):

```
HHTHTTTTHTHHHTTTHTHHTHTHHHTHTTTTTHTHTHT
HTHTHHHHHHHTHTHTHHTHTTTTTTHHTHTHTHHTT
THHHHHTHTHHTHTHTHTHHTHTHHTHTHTTTTTHTHT
HTHTHTHTHHTHTHTHHTHHTHTHHHHHTHTTTTTHTT
HHTHTHTHTHHTTTHTHTHHHHTHTHHTHTHHTHTHTH
HTHTTTHHTHTHTHTHHTTTHTHTTTTHTHTHHTHTTTT
TTHHTHTHTT
```

We can analyze this sequence and discover some interesting patterns that would be completely false, in the sense that these patterns are unlikely to repeat in the future. For example, the above sequence of coin flips may suggest that "every sequence of TTTTTTT is followed by HH," as there are two such occurrences:

HHTHTTTTHTHHHTTTHTHHTHTHHHTHTTTTTHTHTH
THTHTHHHHHHHTHTHTHHTHTTTTTTTHHTHTHTHHT
TTHHHHHTHTHHTHTHTHTHHTHTHHTHTHTTTTTHT
HTHTHTHTHTHHTHTHTHHTHHTHTHHHHHHTHTTTTT
HTTHHHTHTHTHTHHTTTHTHTHHHHHTHTHHTHTHHTH
THTHHTHTTTHHHTHTHTHTHHTTTHTHTTTTHTHTHHTH
TTTTTTTHHTHTHTT

Such conclusions are meaningless and follow the rule: *If we torture the data long enough, the data will confess!* As silly as that sounds, it happens on a daily basis—just turn on the evening news and listen to business commentators "explaining" the stock market moves from that day.

Talking about random sequences and patterns (as the one above), it's hard to resist another observation: Most people tend to think that the number of heads (H) and tails (T) in the long run should be the same. In other words, the deviation of number of heads (H) and tails (T) from the expected values (occurrence of 50%) in the long run is smaller and smaller. They interpret this as a prediction: the larger the deviation at any point, the greater the restoring force toward the expected value. We can see this clearly during a roulette game: after seven "reds" many players bet on "black."

This, however, isn't the case. Although the ratio between the number of heads and tails does get closer to 1 in the long run, it doesn't in absolute numbers. For example, after 100 flips we may get 53 heads to 47 tails, and after 1,000 flips: 523 heads to 477 tails. Then:

53/47 ≈ 1.1277 and 523/477 ≈ 1.0964

so the ratio is indeed getting smaller, whereas the difference in absolute numbers:

53 − 47 = 6 and 523 − 477 = 46

is getting larger! It's important not to confuse these two phenomena.

There's another surprising phenomenon connected with pattern recognition and the number of consecutive heads or tails: There are well-known cases of mathematicians who entertain students by asking someone to generate a made-up sequence of 100 coin flips and a different student is asked to flip a real coin 100 times and record the results. The teacher then looks at both results and correctly identifies the real-life coin flips. Mathematicians manage this by picking the sequence with the longest streak of either heads or tails! The fake results contain much shorter sequences.

The following story illustrates the danger of reasoning on past patterns that are too short to be meaningful: Two men are traveling in the same train compartment, with the younger one being in his mid-twenties, while the older is in early sixties. The younger fellow is deeply involved in a book. His cheeks are pink, his eyes are glued to the text, and he's turning the pages carefully and intently. Every now and then, he expresses his astonishment aloud: "I can't believe it!" he would say, or "This can't be true …"

He finally slams the book shut on his knee and shouts: "Impossible!"

His behavior piques the curiosity of the older man, who asks: "Excuse me, young man. May I ask what book you're reading?"

"Of course," replies the other, "I'm reading the biography of Albert Einstein."

"And what's so *impossible* about it?"

"Listen," said the young man, "I read that Albert Einstein's father was a salesman, and see, my father was also a salesman! I read that Albert had speech difficulties in his early childhood, and you won't believe this, but I also had problems with early speech! Then a few pages later I read that he took violin lessons for seven years while attending Catholic school. This was precisely my situation—I was also sent to a Catholic school and my mother forced me to take violin classes! Albert then met his future wife when he was just seventeen and this also happened to me! But that's not the best part! His most significant contribution to science—the publication of scientific papers for which he was later awarded the Nobel Prize in Physics—happened when he was 26 years old! And listen here … tomorrow is my 26th birthday! I'm so excited!"

This story emphasizes the point that discovering patterns in data is relatively easy, whereas discovering useful patterns that can help us make predictions or solve problems is much harder. In other words, when searching for patterns, we should keep the following questions in mind: How will this help us make an accurate prediction? How will this help us solve the problem? Sherlock Holmes summarized this very well in the story *The Adventure of the Speckled Band*, where he made the following remark: "I came to an entirely erroneous conclusion which shows, my dear Watson, how dangerous it is to reason from insufficient data."

For more information on the material covered in this chapter, please watch the supplementary video at: www.Complexica.com/RiseofAI/Chapter5.

CHAPTER 6

Optimization

"True optimization is the revolutionary contribution of modern research to decision processes."

George Dantzig, *mathematical scientist*

Whether in banking, manufacturing, or retail, there's scarcely an industry where the term "optimization" isn't bandied about. This is because every industry strives for excellence amid continual pressures to reduce costs and increase efficiency, and so over the years many optimization methods have emerged to help organizations find better solutions to their business problems. The field of operations research, in particular, developed many methods to address the complexity of scheduling people, machines, and materials. We often refer to these optimization methods as being "classical," with the best examples being linear programming, branch and bound, dynamic programming, and network flow programming.

However, during the last few decades we've witnessed the emergence of a new class of AI optimization methods, sometimes referred to as "modern heuristics." These methods include (among others) simulated annealing, tabu search, and evolutionary algorithms, and are the main focus of this chapter. But irrespective of the optimization method used, three things always need to be specified: (1) the variables of the problem that represent a solution, (2) the problem-specific constraints, and (3) the objective, which has to be converted into a formula called the *evaluation function*. This function is used to evaluate how "good" or "bad" each solution is by calculating its *quality measure score,* which allows us to compare one solution with any another (in Chapter 3, for example, the quality measure score for *Plan A* was the predicted volume of the promotional plan). These three components constitute the model of an optimization problem. Let's consider each of these points in turn.

The representation of a solution, given in terms of the variables of the problem we're trying to solve or optimize, will determine the *search space* and its size (which represents the total number of possible solutions to any given problem, as we discussed in Chapter 2.1). This is important, because the size of the search space isn't determined by the problem itself, but by its *representation*. Consequently, choosing the right search space is of paramount importance.

The search space for an optimization problem is defined as a collection of *all* potential solutions to the problem, regardless of their quality measure score (i.e. irrespective of how "good" or "bad" they are). And these solutions, on the other hand, are defined by all the variable values we can change during the optimization process. For example, referring back to traveling salesman problem introduced in Chapter 1.3 and discussed further in Chapter 2.1, the search space for a problem with seven cities consists of 360 possible solutions (which is the search space for this problem). Each solution is defined by variable values (which city to visit next), and these values can be changed during the optimization process as we search for the best solution.

Once we've defined the search space, we need to decide what we're looking for, which is our objective. In the traveling salesman problem, our objective is to minimize the overall distance the salesman must travel when visiting all seven cities. After the objective has been clearly defined, the next thing to do is create an *evaluation function* (quality measure score) that allows us to compare the quality of one solution against any other solutions. Some evaluation functions produce a ranking for various solutions (called *ordinal* evaluation functions), while others are *numeric* and provide a ranking and a quality measure score as well.

Because the evaluation function is not provided with a problem, how can we choose the correct evaluation function? Oftentimes, the objective can suggest a particular evaluation function for finding the best solution. In the traveling salesman problem, the evaluation function is intuitive and straightforward, being the total distance traveled for any particular solution. On the other hand, the construction of an evaluation function might be quite tricky, requiring us to combine various objectives into one quality measure score, as well as addressing possible violations of soft business rules and constraints. When designing the evaluation function, it's also essential to keep in mind that most of the solutions we're interested in will be in a relatively small subset of the search space, because we're only interested in *feasible* solutions—those that satisfy our business rules and constraints.

Once these steps are complete, we can begin searching for a solution. Note, however, that the optimization method we use doesn't know what problem we're trying to solve! All it "knows" is the representation of the solution and the evaluation function. If our evaluation function doesn't correspond to the objective, then we'll be searching for the right answer to the wrong problem! Let's start our discussion on optimization with two puzzles that illustrate these concepts, before moving onto business examples.

6.1 Two Optimization Puzzles

The first puzzle requires that four travelers (**A**, **B**, **C**, and **D**) cross a bridge over a deep ravine. It's dark outside and the travelers only have one lamp—which is needed for crossing the ravine—and because the bridge is old with plenty of holes and loose boards, it can only support two travelers at any one time.

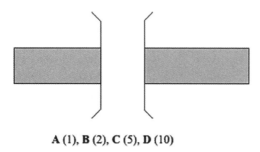

A (1), B (2), C (5), D (10)

It turns out that each traveler needs a different amount of time to cross the bridge: Traveler **A** is young and fast, and only needs a minute; Traveler **D**, on the other hand, is much older and requires ten minutes to get across, while travelers **B** and **C** require two and five minutes, respectively. When the travelers cross in pairs, the crossing time is set by the slower of the two travelers. The question is how to schedule these four travelers, so they cross the bridge in the shortest possible time?

This puzzle is a genuine scheduling problem, without any tricks such as throwing the lamp across the bridge or having traveler A carry traveler **D** on his back. Only two travelers can be on the bridge at any one time, and they can either cross individually or in pairs; but since the lamp is needed to cross from one side to the other, we know that it would be impossible for only one traveler to cross at the beginning because there would be no one available to take the lamp back. So, no matter what solution we come up with, it has to involve a sequence of pairs traveling across the bridge. But which pairs? From the problem description, we know that the only way these four travelers can cross the bridge is by following this sequence of five steps:

1. Two travelers cross the bridge (with the lamp)
2. One returns (with the lamp)
3. Another pair of travelers cross the bridge (with the lamp)
4. One returns (with the lamp)
5. The final pair of travelers cross the bridge (with the lamp).

Of course, we can ignore some unproductive sequences, for example, where traveler **A** wanders aimlessly back and forth, or two travelers cross the bridge and then cross back together again. The above sequence of five steps defines the structure (representation) of the solution that takes into account the constraints, variables (who exactly is traveling at what stage), together with the evaluation function (the total crossing time). These elements represent a model of the problem. Our only task is to specify *who* is crossing the bridge at what stage ...

Most people follow their intuition that the fastest traveler (**A**) should walk back and forth with the lamp, and they come up with the following solution:

1.	Travelers **A** and **B** cross the bridge	time:	2 minutes
2.	Traveler **A** returns	time:	1 minute
3.	Travelers **A** and **C** cross the bridge	time:	5 minutes
4.	Traveler **A** returns	time:	1 minute
5.	Travelers **A** and **D** cross the bridge	time:	10 minutes

Total time: 19 minutes.

In other words, the fastest traveler (**A**) is sent across the bridge with each traveler in turn, so that travelers **A** and **B** cross together (which takes two minutes), then traveler **A** returns with the lamp (which takes an additional minute). Finally, travelers **A** and **C** cross and then traveler A returns to do the same thing with traveler **D**. In all, this schedule takes nineteen minutes to execute, and is based on our intuitive assumption that the fastest traveler should carry the lamp back and forth to minimize the overall crossing time.

But is there is a better way for them to accomplish their task? Is the above solution "optimal"? Well, to answer this question, we should consider other possibilities. Another appealing option is to send the two slowest travelers (**C** and **D**) together. We might have thought about this possibility; however, it's no good, as one of these slow travelers would have to return, and the time gained by pairing them together at the beginning would soon be lost. This is true, but is it really necessary to send one of them back? Indeed, on second thought, we can avoid this by scheduling the travelers as follows:

1.	Travelers **A** and **B** cross the bridge	time:	2 minutes
2.	Traveler **A** returns	time:	1 minute
3.	Travelers **C** and **D** cross the bridge	time:	10 minutes
4.	Traveler **B** returns	time:	2 minutes
5.	**A** and **B** cross the bridge	time:	2 minutes

Total time: 17 minutes.

In the end, it was possible to cut two minutes from our first (intuitive) solution, which represents more than a 10% improvement. If we could reduce our logistics or production costs by more than a 10% through optimized scheduling, this would be something! The above solution is also optimal, as the travelers cannot cross the bridge in less than seventeen minutes.

Interestingly, despite the puzzle being very simple, with only a few variables and constraints, we can easily end up with a suboptimal answer by following our intuition. And if that's the case with simple problems, then what chance do we have of manually finding optimal solutions to more complex problems, such as promotional planning or scheduling supply chain activities? We can see how finding an optimal solution becomes more difficult even in this puzzle, by changing the number of travelers and the constraints. For example, we could have six travelers approach the same bridge and their respective crossing times might be 1, 3, 4, 6, 8, and 9 minutes. Again, what is the best way to schedule them to minimize the crossing time? Or suppose we have seven travelers with crossing times of 1, 2, 6, 7, 8, 9, and 10 minutes, but the bridge is sturdier and can handle three travelers at a time, and so on.

These bridge-crossing puzzles illustrate the nature of optimization problems, where we search for a solution that maximizes or minimizes some measure. As indicated earlier, there usually are *many* possible solutions for any given optimization problem, and the set of all these possible solutions is called the search space, which is further divided into feasible and infeasible solutions (i.e. solutions that satisfy or don't satisfy our business rules and constraints). Again, each solution has a quality measure score (*evaluation function*) that allows us to compare one solution against another to determine which one is "better." The main challenge for any optimization method is how to search through this very large set of possible solutions to find the best one (in terms of the quality measure score) in the shortest number of steps?

As we've already pointed out in Chapter 2.1, a defining characteristic of complex business problems is the astronomical number of possible solutions. To appreciate this, recall that:

- The number of seconds since the Big Bang is around 435,196,800,000,000,000
- The estimated number of planets in the known universe is 200,000,000,000,000,000,000,000
- The number of ways that you can arrange a standard 52-card deck is 80,658,175,000,000,000,000,000,000,000,000,000,000,000, 000,000,000,000,000,000,000,000!

This number is so large that calling it "astronomical" is an understatement. And there's no need to convince anyone that most real-world business problems are more complex than arranging a 52-card deck—as an example, imagine finding the best order of 52 travelers crossing that bridge! So this is the main challenge for all optimization methods: how to find the best solution while testing only a limited subset of the search space.

Let's return to the bridge-crossing puzzle and discuss it from the perspective of the search space and evaluation function. Remember that after analyzing this puzzle, we represented the solution in the following way:

1. Two of them cross the bridge (with the lamp)
2. One returns (with the lamp)
3. Another pair crosses the bridge (with the lamp)
4. One returns (with the lamp)
5. The final pair crosses the bridge (with the lamp).

Note that the constraints are already incorporated into this representation, as we require that only a "legitimate" traveler can cross the bridge. For example, the solution:

1. Travelers **A** and **C** cross the bridge
2. Traveler **B** returns
3. Travelers **B** and **D** cross the bridge
4. Traveler **B** returns
5. Travelers **A** and **D** cross the bridge

doesn't make any sense, as it violates the constraints and is therefore infeasible. The above representation allows us to exclude such solutions, so we restrict the search space to only feasible solutions (but the process of separating feasible and infeasible solutions isn't always easy).

The above representation also dictates the size of the search space, as we can enumerate all the feasible solutions. Six possible pairs can be sent across the bridge in the first step (travelers **A** and **B**, travelers **A** and **C**, travelers **A** and **D**, travelers **B** and **C**, travelers **B** and **D**, travelers **C** and **D**). In each of these cases, there would be two possible choices for a traveler to return with the lamp in step 2 (e.g. if travelers **A** and **B** cross the bridge, then either traveler **A** or **B** would return with the lamp). At this stage, another pair is selected from the three remaining travelers to cross the bridge (step 3), and there are three possible pairs. One of the three travelers on the other side would return with the lamp—again, this would give us three possibilities (step 4). Then the final pair (no choice here) crosses the bridge for the last time (step 5).

Thus, the total number of possible (feasible) solutions is:

$$6 \times 2 \times 3 \times 3 \times 1 = 108$$

where each number corresponds to the number of possible choices at each step (i.e. six choices at step 1, two choices at step 2, etc.). In other words, there are

108 possible combinations here. Also, each possible schedule (i.e. solution), such as:

1. Travelers **A** and **B** cross the bridge
2. Traveler **A** returns
3. Travelers **A** and **C** cross the bridge
4. Traveler **A** returns
5. Travelers **A** and **D** cross the bridge

or:

1. Travelers **B** and **C** cross the bridge
2. Traveler **B** returns
3. Travelers **A** and **D** cross the bridge
4. Traveler **A** returns
5. Travelers **A** and **B** cross the bridge

represents just one solution out of 108 possible feasible solutions. The quality measure score of each solution is the total crossing time (which the evaluation function calculates by adding all the crossing times together), so the two solutions above have quality measure scores of 19 and 20, respectively, making the first solution better than the second. From this perspective, the bridge-crossing puzzle is simple to solve because we can list all 108 possible solutions, calculate their quality measure scores, and select the best one!

This discussion leads us to the following observation: When modeling an optimization problem, it's worthwhile to think about what the solution looks like, and how it can be represented. For example, is the solution a schedule (as was the case for the four travelers), or does the optimization problem call for the best number, best sequence, best arrangement, or best strategy? It's important to keep in mind how the solution is represented in our model, because this representation will define the size of the search space.

Now let's move to a harder optimization puzzle: A manufacturer sells chairs and tables, and the profit per chair is $20, whereas the profit per table is $30. A single unit of wood is needed to make a chair, along with three hours of labor, whereas six units of wood are needed to make a table, along with one hour of labor. The production process also has two constraints: the maximum number of units that can be processed each day is 288, and the maximum available labor each day is 99 man-hours. The question is, how many chairs and tables should the manufacturer build to maximize profit?

Well, as with the previous example, we need a model that would represent the problem. In the case of this manufacturer, we should construct a model of the problem by specifying the following:

- *Variables*: There are only two variables, *x* and *y*, with each variable corresponding to the number of items (chairs and tables, respectively) to be produced
- *Constraints*: There are only two constraints: the 288 available units of wood, and the 99 available man-hours of labor
- *Objective*: There is one objective: to maximize the daily profit

The formulation of the objective (i.e. expressing the objective as an evaluation function) is straightforward. As the profit per chair is $20 and the profit per table is $30, the total profit for a day's production where *x* chairs and *y* tables are produced is:

$20x + $30y

So, if we produce 10 chairs (i.e. *x* = 10) and 15 tables (i.e. *y* = 15), the daily profit would be:

$20 × 10 + $30 × 15 = $200 + $450 = $650

Of course, the larger the number of chairs and tables we produce, the higher the profit. If we produce 20 tables (instead of 15), the profit would be:

$20 × 10 + $30 × 20 = $200 + $600 = $800

Ideally, we'd like to produce millions of chairs and millions of tables to generate tens of millions in profit. Unfortunately, the constraints prevent us from doing this, as the number of available units of wood and man-hours is limited.

What resources do we need to build a chair? We know that "a single unit of wood is required to make a chair, along with three hours of labor." So, what does it take to build *x* chairs? Simple: we need 1*x* units of wood and 3*x* man-hours. And what does it take to build *y* tables? Simple again: we need 6*y* units of wood and 1*y* man-hours. But we only have 288 available units of wood, with each chair requiring one unit of wood, and each table requiring six units of wood. In our model, we can represent this constraint as:

$x + 6y \leq 288$

And we only have 99 available man-hours. Each chair requires three man-hours, while each table needs one man-hour. In our model we can represent this constraint as:

$3x + y \leq 99$

Our modeling is almost complete, although we can be more specific about the domain of values for variables *x* and *y*. In this case, these variables can only take non-negative integer values, such as 0, 1, 2, ... because we cannot produce a negative number of chairs or tables! To summarize, the mathematical model for the profit maximization problem of the manufacturing company can be formulated as follows:

maximize $20x + $30y

subject to:
$$x + 6y \leq 288$$
$$3x + y \leq 99$$

where $x \geq 0$ and $y \geq 0$, and where both of the variables x and y can only take on integer values. It might not be obvious that the solution to this profit maximization problem is:

$x = 18$ and $y = 45$

which generates a profit of $1,710. This is the best we can do: in other words, it's impossible to achieve a higher profit by producing a different number of chairs and tables (while keeping within the constraints of available units of wood and labor). However, there are some additional questions we should ask: Is the model adequate for the problem? Did we include all relevant information? Is the model precise enough?

Again, there are no easy answers to these questions. Note that we've simplified the real problem by making a few "silent" assumptions. We assumed, for example, that the profit per chair is $20 regardless of the number of chairs produced (which also applies to the tables). However, it's doubtful that the manufacturer can make the same profit per produced unit, regardless of the number of produced units. It might be more realistic to assume that the profit per unit goes down as the number of units increases, due to supply and demand factors—the more units produced, the cheaper they become in the marketplace. To illustrate this point, we can assume that the profit per chair is actually:

- $20, if the number of produced chairs stays below 5.
- $18, if the number of produced chairs is 5 or more, but not more than 12.
- $17, if the number of produced chairs is 13 or more, but not more than 20, etc.

These profits are illustrated in the following graph:

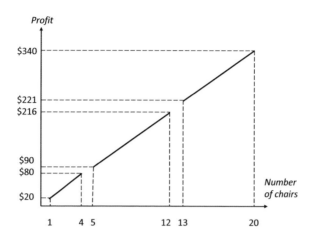

In this model, the profit is zero when no chairs or tables are produced. If up to four chairs are produced, then the profit is $20 per chair. However, if we produce five chairs or more (but not more than twelve), the profit drops to $18 per chair. Thus, the total profit for producing four chairs is 4 × $20 = $80, while the total profit for producing five chairs is only $10 more (5 × $18 = $90). For each additional chair (up to twelve chairs) the profit increases by an additional $18 per chair; for twelve chairs the profit is 12 × $18 = $216. If we produce thirteen chairs or more (but not more than twenty), then the profit drops again, this time to $17 per chair. So, if we produce thirteen chairs, the total profit is $221—only $5 more than for producing twelve chairs.

A similar graph can be created for tables, where the profit per table is a function of the number of tables produced. For example, we can assume that if the number of produced tables stays below ten, the profit per table is $30. If the number of produced tables is greater than ten (but not greater than twenty), the profit per table is $28, and so on. Note that this new (and more precise) model is nearly identical to the previous one, which was:

maximize $20x + $30y

subject to:
$$x + 6y \leq 288$$
$$3x + y \leq 99$$

where x and y are integers, $x \geq 0$ and $y \geq 0$, except that the objective (i.e. the profit formula) is more realistic and consequently more complex. Instead of the simple objective:

maximize $20x + $30y

we now have a more complex expression:

maximize $f(x, y)$

$$\text{where } f(x, y) \begin{cases} \$20\,x + \$30\,y, & \text{if } 0 \le x \le 4 \text{ and } 0 \le y \le 10 \\ \$18\,x + \$30\,y, & \text{if } 5 \le x \le 12 \text{ and } 0 \le y \le 10 \\ \$17\,x + \$30\,y, & \text{if } 13 \le x \le 20 \text{ and } 0 \le y \le 10 \\ \$20\,x + \$28\,y, & \text{if } 0 \le x \le 4 \text{ and } 11 \le y \le 20 \end{cases}$$

We can also make the model more realistic by including data about the skills of particular employees, as some of them might be more skilled at building chairs than tables and vice versa. Thus, the total number of man-hours required to make a table or chair would depend on the employee assigned to the task. We may also consider some labor-saving methods, such as producing chairs or tables in batches (e.g. building three chairs one after another in a batch may only require eight man-hours, and not $3 \times 3 = 9$ man-hours, because we gain some tooling and changeover efficiencies by producing the same product back-to-back). All these considerations would make the model more precise and more representative of the real-word problem, but also more complicated, as we can already see.

Let's move on from our two puzzles and summarize our earlier points. First, every problem-solving activity is a two-step process: (1) building a model of the problem, and (2) solving the model:

This was the case when building predictive models (Chapter 5) and is also the case when building optimization models. Because of this two-step process, we must realize that we're only finding a solution to the *model* of a problem. If the model accurately represents the problem, then the solution will be meaningful. However, it might be more difficult to use a complex model, and consequently, more difficult to find the solution without the use of more advanced optimization methods (i.e. AI algorithms). On the other hand, if the model is inaccurate or has many vague assumptions and approximations, then the solution may be useless.

Let's illustrate the above points with another example: A logistics manager is trying to model the transportation cost of the overall cost of goods sold, and every possible route between each warehouse and a distribution center has a measurable cost. The shape of this cost function depends on a variety of factors, and the transportation model between any given manufacturing site and a specific distribution center might be:

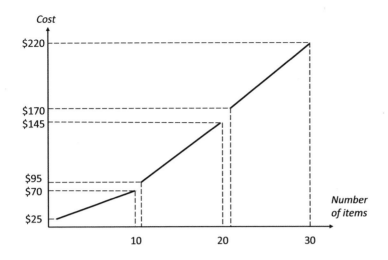

In this model, the cost is zero when there are no deliveries. If up to ten pallets are delivered, then we incur a fixed cost of $25 for the use of a single container and an additional cost of $5 per pallet shipped (thus, the cost for shipping six items would be $25 + (6 × $5) = $55). However, if we transport eleven or more pallets (but not more than twenty), we have to use two containers. In this case, the cost is $70 for ten items, an additional $25 for the second container, and $5 per each additional item.

If the above model accurately describes the real-world situation, we can construct similar models for the other warehouses and distribution centers. Note, however, that the cost functions in such models are discontinuous,[1] and discontinuities usually present severe difficulties for many traditional, non-AI optimization methods. So, what options do we have? Well, there are two ways to go: Either we simplify the model and apply traditional, non-AI optimization methods, or we can keep the model unchanged and apply AI-based optimization methods. Put a different way, the first approach relies on a simplified model of a problem and uses non-AI algorithms to find a solution, whereas the second approach relies on a precise model of the problem and AI algorithms to find a solution.

The first approach is quite tempting. For example, to simplify the original model, we can *approximate* the transportation model as follows:

---

1   In this context, by "discontinuity" we mean the situation where small changes in the number of items (e.g. the change from 10 items to 11 items) cause bigger "jumps" in the cost function (e.g. from $70 to $90).

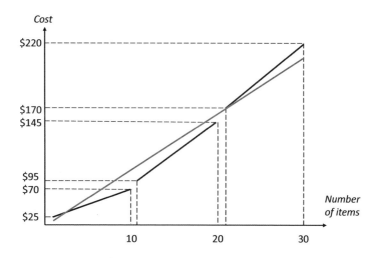

We can simplify the other models in a similar manner, and by making them linear we can use a linear programming method[2] to find a solution. However, this solution would be for the simplified model, and not for the real problem! The second approach is to leave the precise model unchanged—with all discontinuities and irregularities—and use AI optimization algorithms to find a solution. Of these two approaches, the latter is often more difficult to implement because it requires the tuning of AI algorithms—a process that requires significant scientific expertise, which, until recently, had been a barrier for many organizations—but produces superior results (because a solution for a precise model is usually is much "better" than one to a simplified model, which doesn't accurately represent the problem).

To understand why, let's look at this situation in the following way: A simplified model usually hides the "irregularities" of a problem, thus allowing some optimization methods (such as linear programming) to provide a precise solution. However, by hiding the irregularities of a problem, we lose much of the information needed to find the optimal solution! For example, in our simplified model above, note that the difference in transportation cost between 20 and 21 items is now the same as the difference between 19 and 20 items (which isn't the case in the precise model). Consequently, the simplified model does not "see" the thresholds that play a major role in identifying the optimal solution. Thus, the "optimal" solution for a simplified model is usually more appropriate for the wastebasket than for implementation!

As we can now see, the process of modeling a real-world problem is far from trivial! For additional information on some of the concepts presented

---

2   *Linear programming* is a problem-solving method in which a linear function of many variables is subject to a number of constraints in the form of linear inequalities.

in this section, we encourage you to watch the supplementary video for this chapter at: www.Complexica.com/RiseofAI/Chapter6.

6.2 Promotional Planning Optimization

Having worked our way through the puzzles and examples in the previous section, let's revisit the problem of promotional planning. In Chapter 3 we discussed the structure of a solution for this problem: a slotting board matrix where different rows correspond to different products, and columns correspond to different weeks of the year. This solution might be a part of a larger, overall solution, because we might need an overall plan for Mary's Market, whereas this solution only represents one product category (e.g. snacks) in one state (e.g. NSW). Furthermore, apart from the Yes/No decisions on whether particular products should be promoted during particular weeks, there are two additional variables the optimization method must consider: the promotion type and promotional price.

Let's consider one particular promotional plan, *Plan A*, as displayed below:

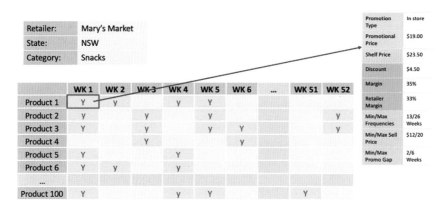

As we have already discussed in Chapter 3, planning just 100 products across a 52-week planning horizon requires 5,200 individual Yes/No decisions (52 × 100), and the number of all possible combinations of these Yes/No values is 1 followed by 1,565 zeros! Although many of combinations wouldn't be feasible, as they would violate some business rule and/or constraints, the number of possible promotional plans is still beyond astronomical. Even if we assume that 99.9999% of all possible combinations of Yes/No values would violate some (hard) business rule or constraint and shouldn't be taken into account when searching for the best plan, the remaining number of feasible solutions would still be 1 followed by 1,559 zeros! Hence, even such a large reduction of the search space wouldn't help us find the optimal solution.

Furthermore, note also that every entry of "Yes" value (a particular product, for a particular week) generates a further number of additional possibilities, because there are now two other variables associated with the promotional plan: the promotion type and promotional price, both of which can take on many different values. Thus, for two plans with an identical distribution of Yes/No values across all products and all weeks, they would still differ in their promotional types and promotional prices in their "Yes" cells, resulting in differences in their predicted outcome. So, even if we only consider feasible solutions (i.e. solutions that satisfy all hard constraints), the complexity, in terms of the size of the search space, would grow again. For five different promotional types and ten different price points for some of these types, we may have more than twenty possible combinations for a single cell. And with 2,000 product/weeks on promotion (out of 5,200 available product/weeks), for each distribution of Yes/No decisions on whether a particular product should be promoted during a particular week, there would be more than $20 \times 20 \times \ldots \times 20$ possible combinations for building a promotional plan (2,000 multiplications)—a number much larger than the number of possible combinations we started with for individual Yes/No decisions for each product and each week!

In Section 6.1, we pointed out that to construct an optimization model of the problem, we need to specify:

- *Variables*: that define the "moving parts" the optimization algorithm can change
- *Constraints*: that define the "feasibility" of a solution
- *Objectives*: that define the optimization goal, which is converted into *evaluation function*

We have already discussed the variables, as there are 5,200 binary (Yes/No) variables for 100 products and 52 weeks. Furthermore, for every product and every week combination, there are two more variables: the promotion type and promotional price, both of which can take on many different values (of course, these two additional variables appear only for corresponding product/week with "Yes" value).

In Chapter 3.2, we presented some typical business rules and constraints related to promotional planning, and as we already explained, some of these business rules and constraints might be for individual products, while others for the entire plan. For example, in the snacks category, we may have a business rule that says that no less than 30 products and no more than 60 products should be on promotion in any given promotional period, and this might be a soft rule. We could also consider a few hard business rules, such as the minimum net revenue of the entire plan should be $1,000,000, and the minimum retailer margin growth over last year should be 2%.

In addition to these business rules, we might also have some product-specific business rules. For instance, the minimum and maximum promotional price of Product 43 should be $4.00 and $7.00, respectively, and the minimum and maximum promotional frequencies should be 5 and 8, respectively. We can classify these business rules and constraints as either hard (must be satisfied) or soft (it would be good if they are satisfied), and then the hard rules and constraints define the concept of feasibility: solutions that violate at least one hard rule or constraint are not feasible.

Now we can move to the third component of the optimization model: the evaluation function. Let's consider the following two plans,[3] A and B, and the question of which is the "better" promotional plan?

Plan A:

Retailer:	Mary's Market
State:	NSW
Category:	Snacks

Promotion Type	In store
Promotional Price	$19.00
Shelf Price	$23.50
Discount	$4.50
Margin	35%
Retailer Margin	33%
Min/Max Frequencies	13/26 Weeks
Min/Max Sell Price	$12/20
Min/Max Promo Gap	2/6 Weeks

	WK 1	WK 2	WK 3	WK 4	WK 5	WK 6	...	WK 51	WK 52
Product 1	Y	y		y	Y				
Product 2	y		y		y				y
Product 3	Y		y		y	Y			y
Product 4			Y			y			
Product 5	Y			Y					
Product 6	Y	y		y					
...									
Product 100	Y			y	Y			Y	

Plan B:

Retailer:	Mary's Market
State:	NSW
Category:	Snacks

Promotion Type	In store
Promotional Price	$19.00
Shelf Price	$23.50
Discount	$4.50
Margin	35%
Retailer Margin	33%
Min/Max Frequencies	13/26 Weeks
Min/Max Sell Price	$12/20
Min/Max Promo Gap	2/6 Weeks

	WK 1	WK 2	WK 3	WK 4	WK 5	WK 6	...	WK 51	WK 52
Product 1	Y	y		y	Y				
Product 2	y		y		y			y	y
Product 3					y	Y			y
Product 4			Y			y			
Product 5	Y			Y				Y	
Product 6	Y			y					
...									
Product 100	Y			Y	Y			Y	Y

---

3   These are the same plans that were introduced in Chapter 3.

There are some differences between these two plans on when some products are promoted; for example, Product 3 is promoted during Week 3 in *Plan A*, but not in *Plan B*. Similarly, Product 2 isn't promoted during Week 51 in *Plan A*, but is in *Plan B*. Clearly, we need an objective measure score by which to compare these two plans; in other words, we need an evaluation function. Because promotional plans are executed in the future, the only way we can evaluate their merit is by accurately predicting their performance. In Chapter 3, we provided the predicted values for the volume, net revenue, retail gross profit, and our gross profit for *Plan A* and *Plan B*:

	Plan A	Plan B
Volume	62,685	59,633
Net revenue	$1,034,303	$1,038,027
Retailer gross profit	$227,547	$342,549
Manufacturer gross profit	$268,919	$363,309

The thing to remember is that the main purpose of an evaluation function is to provide the optimization algorithm "guidance" by facilitating comparisons between different solutions. In particular, the evaluation function should assist in answering the following question: which of these two plans, *Plan A* or *Plan B*, is better?

If we just consider one objective (e.g. volume), then everything is much simpler: *Plan A* had a predicted volume of 62,685 units, and *Plan B* had a predicted volume of 59,633 units, then because *Plan A* had a higher predicted volume, that plan is "better" than *Plan B*, assuming, of course, that both plans are feasible (i.e. that they satisfy all hard business rules and constraints). So, the evaluation function for these plans is just the predicted volume of each plan—nothing more and nothing less.

However, this "simplicity" of the evaluation function implies that:

- The other objectives (net revenue, retail gross profit, and manufacturer gross profit) don't matter
- The number of violated soft constraints doesn't matter
- The degree of violation of each soft constraint doesn't matter.

Besides generating different (better or worse) results on all four objectives (as the table above illustrates), these two plans might also violate some soft business rules and constraints, with one plan violating more than the other. Lastly, the degree of violation might be important; for example, if 75 products are on promotion during Week 10, whereas one of the soft business rules states that no more than 60 products should be on promotion during any given period, then clearly this soft rule is violated. But if just

61 products are on promotion during Week 10, then the same business rule is violated, begging the question of whether these two violations are "equivalent" or not?

These questions are important, as the evaluation function needs to accurately represent the objective of the problem we're solving. After all, the optimization algorithm only understands numbers, so any evaluation function that doesn't accurately reflect our objective may lead the optimization process in the wrong direction. Thus, it might be necessary to create a more complex evaluation function, one that considers the other objectives along with the primary one (volume), or deals with soft business rules and soft constraints directly (as discussed in Section 6.8).

6.3 Local Optimization Methods

Because the evaluation function defines the quality of a solution, it therefore defines the *quality measure score landscape* (also known as a *response surface* or *fitness landscape*). Within this three-dimensional landscape—which resembles a topography of hills and valleys—the problem of finding a solution with the highest quality measure score is similar to searching for a peak in a foggy mountain range. Because our visibility is limited, we can only make local decisions about where to go next. If we always walk uphill, we will eventually reach a peak, but this peak might not be the highest peak in the mountain range; it might just be a "local" optimum. We may have to walk downhill for some time to find a path that will eventually lead to the highest peak (i.e. the "global" optimum).

Graphically, let us consider some abstract search space with a single solution **s**:

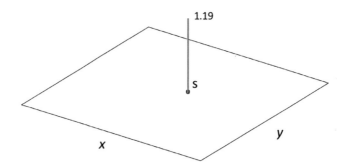

The search space above is presented here as a parallelogram with two variables, x and y, that correspond to the parallelogram's two sides. A solution **s** is interpreted as a pair of numbers (x, y)—for example, (3.1, 0.7). Also, a quality measure score for solution **s** is recorded as 1.19.

In any search space, the goal is to find a solution that is feasible *and* better than any other solution present in the entire search space. The solution that satisfies these two conditions is called the *global optimum*. Because finding the global optimum is extremely difficult, a much easier approach is to find the best solution in a subset of the search space.[4] If we can concentrate on a region of the search space that is "near" some particular solution, we can describe this as looking at the "neighborhood" of that solution (the shaded area around **s**):

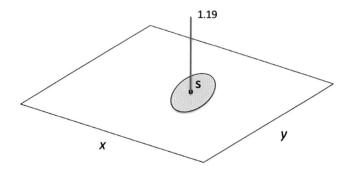

Our intuition might tell us that solution **s** lies within a part of the search space where all solutions are very similar to one another. Consequently, we can use a "local" optimization method to find the best solution in this neighborhood. Such a solution is called a *local optimum*: a solution that's feasible *and* better than any other solution present in its neighborhood. The sequence of solutions these optimization methods generate while searching for the best possible solution relies on *local* information at each step of the way.

Local optimization methods present an interesting trade-off between the size of the neighborhood and the efficiency of the search. If the neighborhood's size is relatively small, then the algorithm may be able to search the entire neighborhood quickly. Only a few potential solutions may need to be evaluated before a decision is made on which new solution should be considered next. However, such a small neighborhood increases the chance of becoming trapped in a local optimum! This suggests using larger neighborhoods, as a greater range of visibility makes it easier for the algorithm to decide where to search next. In particular, if the visibility were unrestricted (i.e. the size of the neighborhood was the same as the size of the whole search space), then eventually we would find the best series of steps to take. However, the number of evaluations might become overwhelming and impossible to compute.

All optimization methods generate new solutions from existing solutions. The main difference between these methods—some based on Artificial

---

4   This observation forms the fundamental basis of many optimization methods.

Intelligence algorithms and others not—lies in how these new solutions are generated. Because we can only investigate a small fraction of the search space (otherwise the computation time would be billions of years!), we should be economical in the process of generating and evaluating new solutions.

The whole landscape of all the quality measure scores for such a two-variable function is illustrated below. The graph displays the quality measure score for every pair of values for the first and second variable, which allows us to visualize the mountain ranges, valleys, and peaks of differing heights:

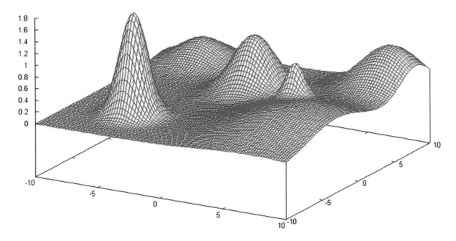

There's usually only a single highest peak, which represents the global optimum, and possibly many lower peaks, representing local optima. It's important to make this distinction when discussing various optimization methods, because the goal of local optimization methods is to search for local optima, whereas the goal of global optimization methods is to search for the global optimum.

Keeping this in mind, let's examine a local optimization method called *hill climbing*,[5] and its connection with the concept of exploring a particular neighborhood of the search space. Hill climbing is a traditional, non-AI optimization method, and by discussing it in more detail, we can then contrast it with AI-based methods. The hill climbing algorithm begins with a single solution in the search space (i.e. the current solution) and uses iterative improvements to find the local optima. During each iteration, a new solution is selected from the neighborhood of the current solution. If that new solution has a better quality measure score, then the new solution becomes the current solution. Otherwise, some other neighbor is selected and tested against the

---

5   The term *hill climbing* implies a maximization problem, but the equivalent *descent* method is easily envisioned for minimization problems. For convenience, the term will be used to describe both methods without any implied loss of generality.

current solution. The algorithm terminates if no further improvements are possible, or when the allotted time runs out.

A simple flowchart of a hill-climbing algorithm is given below:

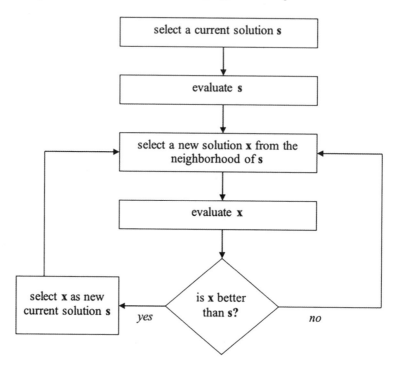

Note that this flowchart expresses only the general principle of hill climbing without any termination conditions. We have to start with some (possibly randomly generated) solution **s**, evaluate it, and then generate a new solution **x** from a neighborhood of **s**. If the new solution **x** is better than **s**, then we take an uphill step, meaning that we accept this new solution as the current solution, and try to improve it further by generating yet another new solution from the neighborhood of the current one. On the other hand, if the new solution **x** is not better than **s**, we generate another new solution and repeat this process several times until either (1) the whole neighborhood has been searched, or (2) we have exceeded the threshold of allowed attempts (which is missing from the flowchart above). At this stage, we can exit the loop and confirm that the current solution is the best solution, or we can store the current solution in "memory" and restart the whole process, hoping that the next hill-climbing iteration (which starts from a new solution) may produce a better overall solution (a process called *iterated hill-climbing*).

Hill-climbing algorithms can only provide locally optimal solutions that are highly dependent on the starting solution. Moreover, there's no general procedure for measuring the relative error with respect to the global optimum because it remains unknown. As this method returns locally optimal solutions, we often have to start the hill-climbing algorithm from a large variety of different solutions. The hope is that at least some of these initial solutions will lead us to the global optimum. We might choose the initial solutions at random, or base them on some grid, regular pattern, or other information (perhaps using the search results from somebody else's effort to solve the same problem).

The figure below continues our example of the search space where solutions are represented by two variables, x and y, that correspond to two sides of the parallelogram. This figure also visualizes a case of random search, where the search spaced is sampled 15 times (i.e. 15 solutions were generated and evaluated) in our search for the best solution. Such random sampling is then followed by a local optimization method such as hill-climbing to improve the best solution found by searching its neighborhood.

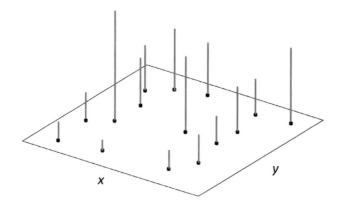

The success or failure of a single hill-climbing iteration (i.e. one complete climb) is entirely dependent on the initial solution. For problems with many local optima, it is often very difficult to find the global optimum. Consequently, hill-climbing methods have several weaknesses:

- They usually terminate at solutions that are only locally optimal.
- They don't provide information on how much the discovered local optimum deviates from the global optimum, or even from other local optima.
- The optimum they obtain depends on the initial solution.
- In general, it's *not* possible to provide an upper bound for the computation time.

On the other hand, there's a tempting advantage to hill-climbing algorithms: they are very easy to use! All we need is a representation of the solution, an evaluation function, and a measure that defines the neighborhood around a given solution.

Effective optimization algorithms provide a mechanism for balancing two conflicting objectives: *exploiting* the best solutions found so far and *exploring* the search space. Hill-climbing algorithms exploit the best available solution for possible improvement, but they neglect exploring a large portion of the search space. In contrast, a random search (where various solutions are sampled from the entire search space with equal probability) explores the search space thoroughly but foregoes exploiting promising regions of the search space. Different optimization algorithms handle this balance between exploitation and exploration of the search space differently, which is why different algorithms return different results (better or worse) for different problems. As a result, there's no way to choose a single optimization algorithm that performs well in every case (more on this topic in Chapter 7.3). However, it seems that AI optimization algorithms address the issue of balancing between exploitation and exploration directly, thus providing better results than traditional methods when the global optimum is required.

Let's now apply the hill-climbing method to promotional planning. Looking at the representation of a solution (i.e. a slotting board matrix) we may define the neighborhood for a given solution (say, *Plan A*) as any solution that differs at most by one value from *Plan A* (in terms of one product for one particular week). For example, we can take *Plan A*:

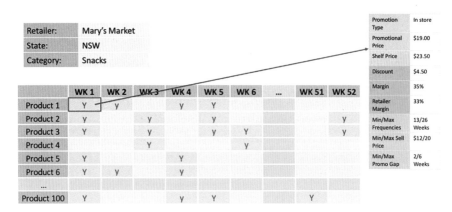

and create two new solutions that are neighbors of *Plan A*, which are *Plan A₁*:

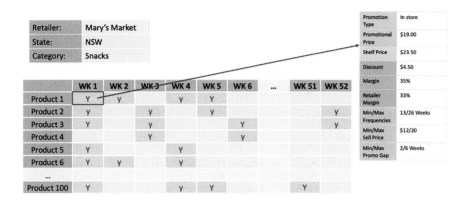

and *Plan A$_2$*:

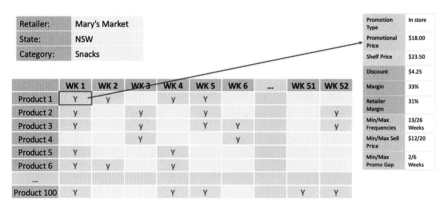

The difference between *Plan A* and *Plan A$_1$* is in Product 3 for Week 5, whereas the difference between *Plan A* and *Plan A$_2$* is in the promotional price for Product 1 in Week 1. This means there are two different types of neighborhoods: one is defined by the distribution of Yes/No decisions on the slotting board (i.e. when each product is promoted), whereas the other is defined by the values of other variables (i.e. the promotion type and promotional price).

For any single objective, such as volume, the hill-climbing algorithm would work in the following way: The algorithm begins with a starting solution, which might be generated randomly (most likely resulting in an infeasible solution that violates a few hard business rules or constraints) or based upon a human-generated solution, such as last year's plan. Either way, let's call that initial solution *Plan A*. The predictive model then evaluates this plan and assigns a quality measure score. For this example, let's assume the above solution generates a quality measure score (volume) of 62,685 units. Now we're ready to do some "hill-climbing"! First, the algorithm generates a neighbor solution by selecting a random location on the slotting board (anywhere between row

1 to 100 and column 1 to 52) and either makes a change to this value from "Yes" to "No" or vice versa, or makes a change to one of the other variables in this location, such as the promotion type or promotional price. We can call this new solution *Plan A₁* and the next step is to evaluate it. If the evaluation produces a quality measure score higher than *Plan A* (e.g. 62,691), then the algorithm will accept this new solution as the current solution and continue the process.

We should note that this new solution (*Plan A₁* with a higher quality measure score) has its own new neighborhood, and the subsequent new solution is drawn from this new neighborhood. Any acceptance of a new solution means that the algorithm has found a better solution and made an uphill step. However, the quality measure score of the new solution might be lower than *Plan A* (e.g. 62,679), in which case the algorithm will discard this solution and generate another solution from the neighborhood of *Plan A* (because we're not interested in an inferior promotional plan). And again, if there is an improvement in the quality measure score, then the algorithm will accept this solution and continue. If not, the algorithm will generate another solution from the original neighborhood …

Note also that a hill-climbing algorithm can (a) accept the first solution found that is better than the current one (as presented above), or (b) accept the best solution found in the whole neighborhood. These two possibilities represent two extremes, with plenty of "in-between" possibilities, such as accepting the best solution from 100 generated solutions in the neighborhood.

The question is, how long should the hill-climbing algorithm generate random solutions before giving up? We usually have a counter responsible for counting the algorithm's attempts to improve the current solution. Each time the algorithm finds an improvement, the counter is reset to zero. However, if the hill-climbing algorithm experiences a long sequence of unsuccessful attempts, it will stop searching when the counter reaches a predefined threshold. What this threshold should be depends on a few factors, with the neighborhood's size being the most important. It's difficult to claim that we've found the "local optimum" without searching the entire neighborhood, but the neighborhood's size might preclude the algorithm from evaluating all neighbors! This problem can be resolved by defining a neighborhood differently, and/or imposing stricter conditions for what constitutes a "neighbor."

If the algorithm has time to search the whole neighborhood before arriving at the local optimum, then we don't need a counter for controlling the number of unsuccessful attempts because *all* solutions in the neighborhood will be searched. However, if it's not feasible to search the entire neighborhood, we have to settle for a counter and stop our search after some number of unsuccessful attempts. Let's say we instruct the algorithm to stop after 100,000 unsuccessful attempts, and we arrive at the following solution after many, many iterations and improvements (and all attempts to improve it have failed):

Retailer:	Mary's Market
State:	NSW
Category:	Snacks

	WK 1	WK 2	WK 3	WK 4	WK 5	WK 6	...	WK 51	WK 52
Product 1	Y			Y	Y			Y	
Product 2				y					y
Product 3	Y	y			y	Y			
Product 4				Y		y			
Product 5	Y				Y				
Product 6	Y	y			Y	y			Y
...									
Product 100			y	Y				Y	y

Promotion Type	In store
Promotional Price	$19.00
Shelf Price	$23.50
Discount	$4.50
Margin	35%
Retailer Margin	33%
Min/Max Frequencies	13/26 Weeks
Min/Max Sell Price	$12/20
Min/Max Promo Gap	2/6 Weeks

Note the significant number of changes the algorithm went through: the original length of promoting Product 1 changed from two weeks to one, and some regularity emerged in promoting products 2, 3, 6, and 100. In all likelihood, we have arrived at the local optimum, and this solution is the outcome of our hill-climbing exercise. The quality measure score is now 65,117 and we are confident about the solution's quality. After all, 100,000 neighboring solutions failed to produce any improvement!

However, we're still not sure if this is the *best* solution. If we had started our hill-climbing from a different initial solution, which might be located in a very different part of the search space, we might have finished up with a local optimum solution that looks like:

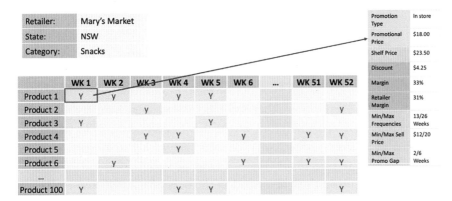

Retailer:	Mary's Market
State:	NSW
Category:	Snacks

	WK 1	WK 2	WK 3	WK 4	WK 5	WK 6	...	WK 51	WK 52
Product 1	Y	y		y	Y				
Product 2			y						y
Product 3	Y				Y				
Product 4			Y	Y		y		Y	Y
Product 5				Y					
Product 6		y				Y		Y	Y
...									
Product 100	Y			Y	Y				Y

Promotion Type	In store
Promotional Price	$18.00
Shelf Price	$23.50
Discount	$4.25
Margin	33%
Retailer Margin	31%
Min/Max Frequencies	13/26 Weeks
Min/Max Sell Price	$12/20
Min/Max Promo Gap	2/6 Weeks

which has a quality measure score of 67,089 and is *much* better than the solution we discovered earlier!

Recall our earlier discussion of the "hills and valleys" in the fitness landscape (i.e. the quality measure score landscape). Because there are many hills (and many locally optimum solutions), the hill-climbing algorithm will produce a solution that represents one of these hills. However, we don't know whether there are other (possibly much higher) hills somewhere else! And the neighborhood's size corresponds to our "visibility" during the search: the larger the neighborhood, the better the visibility, and the better our chances of discovering the highest peak! However, it might not be feasible to search the whole neighborhood if it's too large.

So, what can we do? We can restart the hill-climbing algorithm several times, choosing a different (possibly random) location in the search space and hope that one of these runs will provide us with the global optimum solution (which may or may not happen). This is referred to as *iterated hill climbing*. Alternatively, we can try some other approaches, which we'll discuss in the next section.

6.4 Stochastic Hill Climber

Getting stuck in local optima is a serious problem for optimization algorithms, and is one of the main deficiencies of traditional methods, as the solutions they produce to real-world problems such as production planning, labor scheduling, journey planning, and so forth, are only locally optimal at best.

So what can we do? How can we escape local optima by balancing exploration and exploitation, and making the search independent from the initial solution? There are a few possibilities, and we'll discuss some of them here, but we need to keep in mind that the proper choice is always dependent on the problem. One option, as we discussed earlier, is to execute a large number of initial solutions for the chosen algorithm. Moreover, it's often possible to use the results of previous attempts to improve the initial solution for the next attempt. One such approach is called iterated hill climbing, where after reaching the local optimum the algorithm restarts its search from a different starting solution (i.e. in effect, starting in a different neighborhood in a different part of the search space). Although we can apply this approach to other algorithms, let's discuss some other possibilities of escaping local optima within a single run of an algorithm. One way of accomplishing this is by modifying the criteria for accepting new solutions that correspond to a *negative* change in the quality measure score. That is, we might want to accept an inferior solution from the local neighborhood in the hope that it will eventually lead to something better.

Turning an ordinary hill climber into such an algorithm requires a few modifications. First, let's recall the structure of an ordinary hill climber:

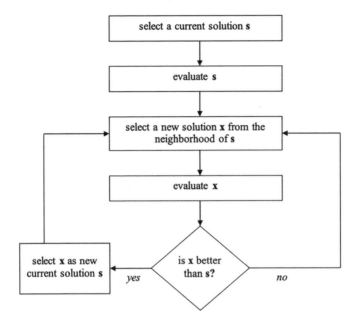

Note again that the inner loop *always* returns the local optimum. The only way for this algorithm to "escape" local optima is by starting a new search (outer loop) from a new (random) location. After some maximum number of attempts, the best overall solution is the final outcome.

By modifying this procedure so that the acceptance of a new solution depends on some probability based on the difference between the quality measure score for these two solutions, we move into a new method called the *stochastic hill climber.*

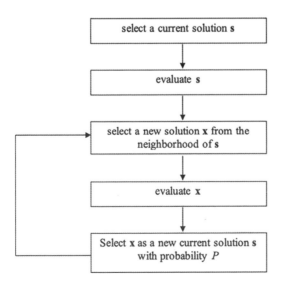

The slight (but significant) difference between an ordinary and stochastic hill climber lies in a single box inserted in the flowchart that replaces the condition box. During the execution of an ordinary hill climber's internal loop (where the hill-climbing searches for a better solution in the neighborhood of the current one), only a superior solution can be accepted as a new current solution, whereas in a stochastic hill climber, the same internal loop may accept an inferior solution as a new current solution. This feature doesn't appear in local optimization methods, and represents a probabilistic decision on the acceptance of a new solution (as opposed to a deterministic decision in ordinary hill climbers) in order to escape local optima …

Let's discuss this feature carefully. A new solution **x** is accepted with some probability P, which means that moving from the current solution to a new neighbor is probabilistic. Consequently, the newly accepted solution **x** can be *inferior* to the current solution s, and it's also possible that a superior solution will *not* be accepted! This probability of acceptance depends on the quality measure score difference between these two solutions, as well as the value of an additional parameter T (which remains constant during the execution of the algorithm).

Rather than providing a mathematical function for calculating the values of probability P (which is based on a constant value of parameter T), we will instead explain how this function works. In general terms, the probability function is constructed in a such way that:

- If the new solution **x** has the same quality measure score as the current solution s, then the probability of acceptance is 50% (it doesn't matter which one is chosen, because they're of equal quality).
- If the new solution **x** is superior, then the probability of acceptance is greater than 50%. Moreover, the probability of acceptance increases together with the (negative) difference between these two quality measure scores.
- If the new solution **x** is inferior, then the probability of acceptance is smaller than 50%. Moreover, the probability of acceptance shrinks together with the (positive) difference between these two quality measure scores.

The probability of accepting a new solution **x** also depends on the value of parameter T, and the general principle is as follows:

- If the new solution **x** is superior, then the probability of acceptance is closer to 50% for *high* values of parameter T, or closer to 100% for *low* values of parameter T.

- If the new solution **x** is inferior, then the probability of acceptance is closer to 50% for *high* values of parameter T, or closer to 0% for *low* values of parameter T.

This is interesting, because it means that a superior solution **x** would have a probability of acceptance of *at least* 50% (regardless of the value of parameter T). Likewise, an inferior solution would have a probability of acceptance of *at most* 50% (varying between 0% for low values of T and 50% for high values of T). The general conclusion is clear: The lower the value of T, the more the algorithm behaves like an ordinary hill climber that rejects inferior solutions and accepts superior ones. On the other hand, if the value of T is very high, then the algorithm resembles a random search, because the probability of accepting inferior or superior solutions is close to 50%. Thus, we have to find a value for parameter T that is neither too low nor too high for any given problem.

The stochastic hill climber method is also a forerunner to another optimization method called *simulated annealing*, which we'll cover next.

6.5 Simulated Annealing

Simulated annealing[6] is based on an analogy taken from the thermodynamics of heating metal or glass and then cooling it slowly to toughen it, and is similar to a stochastic hill climber in that it may accept an inferior solution as a new current solution, and the acceptance decision is based on the value of parameter T. However, unlike the stochastic hill climber (which has a fixed value for parameter T), simulated annealing changes the value of parameter T (commonly referred to as *temperature*) during the run. Simulated annealing starts with high values of parameter T—making the process similar to a random search—and then gradually decreases this value during the run. The value of parameter T is quite small toward the end of the run, so the final stages of simulated annealing resemble an ordinary hill climber.

Another difference between the stochastic hill climber and simulated annealing is that the latter always accepts superior solutions. Recall from the previous section that the stochastic hill climber used some probability for accepting both inferior *and* superior solutions, which is not the case in simulated annealing.

---

6 Simulated annealing is known as *Monte Carlo annealing, statistical cooling, probabilistic hill-climbing, stochastic relaxation,* and *the probabilistic exchange algorithm.*

The following flowchart represents a simulated annealing algorithm:

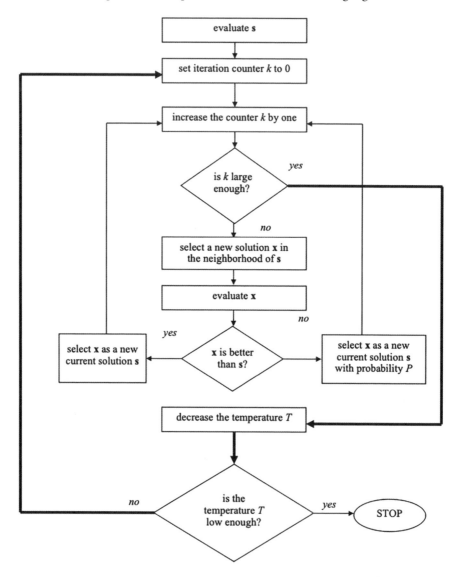

The internal loops (represented by the thin lines) generate a new solution in the neighborhood of the current solution, accept superior solutions, accept or decline inferior solutions (according to probability P, which depends on the quality measure score difference between two solutions and the current value of parameter T), and then repeat this process many times. When this iterative cycle is complete, the algorithm drops the temperature a bit (outer loop, represented by the thick line) and then switches back to the internal loop, repeating the process of creating, evaluating, and possibly accepting the neighboring solutions. If the temperature is low enough (i.e. when it reaches the freezing point), the algorithm stops.

As this flowchart illustrates, simulated annealing applies a modified version of the stochastic hill climber where the value of parameter T is gradually decreased during a run. As mentioned earlier, the behavior of this algorithmic method resembles random search at higher temperatures (i.e. at the beginning of the run) and an ordinary hill climber at lower temperatures (i.e. toward the end of the run). This process can be humorously compared to that of a drunken explorer searching for the highest peak on a quality measure score landscape. Initially, the explorer doesn't care whether he goes up or down, and this directional indifference corresponds to the early stages of simulated annealing when the temperature is high and the probability of accepting or rejecting an inferior solution is close to 50%. However, as the explorer sobers up—which corresponds to the temperature dropping—he makes more "reasonable" decisions on where to go next. Towards the end of the search, the explorer is completely sober and is always walking uphill.

Using simulated annealing would require us to answer a few general questions applicable to any optimization method, such as: *What is the representation of a solution? How are neighbors defined? What is the evaluation function?* And so on. Let's assume the answers are the same as those for the hill climbing example discussed in Section 6.3, allowing us to move onto questions specific to simulated annealing:

- How to determine the initial value of parameter T?
- How to determine the number of iterations (i.e. how to define the "is k large enough?" statement in the flowchart)?
- How to "cool" the algorithm (i.e. how to define the "decrease the temperature T" statement in the flowchart)?
- When to stop the algorithm (i.e. how to define the "is the temperature T low enough?" statement in the flowchart)?

The first question deals with the temperature parameter, which must be set before starting the algorithm. Should we start with $T = 100$, $T = 500$, or something else? Well, at the beginning of the simulated annealing run, we want to make almost random decisions for accepting or rejecting inferior solutions (remember that superior decisions are always accepted), which means that the answer to "$T = ?$" depends on the "average difference" in quality measure scores between two randomly generated solutions from the same neighborhood. This is important, as the probability of accepting a new solution is based on this difference and the value of the parameter T. The initial temperature should be high enough to generate probabilities of acceptance close to 100% for inferior (new) solutions that have an average difference in the quality measure score. For example, if we generate 1,000 random pairs of solutions plus their neighbors, evaluate them, and then discover that the average difference in the quality measure score is 5, we should start with an initial temperature of $T = 100$, because at this temperature the probability of accepting an inferior solution (worse than the current solution by 5) is over 95% (whereas for temperature $T = 50$, this probability would only be around 90%).[7]

The next two questions (*How to determine the number of iterations?* and *How to cool the system?*) are really about the number of temperature levels and the number of iterations performed at each level. Assuming limited computing resources (which is usually the case), these questions represent a trade-off in most implementations of simulated annealing: Is it better to have more temperature levels or to search each level more thoroughly? Unfortunately, there's no easy answer here, as these questions are problem dependent. For some problem types, a superior solution can be obtained through the implementation of more temperature levels, while for other problem types, allowing the algorithm to spend more time searching each level produces a better result.

The final question (*When to stop?*) can be answered as follows: Think about a freezing temperature for which it would be almost impossible to accept any inferior solution. For instance, if the temperature is $T = 0.001$, then a new solution that is inferior to the current solution by just 0.01 would have a probability of less than 0.005% of being accepted. Using these numbers, there would be 225 different temperature levels: $T = 100$ at the first iteration, $T = 95$ at the second (100×0.95), $T = 90.25$ at the third (95×0.95), and so forth, until iteration 225, where the temperature value drops for the last time to 0.001. This happens for $T = 0.001023$, where the next drop, from $T = 0.001023$ to $T = 0.000972$ (0.001023×0.95), would trigger the termination condition of the algorithm. In total, the simulated annealing algorithm would generate and evaluate 2,250,000 solutions, as 10,000 solutions are generated and evaluated at each temperature level.

Implementing the rest of the simulated annealing algorithm is straightforward. Note that the algorithm would start with some initial solution **s**. The solution is evaluated and draws a quality measure score of 62,685, and then a new solution **x** is generated from its neighborhood. If the new solution **x** is better than solution **s**, then it's selected as the new current solution. If solution **x** is worse, it might still be selected—everything depends on the quality measure score difference between the current solution s and the new

---

7   These numbers are derived from standard functions used in simulated annealing for calculating such probabilities.

solution **x**, as well as the current value of parameter T. Since the temperature is relatively high at the beginning of the run, the probability of accepting an inferior solution would be also high. For instance, if solution **x** generated a quality measure score of 62,680, then the probability of acceptance would be around 95% (recall that parameter $T = 100$ at this stage, and solution **x** is worse than solution s by 5, which is the exact "average difference" in quality measure scores computed earlier in this section).

This process continues for 10,000 iterations, with the algorithm generating, evaluating, and accepting or rejecting a neighboring solution at each iteration. Because the temperature stays very high ($T = 100$) during these 10,000 iterations, the probability of accepting inferior solutions is also very high. Thus, this phase of the search resembles a random search.

Let's assume that after 10,000 iterations, the current solution has a quality measure score of 62,845. We then lower the temperature slightly (from 100 to 95) and repeat the next sequence of 10,000 iterations. However, we start with the current solution and temperature of $T = 95$. Although the probability of accepting an inferior solution is not as high as before (because the value of parameter T is lower), it's still relatively high. Again, let's assume that 10,000 iterations later we arrive at another solution, which has a quality measure score of 63,711. We then repeat the process by lowering the temperature from 95 to 90.25 and continue for another 10,000 iterations. The probability of accepting an inferior solution is lower now, but still relatively high.

After some time, we'll arrive at the final temperature level, where the value of parameter $T = 0.001023$. Accepting an inferior solution is very unlikely at this low temperature, and the algorithm (for its final 10,000 iterations) acts like an ordinary hill climber. It wouldn't be surprising to get a good solution, with a quality measure score of 68,428 (as an example). Clearly, during these 2,250,000 total iterations at different temperature levels, the algorithm climbed and escaped from many "local" hills. Finally, during the hill-climbing stage in the end, it made the final climb hoping to arrive at the global optimum!

6.6 Tabu Search

The main idea behind *tabu search* is very simple: "memory" forces the algorithm to explore new areas of the search space in order to escape from local optima. We can memorize some recently examined solutions and these become "tabu" (forbidden) when selecting the next solution. Note that tabu search is deterministic (as opposed to simulated annealing, which is probabilistic), but it's possible to add some probabilistic elements.

The following flowchart outlines the basic steps of tabu search (without a termination condition):

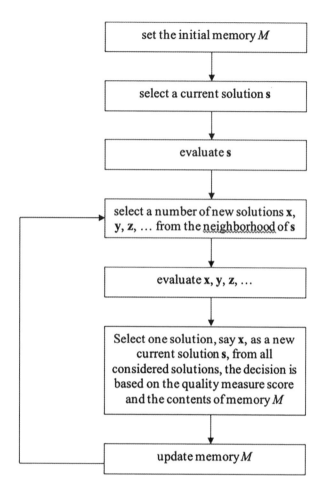

So, what's so special about tabu search? Well, there are a few exciting features that deserve a more detailed explanation. Most of the boxes in the above flow-chart are self-explanatory, and after discussing hill climbers and simulated annealing, we now know how to define a neighbor and select and evaluate a solution (or several solutions) from a neighborhood. What's new in tabu search is the "memory" component.

The best way to explain this feature is by describing some steps of the tabu search algorithm within the context of promotional planning, starting with our usual initial solution s, which is *Plan A*. From the neighborhood of this solution, we generate several solutions (solutions **x**, **y**, **z**, etc.) and select the best one[8] (say it was solution **x**, *Plan A₁*). Now we'll introduce the special feature of tabu search: *memory*. To differentiate between older and more recent

---

8   Note that current solution **s** and the new solution **x** differ in several positions. Later, choosing the "best" solution is influenced by memory, which is discussed later in this section.

changes in the solution, we need to remember the index of variables that were changed, as well as the "time" when these changes were made. In the case of a promotional plan, we need to keep a time stamp for each position in the slotting board matrix that provides information on the recency of the change. A matrix M (of dimensions 100×52) will serve as our memory. This matrix is initially set to all 0s, and if some entry of this matrix has a value j at any stage of the search, it means that j was the most recent iteration when that cell of the solution matrix was changed.[9] Hence, our memory matrix (after selecting solution **x** at the first iteration, which replaces **s** as a current solution) is:

	WK 1	WK 2	WK 3	WK 4	WK 5	WK 6	...	WK 51	WK 52
Product 1	0	0	0	0	0	0		0	0
Product 2	0	0	0	0	0	0		0	0
Product 3	0	0	0	0	1	0		0	0
Product 4	0	0	0	0	0	0		0	0
Product 5	0	0	0	0	0	0		0	0
Product 6	0	0	0	0	0	0		0	0
...									
Product 100	0	0	0	0	0	0		0	0

as the new solution **x** (*Plan A₁*) introduced a change for Product 3 in Week 5. Note that the entries in the solution that did not change have corresponding values of 0 in this memory matrix. Now let's assume that a new solution selected in the second iteration made a change for Product 6 in Week 51. In that case, the memory matrix would have the following values:

	WK 1	WK 2	WK 3	WK 4	WK 5	WK 6	...	WK 51	WK 52
Product 1	0	0	0	0	0	0		0	0
Product 2	0	0	0	0	0	0		0	0
Product 3	0	0	0	0	1	0		0	0
Product 4	0	0	0	0	0	0		0	0
Product 5	0	0	0	0	0	0		0	0
Product 6	0	0	0	0	0	0		2	0
...									
Product 100	0	0	0	0	0	0		0	0

The general idea behind memory is that if some positions in the solution have changed, then the algorithm should leave these positions alone for some number of future iterations (i.e. they would be tabu for some number of iterations). This forces the algorithm to explore other parts of the search space,

---

9   Of course, $j = 0$ implies that the corresponding position of the solution matrix hasn't been changed. Only for $j > 0$ does the value of j indicate the iteration number when a change was made.

and after a predefined number of iterations has elapsed, these positions would become available again.

It might also be useful to alter the definition of memory so that the data stored in memory is erased after some number of iterations. If the data can stay in memory for 50 iterations, and then some entry of matrix M has a value j, a new interpretation of this can be that this cell in the solution was changed j iterations ago. Under this interpretation, matrix M might have the following values after several iterations:

	WK 1	WK 2	WK 3	WK 4	WK 5	WK 6	...	WK 51	WK 52
Product 1	25	0	50	0	0	0		0	0
Product 2	0	17	0	0	0	0		7	0
Product 3	0	0	0	0	19	0		0	0
Product 4	0	0	0	0	0	0		0	47
Product 5	0	33	0	0	23	0		0	0
Product 6	0	0	0	0	0	0		20	0
...									
Product 100	0	0	40	0	0	0		0	0

The numbers in this memory matrix provide the following data:

- The cell for Product 1 and Week 1 isn't available for the next 25 iterations
- The cell for Product 2 and Week 2 isn't available for the next 17 iterations
- The cell for Product 1 and Week 3 isn't available for the next 50 iterations (i.e. this cell in the solution has just changed)
- The cell for Product 2 and Week 4 is available for a change
- The cell for Product 100 and Week 3 isn't available for the next 40 iterations
- Etc.

In other words, the most recent change took place in the first row and third column (Product 1 and Week 3), and all non-zero positions in the memory matrix are considered tabu.

It's interesting to point out that the main difference between these two interpretations of memory is simply a matter of implementation. The latter approach interprets the values as the number of iterations for which a given position is not available, while the former interpretation simply stores the iteration number of the most recent change at a particular position. In the above example, if the difference between the iteration counter and the memory value is greater than 50 (our memory horizon), it should be forgotten. Hence, this interpretation only requires updating a single memory position per iteration and then increasing the iteration counter. In either case, tabu search utilizes the memory matrix to force the algorithm to explore new areas of the search space in an effort to escape from local optima. The recent changes in the solution are tabu for the next iteration (i.e. entries with corresponding non-zero values in the memory matrix).

Suppose that at some stage of the tabu search process, the quality measure score of the current solution **s** is 65,735 and the *best available* neighbor is solution **y**, with a quality measure score of 65,511. Note that this value represents a *decrease* in quality between the current and new solution, and the available neighborhood is much smaller than the entire neighborhood, as many positions in the solution are tabu (i.e. their corresponding values in the memory matrix are non-zero). On the other hand, imagine that a tabu neighbor, solution **q**, yields a quality measure score of 65,997. Assume further that the score of 65,997 is the best score from the beginning of the search, but because solution **q** is tabu we must ignore it!

Upon reflection, this policy might be too restrictive. One of the tabu neighbors of current solution **s** might produce a quality measure score that is *much* better than that of any previous solution. Perhaps we should make the search more flexible and bend the rules somewhat if we find an outstanding solution? Under normal circumstances, the tabu search algorithm should evaluate the *entire* neighborhood, and select a non-tabu solution as the next current solution, whether or not this non-tabu solution has a better-quality measure score than the current solution. But in other circumstances, where an outstanding tabu solution is found in the neighborhood, the superior solution should be selected as the next current solution.

Of course, there are other possibilities for increasing the flexibility of the search. For example, we could change the previous deterministic selection procedure into a probabilistic method, where better solutions have an increased chance of being selected. In addition, we could change the memory horizon during the search: sometimes it might be worthwhile to remember "more," and at other times, "less" (e.g. when the algorithm climbs a promising hill in the search space).

Another option is even more interesting: The memory structure discussed so far can be labeled as *recency-based* memory, because it only records the last few iterations. This structure might be extended by a *frequency-based* memory, which operates over a much longer time horizon h (by time horizon, we mean the number of past iterations taken into account) and measures the frequency of change at each position. For example, an additional matrix H may serve as long-term memory. This matrix is initially set to all 0s, and at any stage of the search the value j at any cell of this matrix is interpreted as "during the last h iterations of the algorithm, this cell of the solution was changed j number of times." Usually, the value of time horizon h (i.e. the number of past iterations we consider) is quite large, at least in comparison with the horizon of the recency-based memory. Thus, after many iterations with $h = 100,000$, the long-term memory H might have the following values:

	WK 1	WK 2	WK 3	WK 4	WK 5	WK 6	...	WK 51	WK 52
Product 1	0	0	0	0	0	0		0	0
Product 2	0	0	0	0	0	0		0	0
Product 3	0	0	0	0	1	0		0	0
Product 4	0	0	0	0	0	0		0	0
Product 5	0	0	0	0	0	0		0	0
Product 6	0	0	0	0	0	0		0	0
...									
Product 100	0	0	0	0	0	0		0	0

These frequencies (the total of which should equal 100,000 after 100,000 iterations) show which changes happened in the plan during the last 100,000 iterations. The principles of tabu search indicate that this type of memory might be useful for diversifying the search. For example, the frequency-based memory provides data on changes in the solution that have been infrequent, and we can diversify the search by exploring these positions.

The use of long-term memory in tabu search is usually restricted to special circumstances. For example, we might encounter a situation where all non-tabu neighbors produce inferior quality measure scores. Thus, to make a meaningful decision about which direction to explore next, it might be worthwhile to consult the long-term memory. There are many possibilities for incorporating this information into the decision-making process, but the most typical approach makes the most frequent changes less attractive by penalizing the quality measure score. As an example, let's assume that the quality measure score of the current solution s is 63,345 and all non-tabu neighbors produce inferior values (63,135, 63,019, 62,912, etc.), while none of the tabu neighbors provide a value greater than 63,720 (the highest value found so far), so we cannot apply the aspiration criterion by objective.[10] This is typical when consulting the frequency-based memory, as the evaluation function used in such circumstances (for a new solution $\mathbf{x}$) is the original evaluation function minus some penalty. This penalty, on the other hand, is calculated as a product

---

10 *Aspiration criteria* are used to override a solution's tabu state, thereby including the otherwise-excluded solution in the allowed set (provided that the solution is "good enough" according to its quality measure score or diversity). A simple and commonly used aspiration criterion is to allow solutions that are better than the current best solution (and this criterion is called: *aspiration by objective*).

of some parameter (say it is 0.1) and the total of all entries in the memory matrix that correspond to changed entries of the solution. Let's explore this through an example.

Assume that an inferior neighbor solution **x** (with a quality measure score of 63,135) differs from the current solution **s** in several places on the slotting board (for Product 3 in Week 4, for Product 6 in Week 3, for Product 5 in Week 51, and so on). By referring to memory matrix H and adding together all values in these positions, we arrive at a total of 3,704, which is multiplied by the penalty parameter 0.1 to produce a penalty of 370.4. After repeating this procedure for all other neighbors under consideration, let's assume that neighbor solution **y** (with a quality measure score of 63,019) generates a penalty of 364.1 and neighbor solution **z** (with a quality measure score of 62,912) generates a penalty of 130.9. In this case, it's clear we should select neighbor solution **z**, because it has the highest final quality measure score:

- Neighbor solution **x** has a final evaluation of 62,764.6 (63,135 – 370.4)
- Neighbor solution **y** has a final evaluation of 62,654.9 (63,019 – 364.1)
- Neighbor solution **z** has a final evaluation of 62,781.1 (62,912 – 130.9)

Although using frequency values to create a penalty measure diversifies the search, we can also consider some other options.

Over the years, tabu search has become increasingly complex as different researchers have modified the original method by incorporating additional rules. We have already seen one such rule, called *aspiration by objective*, which overrides the tabu search when a neighbor yields a solution that's the best so far. We can also use an additional rule, called *aspiration by default*, to select a neighbor that is the "oldest" of all those considered. It might also be a good idea to memorize not only the recent neighbors, but also whether or not these neighbors generated any improvement. This information can be incorporated into search decisions, called *aspiration by search direction*. We can also apply the concept of "influence," which measures the degree of change of a solution, either in terms of distance between the current and new solution, or the change in the solution's feasibility if we are dealing with a constrained problem. A neighbor has more influence if a "larger" step was made from the current solution to the new, and this information can be incorporated into the search (so-called *aspiration by influence*). Apart from these approaches, there are many other ways of implementing memory structures, aspiration criteria, and other features.

6.7 Evolutionary Algorithms

In previous sections of this chapter, we discussed hill climbers, stochastic hill climbers, simulated annealing, and tabu search. All of these optimization methods represent an approach of processing a single solution (i.e. of holding onto the best solution found so far and trying to improve it). This is intuitively sound, remarkably simple, and often quite efficient. The only decision to make during the execution of the algorithm is whether to "accept" or "reject" a newly generated neighbor solution. To make this decision, we can use many different rules. For example, hill climbers use deterministic rules: If a neighbor solution is superior, proceed to that neighbor and continue searching from there; otherwise, continue searching in the current neighborhood. Simulated annealing uses probabilistic rules: If a neighbor solution is superior, accept it as the new current solution; otherwise, either probabilistically accept this new inferior solution anyway or continue to search in the current neighborhood. Whereas tabu search uses the history of the search: Take the best available neighbor, which need not be better than the current solution, but which is not listed in memory as a restricted or "tabu" neighbor.

Rather than processing a single solution, evolutionary algorithms[11] process a "population" of competing solutions. In other words, evolutionary algorithms simulate the evolutionary process of competition and natural selection, where the solutions in the population fight for room in future generations. Additionally, new solutions are generated by means of genetically inspired operators (e.g. mutation or crossover) in a manner similar to natural evolution.

So, how do evolutionary algorithms work? Instead of generating an initial solution as we did for other methods, we start with a population of initial solutions[12] that could be generated at random from the search space, or through some manual process (i.e. starting with last year's plan). The evaluation function determines the quality measure score of each initial solution, and superior solutions are favored to become "parent" solutions for the next generation of "offspring" solutions. As before, new solutions can be generated probabilistically in the neighborhood of old solutions.

However, evolutionary algorithms provide an additional twist: we can also examine the neighborhoods of pairs of solutions. That is, we can use more than one parent solution to generate a new offspring solution. One way to do this is by taking different "parts" of two parent solutions and then

---

11 There are many terms related to evolutionary algorithms. The most popular are *genetic algorithms*, *evolutionary programming*, and *evolution strategies*. In this book, we use just one term, "evolutionary algorithms" without going into deeper details on similarities and differences between these algorithms. Many people use the term "genetic algorithms" in the same manner as we use "evolutionary algorithms," because the term "genetic algorithms" is better known than "evolutionary algorithms" in the business community. In Chapter 5.3, we already introduced a special class of evolutionary algorithms called *genetic programming*.

12 In evolutionary algorithms, these solutions are called *chromosomes* (to emphasize a link with genetics).

splicing them together to form an offspring solution. For example, we might take the first half of one parent together with the second half of another; or take the "middle" segment from one parent and implant it as the new "middle" segment of the second parent; or take the numbers present in the solution vector of both parents and create some (possibly weighted) average of numbers for the offspring solution.[13] With each generation, the individual solutions compete against one another (or also against their parents) for inclusion in the next generation of solutions. After many generations (i.e. iterations), we can often observe a succession of improvements in the quality measure score and convergence toward the neighborhood of a quality solution.

The following flowchart outlines the basic steps of an evolution algorithm:

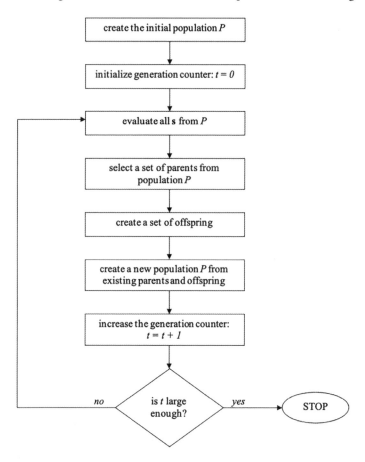

---

13 While designing such recombination operators, it is sometimes difficult to escape the temptation of "improving nature" … For example, should we consider more than two parents to generate offspring? In evolutionary algorithms, this is relatively easy: we can build an offspring solution by taking the first segment of the first parent, splice it with the second segment of the second parent, and combine it with the third segment of the third parent.

Here we generate some number of (possibly random) solutions, and since the population size is one of the parameters of the evolutionary algorithm, let's say we create 100 solutions as members of population *P*. At the same time, we set our generation counter *t* to zero. At this stage, we'll be entering an "evolutionary loop" that's repeated many times, and the process will terminate when *t* is "large enough" (many other termination conditions are possible, such as lack of progress after some number of generations). During this evolutionary loop, a few activities are performed. To start with, all solutions in population *P* are evaluated, as we'd like to know which solutions are superior and which are inferior. Secondly, a subset of parent solutions is selected from the population, favoring superior solutions. Thirdly, the selected parent solutions produce offspring solutions by means of some variation operators—quite often, through terms like *crossover* or *mutation*. Fourthly, a new population is selected (usually of the same size as the original population) from the existing parent and offspring solutions. And lastly, the generation counter *t* is increased, and the loop is repeated.

Without a doubt, this is an appealing approach for solving complex problems! Why should we labor to solve a problem by calculating difficult mathematical expressions or developing complicated computer programs to create approximate models of a problem, when we can discover quality solutions using models of greater granularity and accuracy? However, it's appropriate to ask some important questions: How much work does it take to implement these concepts into an algorithm? Is this cost-effective? And can we solve *real* problems using evolutionary algorithms? Well, let's explore these questions by continuing with the promotional planning example.

Say we started from an initial population *P*, which consists of 100 randomly generated solutions (the first four are displayed below, with each solution being a promotional plan, as before). We have to set our generation counter *t* to zero, and then we're ready to enter the evolutionary loop:

	WK1	WK2	WK3	WK4	WK5	WK6	…	WK51	WK52
Product 1	Y	Y		Y	Y				
Product 2	Y		Y		Y	Y			Y
Product 3	Y		Y		Y	Y			Y
Product 4			Y						
Product 5	Y			Y					
Product 6	Y	Y		Y					
Product 100	Y			Y	Y			Y	Y

	WK1	WK2	WK3	WK4	WK5	WK6	…	WK51	WK52
Product 1		Y		Y	Y				
Product 2	Y					Y			Y
Product 3	Y		Y		Y				
Product 4		Y						Y	
Product 5	Y			Y		Y		Y	Y
Product 6		Y		Y					
Product 100	Y			Y	Y			Y	Y

	WK1	WK2	WK3	WK4	WK5	WK6	…	WK51	WK52
Product 1		Y		Y	Y				
Product 2			Y			Y			Y
Product 3	Y			Y					
Product 4						Y			
Product 5		Y		Y		Y			Y
Product 6		Y			Y				
Product 100	Y			Y	Y	Y		Y	

	WK1	WK2	WK3	WK4	WK5	WK6	…	WK51	WK52
Product 1		Y							
Product 2	Y		Y	Y		Y			Y
Product 3			Y	Y	Y				
Product 4		Y				Y			
Product 5	Y			Y					
Product 6	Y		Y						
Product 100	Y			Y		Y			Y

.
.
.

We begin the evolutionary loop by evaluating all the initial solutions and identifying the parent solutions. From the 100 solutions present in the population, say we select 80 parents. One of the more popular parent selection strategies is called *tournament selection*, where we select two random solutions from our population, compare their quality measure scores, and select the better one as a parent. We repeat this process 80 times (i.e. until we select 80 parents). For example, say we selected:

	WK 1	WK 2	WK 3	WK 4	WK 5	WK 6	...	WK 51	WK 52
Product 1	y	y		y	y				
Product 2	y		y		y				y
Product 3	y		y		y	y			y
Product 4			y			y			
Product 5	y			y					
Product 6	y	y		y					
...									
Product 100	y			y	y			y	

	WK 1	WK 2	WK 3	WK 4	WK 5	WK 6	...	WK 51	WK 52
Product 1		y		y	y				
Product 2	y		y						y
Product 3	y				y	y		y	y
Product 4			y	y		y		y	
Product 5				y					
Product 6	y	y			y				
...									
Product 100	y			y	y			y	

for tournament competition. If the first solution has a quality measure score of 62,845 and the other 62,911, then the latter solution is the "winner" and is placed in a temporary population of parents. Note, however, one interesting twist: As we select pairs of random solutions, it is quite possible that some (relatively strong) solution is selected more than once for a tournament, and it wins such a tournament more than once. In that case, this solution would be represented in the population of parents a few times (e.g. the population of parents may contain several identical solutions). This is a desired side effect, as we would like to promote good "genetic traits" during the evolutionary process. If a solution is capable of winning several tournaments, then the chances are its quality measure score is above average, and it will produce "above average" offspring solutions.

Now we're ready for the most important step of the evolutionary loop: creating offspring solutions. As mentioned earlier in this section, we can consider several possibilities here. For example, we can use a *mutation operator* to change some positions in a parent solution to produce an offspring solution. For example, a parent solution of:

	WK 1	WK 2	WK 3	WK 4	WK 5	WK 6	...	WK 51	WK 52
Product 1	y	y		y	y				
Product 2	y		y		y	y			y
Product 3	y		y		y	y			y
Product 4				y					
Product 5	y			y					
Product 6	y	y		y					
...									
Product 100	y			y	y			y	y

can be "mutated" to produce the following offspring solution:

	WK 1	WK 2	WK 3	WK 4	WK 5	WK 6	...	WK 51	WK 52
Product 1	y	y		y	y				
Product 2	y		y			y			y
Product 3	y	y	y		y	y			y
Product 4				y					
Product 5	y								
Product 6	y	y		y					
...									
Product 100	y		y	y	y				y

Mutations occurred for Product 2 in Week 5, for Product 3 in Week 2, for Product 5 in Week 4, for Product 100 in Weeks 3 and 51) of the parent solution (where the values "Yes" and "No" were transposed). Note that the mutation corresponds to generating a new solution in the broader neighborhood of the current solution as several cells might be affected. When it comes to the process of mutation, we have some flexibility. For example:

- The probability of mutation may be quite low (e.g. 1%). This would mean that on average, 1% of the cells in the parent solution would be changed to produce a new offspring solution. As there are 5,200 cells in the solution, then on average 52 cells would receive new values. Of course, the probability of mutation (whether it be 1%, 0.1% or 10%) is an adjustable parameter of the algorithm.

- Regardless of whether the probability of mutation is 0.1% or 10%, we also have to decide "how" to mutate. One possibility is to switch values from "Yes" to "No" and vice versa. But another possibility would be to replace a selected variable—such as the promotional price—by a random value from the feasible range. However, we may also consider some "intelligent" mutations. For example, the mutation operator can refer to the feasibility status of a solution before deciding on "where to mutate." In particular, the mutation operator may know that Product 15 violates the minimum constraint on the number of promotions, so it would make perfect sense to mutate Product 15 by changing the "No" value into a "Yes" value, which is a step towards satisfying this very constraint.

- We might also have to consider more than one mutation operator. One mutation operator, for example, might be responsible for "small" changes, and the other for "big" changes (to introduce some additional diversity into our search, similar to microevolution and macroevolution in nature). Or one mutation operator might be responsible for playing with "Yes" and "No" values in the solution, whereas other mutation operator—for adjusting the promotional prices and promotion types.

As indicated earlier, we can explore some other operators for creating offspring. For example, we can use a *crossover* operator, which produces two offspring solutions by mixing some positions from two-parent solutions. For example, the parents:

	WK 1	WK 2	WK 3	WK 4	WK 5	WK 6	...	WK 51	WK 52
Product 1	y	y		y	Y				
Product 2			y		y				y
Product 3	Y		y		y	Y			y
Product 4			y			y			
Product 5	Y			y					
Product 6	Y	y		y					
...									
Product 100	Y			y	Y				Y

	WK 1	WK 2	WK 3	WK 4	WK 5	WK 6	...	WK 51	WK 52
Product 1		y		y	Y				
Product 2	y		y						y
Product 3	Y				y	Y		Y	y
Product 4			y	Y		y		Y	
Product 5					Y				
Product 6	Y	y				Y			
...									
Product 100	Y			y	Y			Y	

may produce an offspring that combines values from both parents. In particular, a subset of products might be selected from the first parent (randomly or by some rule, like products with the smallest degree of constraint violations related to these products), whereas the remaining products are selected from the second parent. Such a combination would result in a new solution (i.e. a new promotional plan). Note that crossover operators are fundamentally different from mutation operators, which modify (to a larger or smaller extent) a single solution. Crossovers, on the other hand, allow the combining of good parts of two promotional plans producing an offspring plan with a higher quality measure score than either parent!

Of course, there are a few additional possibilities. First, it's possible for the crossover operator to consider weeks rather than products, thus selecting a subset of weeks from the first parent and complementing these by weeks from the second parent. Secondly, the crossover operator can be assisted by additional knowledge, such as the min/max constraint on the number of products to be promoted every given week. We may also have a few parents contributing to an offspring, with each participating parent "contributing" a few of its best products or weeks—after all, who said that "two parents" is an optimal arrangement? Some AI scientists call this "an orgy," but there's no need to think along such lines. Clearly, the role of the crossover operator is straightforward: It might be that at some stage of the search process, some solutions contain individual values of high quality (e.g. promoting a particular product during a particular week). Crossover can speed up the search process, as these quality building blocks can be spliced together to create a solution with a larger number of quality components.

Anyway, after applying several variation operators (possibly different types of mutations and different types of crossovers), a set of 200 offspring solutions is created. We're now ready for the final step in the evolutionary loop—the creation of the next generation of plans—and we also increase our generation counter t at this stage, so it takes on a value of 1. As the population size is 100, our task is to create a new population of 100 solutions from 80 selected parents and 200 created offspring solutions. Again, there are many ways to accomplish this task. For example, we can:

- Apply a tournament selection process to all 280 plans (80 parents and 200 offspring)
- Exclude the parents and build a new generation from the offspring only.
- Design a tournament selection where more than two individuals compete to be selected
- Rank all plans by their quality measure score and then allocate the probabilities of selection accordingly

In most implementations, however, we use the *elitist* strategy to select the best solutions from one generation to the next. The cycle is then complete. We arrive at $t = 1$ generation with a new population of 100 solutions. We then repeat the evolutionary loop by evaluating these 100 solutions, selecting the parents, applying crossover and mutation operators to create offspring, selecting the next generation, and so on, until generation counter t hits a threshold (e.g. 100,000). At this stage, we expect to have many quality solutions in our final population, and the best one is selected as our final promotional plan.

For additional information on evolutionary algorithms and how they mimic nature, we encourage you to watch the supplementary video for this chapter at: www.Complexica.com/RiseofAI/Chapter6.

6.8 Constraint Handling

Most real-world optimization problems involve many problem-specific business rules and constraints[14] that impose some limitations on variables for the purpose of optimization. Usually they're classified into hard constraints, which an optimizer cannot violate, and soft constraints that can only be violated if necessary. A solution is feasible if it doesn't violate any hard constraints, and infeasible if at least one hard constraint is violated.

There are generally three ways to influence the search toward feasible solutions: through the evaluation function, the operators used, or the representation of the solution (and, of course, through some combination of the three). These three approaches are often independent of the optimization method, and we'll discuss them in turn.

Many constraint-handling methods have been proposed over the last few decades, but the simplest approach is the *death penalty*, where solutions that violate a hard constraint are immediately removed from further consideration. However, in many cases, the death penalty approach doesn't work very well. For instance, for highly constrained problems (like promotional planning), the algorithm generates a new solution, checks its feasibility, and then discards it because it's not feasible. And if 99% of the algorithm's effort is wasted on generating and removing infeasible solutions, then 99% of the algorithm's effort is unproductive! Moreover, the first feasible solution found (regardless of quality) can drive the whole algorithm to that area of the search space, where the final solution might be feasible but of poor quality (some local optimum at best).

---

14 The term "constraint" is often used in a more general sense: to mean either "constraint" or "business rule" (as defined at the beginning of Chapter 3). From an optimization point of view, both constraints and business rules restrict the feasibility of solutions in the same way. In this section, we often follow this convention.

Therefore, some leniency may be warranted toward infeasible solutions, as the death penalty may be too harsh and may prevent us from finding the best solution. A less drastic approach is through the use of various penalty functions. The idea is that if a potential solution (whether in evolutionary algorithms, simulated annealing, or tabu search) violates some problem-specific constraint, then the solution is "penalized" by reducing its quality measure score (i.e. less attractive). In other words, the quality measure score of a solution consists of two parts: the output of the evaluation function and a penalty score for violating constraints. Let's illustrate this approach by continuing the promotional planning example.

As explained in Chapter 3, promotional planning involves many hard and soft business rules and constraints. Some may be specific to individual products, for example, where each product may have minimum and maximum promotional frequencies, and a minimum promotional price or margin. Some other business rules and constraints may relate to the overall plan and are typically based on KPIs and various business objectives. For example, these business rules and constraints may include:

- Minimum and maximum discount, by product (hard)
- Price or discount step by product (hard)
- Minimum and maximum promotional frequencies, by product (soft)
- Maximum consecutive promotional periods, by product (soft)
- Maximum consecutive non-promotional periods, by product (soft)
- Minimum and maximum products on promotion in any given period (soft)

and can be visualized as follows:

	Type
Min freq	Soft
Max freq	Soft
Max promo length	Soft
Min promo gap	Soft
Max promo gap	Soft
Min promo price	Hard
Max promo price	Hard
Price step	Hard
Min promos p/ period	Soft
Max promos p/ period	Soft

together with their corresponding values:

	Shelf Price	Min Freq	Max Freq	Max Promo Length	Min Promo Gap	Max Promo Gap	Min Promo Price	Max Promo Price	Price Step	Allow rounding?
Product 1	$23.50	13	26	2	2	6	$12.00	$20.00	$0.50	Y
Product 2	$14.50	13	26	2	2	6	$10.00	$12.50	$0.50	Y
Product 3	$19.00	13	26	2	2	6	$10.00	$15.00	$1.00	N
Product 4	$22.00	20	39	3	2	4	$11.00	$18.00	$1.00	Y
Product 5	$52.00	20	39	3	2	4	$40.00	$48.00	$1.00	Y
Product 6	$48.00	13	39	3	2	4	$40.00	$45.00	$0.50	Y
...										
Product 100	$65.00	13	13	1	2	6	$55.00	$60.00	$1.00	Y

Additional business rules and constraints may include:

- Minimum and maximum number of products on promotion (soft)
- Overall plan minimum net revenue ($, hard)
- Overall plan minimum retailer margin growth over last year (%, hard)
- Overall gross profit growth over last year (%, hard)

For particular products (or an entire product category), these rules and constraints can be visualized as follows:

	Type
Min number of products on promotion	Soft
Max number of products on promotion	Soft
Min net revenue	Hard
Min retailer margin YOY growth	Hard
Gross profit YOY growth	Hard

Note also, that business rules and constraints can be quite complex. For example, the minimum and maximum numbers of products on promotion (expressed as a ratio of all products in a category) may depend on a particular week (e.g. the percentages may be larger towards public holidays):

	Min Promos	Max Promos
W 1	30%	60%
W 2	40%	70%
W 3	30%	60%
W 4	30%	60%
W 5	35%	65%
W 6	30%	60%
...		
W 52	60%	80%

Let's now assume that the quality measure score (say, volume is the only objective) of a particular solution plan, say *Plan A*, which we've used throughout this chapter:

		WK 1	WK 2	WK 3	WK 4	WK 5	WK 6	...	WK 51	WK 52
Retailer:	Mary's Market									
State:	NSW									
Category:	Snacks									
Product 1		Y	y		y	Y				
Product 2		y		y		y			y	
Product 3		Y		y		y	Y		y	
Product 4				Y			y			
Product 5		Y			Y					
Product 6		Y	y		Y					
...										
Product 100		Y			y	Y			Y	

Promotional Mechanism	In store
Promotional Price	$19.00
Shelf Price	$23.50
Discount	$4.50
Margin	35%
Retailer Margin	33%
Min/Max Frequencies	13/26 Weeks
Min/Max Sell Price	$12/20
Min/Max Promo Gap	2/6 Weeks

is 62,685, but seven products violate the minimum frequency soft constraint. So we should make this score less attractive. One way of setting such penalties is to assign a penalty weight for each violated constraint, which might look as follows (for both business rules and constraints):

	Type	Penalty
Min freq	Soft	60
Max freq	Soft	60
Max promo length	Soft	40
Min promo gap	Soft	40
Max promo gap	Soft	40
Min promo price	Hard	NA
Max promo price	Hard	NA
Price step	Hard	NA
Min promos p/ period	Soft	20
Max promos p/ period	Soft	21
Min net revenue	Soft	80
Min retailer margin YOY growth	Soft	80
Gross profit YOY growth	Soft	80

Now, the quality measure score of *Plan A* is $62,685 - 7 \times 60 = 62,265$ (as seven products have violated the minimum frequency soft constraint with penalty weight of 60). However, the application of these penalty weights might not be that straightforward. For example, Product 5 violated the minimum frequency constraint by one, whereas Product 77 violated this constraint by three; should we apply the same penalty weight of 60 to both cases?

Now we've begun designing a penalty function for each type of constraint violation. For example, we can assign 60 penalty points for every "unit" of violation of minimum frequency. In this case, the penalty for violating the minimum frequency soft constraint for Product 5 would remain 60, whereas the penalty for violating of minimum frequency soft constraint for Product 77 would be $3 \times 60 = 180$. However, such a linear allocation of penalty weights might not work that well, because a significant violation, even of a soft constraint, should result in a significant penalty. So, what about squaring the number of units that represent the degree of violation, before multiplying it by the penalty weight? In our example of Product 77, the penalty for violating the minimum frequency soft constraint would be $3^2 \times 60 = 9 \times 60 = 540$.

But that's not all! There is a variety of other constraints, such as maximum frequency, maximum promo length, or minimum promo gap, to name a few. The above solution may contain additional violations of these types. For example, the maximum promo gap is exceeded by two weeks for Product 43, and we need to assign a penalty weight for this violation. All constraints must be dealt with in a similar fashion, and we can only calculate the final quality measure score after totaling all applicable penalties. During this process of assigning penalty weights for different violations and different degrees of such violations, we should reflect the relative importance of each constraint (or rather, reflect our relative desire of getting these constraints satisfied). If violating the minimum frequency has a penalty weight of 60, what should be the penalty weight for violating the maximum promo length? Also 60? Lower? Higher?

Despite all these small complications, assigning penalty weights is relatively straightforward and easy to implement. However, it has many disadvantages. First of all, it's tricky to assign accurate penalty weights, as it might be difficult to weigh the relative importance of violating different constraints. Secondly, it's even harder to tune all these penalty weights to the original evaluation function: If the solutions are evaluated in the range from 50,000 to 80,000, then the penalty weights of 10 or 20 might be too low. And if the penalty weights are too high, then the effect is similar to that of the death penalty, as heavily penalized solutions wouldn't have much chance of survival. However, if the penalty weights are too low, then the final solution may violate too many constraints.

These issues can be addressed in several ways, one of which is to design *dynamic penalties* that change (usually increase) with each iteration of the optimizer. The purpose of dynamic penalties is to allow early solutions to sample various points in the search space without paying too much attention to feasibility (as the penalty weights are low). Then, as these penalty weights gradually increase, greater emphasis is placed on feasible solutions.[15] The drawback of this approach is that we have to specify the penalty weight changes in advance, and, again, this can be quite tricky. Another approach deals with this difficulty in the following way: For population-based methods, we can assume some "healthy" ratio between the number of feasible and infeasible solutions in the population, and during the algorithm run we maintain this ratio by increasing or decreasing the penalty weights. Thus, there's no need to specify the penalty weight changes in advance; the algorithm can adapt these weights by itself. Again, the drawback lies in defining this "healthy" ratio for a particular problem. Another way of dealing with constraints is based on the idea of smart operators (or *repair algorithms*).

Let's take the same solution as above, which violated the minimum frequency soft constraint for several products. Instead of penalizing this solution, we may try to "repair" it. As the minimum frequency constraint is violated for Product 5, we can try to repair this single row of the promotional plan by inserting an additional promotion. However, we may have to consider several possibilities on how to repair it: We could select several promotional slots at random and check if they satisfy this constraint or consider promotional slots in some predefined order (e.g. based on the length of a gap). Either way, we can repair the solution by adding extra promotions in the appropriate place.

We can handle the other constraint violations in the same way: If a constraint is violated, we can search for a suitable replacement. However, this process isn't so straightforward, because complex constraints might involve many variables, and so repairing one segment of the plan might trigger a constraint violation in some other segment. For example, by adding extra promotions for Product 5 to satisfy the minimum frequency constraint, it would be worthwhile to check whether other soft rules and constraints aren't violated due to the change (e.g. the minimum or maximum number of products on promotion for a particular week). In complicated cases such as these, we should use a problem-dependent repair algorithm with a degree of intelligence.

There is an additional twist in the use of repair algorithms: When a solution is repaired, the repair might be temporary, meaning that the solution is changed only for evaluation, as it's much easier to evaluate a feasible solution than an infeasible one. The original (infeasible) solution stays in the population for further processing. This is called the *Baldwin effect,* where solutions are evaluated based on their potential, rather than their current state.

---

15 Note the similarity of this approach to simulated annealing: the algorithm initially allows up and down movements, but, with time, it begins to favor moves that lead to superior solutions.

The other approach would be to make the repair permanent by changing the solution permanently, which is called *Lamarckian evolution*. This would be equivalent to improving our genes, so they're passed down to our offspring. Both of these approaches—Baldwin effect and Lamarckian evolution—have their advantages and disadvantages. Many researchers apply them in some ratio (e.g. only 10% of repairs are permanent, so that the process is 10% Lamarckian and 90% Baldwinian) and claim that a mixture produces better results than a pure Baldwinian or Lamarckian approach.

Yet another approach for constraint handling is based on the clever representation of a solution and *decoders*. The idea is that data present in a solution is interpreted (i.e. decoded) to correspond to a feasible solution. This would require an indirect representation of solutions and some pre-processing. In the case of promotional planning, this may mean sorting all products by their importance in terms of the main objective (i.e. their ability to generate volume). The representation of a solution may look something like:

Product 71	Product 34	Product 5		Product 13	Product 51

The above structure may indicate that product 71 should be considered first, along with the appropriate arrangement of all promotions for that product. At this stage of the process, when the slotting board representing the promotional plan is empty, it would be relatively easy to find an assignment of promotions that satisfy all product-specific constraints. We would consider Product 34 next, and repeat the process of assigning promotions so that all product-related constraints are satisfied, then Product 5, and so on. Using the decoder approach, we can maintain a dynamic list of *available* (i.e. feasible) slots for each product on the slotting board (based on some other constraints). By doing so, many constraints (e.g. min/max frequencies, minimum gaps, min/max number of products promoted for a particular week) can be satisfied during the pre-processing stage. For example, if Week 17 has already reached the maximum limit of promoted products for some iteration, then we can exclude that week from the list of available weeks for the remaining products. Thus, there's no need to penalize or repair infeasible solutions.

These three main approaches—penalties, repairs, and decoders—aren't the only constraint-handling methods. Sometimes it's worthwhile to develop a hybrid approach that combines elements of all these approaches. For example, some constraints are avoided through the application of a decoder, some constraints are handled by a family of repair algorithms, and other constraints through the application of penalties. The number of possibilities is almost endless, and for this reason, the selection of constraint-handling methods for any given optimization problem is still more art than science, and heavily based on experience.

6.9 Additional Aspects of Optimization

So far we have explained the basic idea behind several non-AI and AI optimization methods: The idea of traveling "up" in hill climbers, the idea of traveling almost randomly at first and gradually switching to uphill moves in simulated annealing, the idea of remembering our previous decisions in tabu search, and the idea of "breeding" the best possible solution in evolutionary algorithms. However, many other methods have emerged during the past few decades, and because real-world problems are usually very complex, some additional issues must be taken into account when applying these methods.

In the following sub-sections, we'll discuss ant systems, which we used in Chapter 1 to demonstrate the difference between AI algorithms and non-AI algorithms, before returning to the topic of multi-objective optimization from Chapter 2.2 and continuing with our exploration of constraints and other interesting features of optimization algorithms.

Other optimization methods

Many AI optimization methods that have emerged during the past few decades are based on natural phenomena, such as *ant systems* (also known as *ant colony optimization*), which were inspired by colonies of real ants that deposit a chemical substance (*pheromone*) on the ground. This substance influences an individual ant's behavior, in the sense that the greater the amount of pheromone deposited on a particular path, the larger the probability that an individual ant will select that path. Artificial ants behave in a similar way. In a nutshell, an ant algorithm is a multi-agent system, where simple interactions between artificial ants result in an overall complex behavior of the entire colony.

Ant systems are another population-based method, much like evolutionary algorithms. In evolutionary algorithms, the parent solutions are modified through some operators (e.g. mutations, crossovers) to create offspring solutions; in ant systems, however, the pheromone levels influence the creation of new solutions. The general idea behind ant systems is provided in the following flowchart:

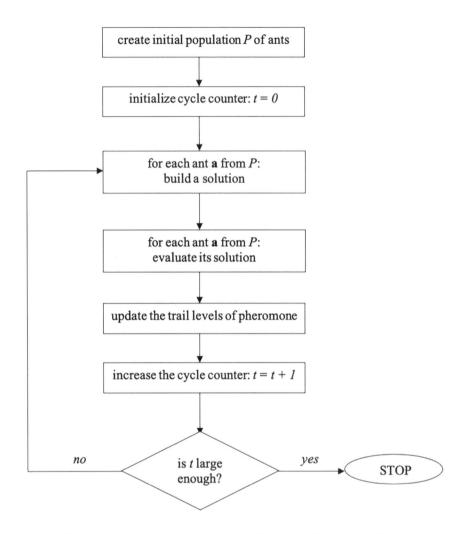

In this flowchart, we can start with a population P of ants, initialize each ant (i.e. set its various behavioral rules) and then set a cycle counter to repeat the process several times. Now, in the loop above there's a central action box where each ant is responsible for *building* a solution. The process of building the solution is usually influenced by two factors: (a) problem-specific knowledge, and (b) decisions made during the previous cycles (these decisions are summarized by the current *pheromone* levels). Once the new solutions are ready, we can evaluate them, update the pheromone trail levels (thus summarizing the "popularity" of the various decisions made by the ants), increase the cycle counter, and then repeat the process. Through this loop, the updated pheromone trail levels would influence the future decisions of the ants, and better decisions made in previous cycles would be reinforced in future cycles!

So, what is an "ant" within an "ant system"? In simple terms, an ant is a computer "agent" responsible for making decisions in the process of building new solutions, and each ant can be programmed with parameters (behavioral rules) that will influence its decisions. For example, when building a new solution, an ant will typically have parameters for controlling the relative importance of the pheromone trail versus problem-specific knowledge. The most obvious application (covered briefly in Chapter 1.3) of ant systems is for route optimization, where each solution represents a particular route traveled by the ant, but this AI-based algorithmic method can also be applied to other optimization problems. In the case of promotional planning, for example, these ant parameters may represent a trade-off between *visibility*, where more important products (in terms of generated volume) should be chosen with a higher probability, and *trail intensity*, which states that if a particular product enjoys a lot of "traffic," then it must be a profitable path to follow. It's also common for each ant to have a memory structure that records earlier decisions. This data structure (sometimes called a *tabu list*) is useful in avoiding the construction of infeasible solutions.

Each ant generates a solution, and each ant makes an independent decision on the promotion for each product and each week. These decisions are based on selection probabilities, which are calculated on the basis of attractiveness (importance of the product and significance of the week) and pheromone levels (related to the popularity of promoting this product that week). At this stage, each ant has built its own plan, and we have a population of many plans. The evaluation process is the same as for any other optimization method discussed in this chapter, with each plan being evaluated with a quality measure score.

Now it's time to update the pheromone levels. Let's consider a particular value p, which represents the pheromone level for some product for a particular week. At the end of the cycle, when the ants have built their individual plans, there are two independent processes for updating the value p: *evaporation* (which decreases the value of p) and *accumulation* (which increases the value of p). The first process is responsible for reducing the pheromone values of less popular slots in the promotional plan, and the second process is responsible for increasing the pheromone values of more popular slots.

We can easily implement the evaporation process by multiplying the original value p by a number smaller than 1 (e.g. assume that this number— also known as the *evaporation rate*—is equal to 0.9). This means that if, at the beginning of some cycle, $p = 0.087945$, then after evaporation, $p = 0.079151$ (0.9 × 0.087945). The accumulation process, on the other hand, is a bit more complex. We have to measure the "popularity" of each slot in the promotional plan across all ants. Moreover, better ants (i.e. ants that found better plans) should be able to exert greater influence in this popularity contest. Without going into too much detail, some product/weeks would be selected several times (getting significant increments in pheromone levels), while others might not be selected at all. The latter ones would experience evaporation, so their chances of being selected in future cycles would be smaller. The updating of pheromone levels is an essential part of the ant system, as better plans influence the probability of selecting the more promising parts of a plan in future cycles.[16] After many cycles, some ants should build a quality promotional plan.

This distribution and updating of pheromone levels represent a "communication process" that is essential for the ant algorithm to converge upon an optimal solution: The ants in the population become more and more similar in their behavior, as they *swarm* toward the optimum solution. Other related methods that use this "swarming" phenomenon include *particle swarm optimization*, which applies some *variation operator* (like mutation and crossover in evolutionary algorithms) to a population of agents, but without any selection process, as all agents are in constant motion and they "live" forever. In particle swarm optimization, the concept of "generation" is replaced with "iteration." Within a search space, each agent has a location and velocity that's updated according to the relationship between the agent's individual parameters (e.g. behavioral rules) and some other global parameters (e.g. the location of the best individual solution found so far). The search is biased toward better regions of the search space, resulting in a sort of "flocking" (i.e. swarming) toward superior solutions. As in ant systems, the agents exchange information through some medium that collects information on the locations and velocities of the particles, processes this information (e.g. selects the best particles in the current iteration, or selects the best locations found so far), and disseminates this information to other particles, influencing their subsequent direction and velocity.

In general, applications of the "social insect metaphor" for solving problems are often called *swarm intelligence systems*. As with ant systems and particle swarm optimization, these AI algorithmic methods assume the presence of many simple agents (e.g. ants, bees, wasps, termites) with direct or indirect interactions and communications that influence their future behavior. There are many possible applications of swarm intelligence for problems in distribution, communication networks, robotics, and so on. In all these cases the underlying principle is the same: Each insect is an "independent individual" performing some "individual activities" (often specialized activities); however, all these activities *seem* very well organized, without any outside organizer or

---

16 This process is similar to the "selection pressure" in evolutionary algorithms.

supervisor. This "self-organized behavior" is the essence of systems based on swarm intelligence.

Multi-objective optimization

Real-world problems usually have multiple considerations, and it's rare to find a problem with only one objective. In promotional planning, we'd like to maximize volume, while at the same time maximizing net revenue. And clearly, there is a trade-off between these two objectives: the larger the volume (triggered by larger discounts and selection of smaller value products), the smaller the net revenue.

Addressing multiple objectives may require methods that are quite different from standard single-objective optimization algorithms. As illustrated above, if there are even two objectives, it might be possible to find the best solution with respect to the first objective but not the second, and another solution that is best with respect to the second objective but not the first. If we go back to *Plan A* and *Plan B* from Chapter 3, *Plan A* had a predicted volume of 62,685 units and net revenue of $1,034,303, whereas *Plan B* had a predicted volume of 59,633 units and net revenue of $1,038,027:[17]

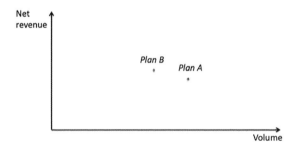

From these two solutions, *Plan A* is better with respect to volume and *Plan B* is better with respect to the net revenue, and neither is better than the other with respect to both objectives. Clearly, we can't directly compare *Plan A* with *Plan B* without creating some way of "translating" between net revenue and volume—in their current form, it's impossible to state which plan is better. But is one or the other of these two plans "optimal" in some sense?

It's clear that *Plan B* "dominates" an area of the search space, in the sense that any *Plan Q* from this area will be inferior to *Plan B* with respect to both objectives:

---

17 Note that we'd like to maximize our aggregate volume and aggregate net revenue across all products.

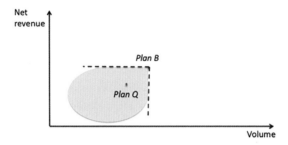

As this diagram clearly illustrates, *Plan Q* can't be "optimal" when compared with *Plan B*. On the other hand, if we find that there's no other plan that's better with respect to *all* objectives of the problem, then *Plan B* is "optimal" in some sense.

It might be convenient to classify all possible solutions of a multi-objective optimization problem into *dominated* solutions and *non-dominated* solutions. *Plan Q* is dominated if there exists a feasible *Plan B* that is at least as good as *Plan Q* with respect to every objective, and strictly better than *Plan Q* on at least one objective. Furthermore, any plan that's not dominated by any other feasible plan is called a non-dominated solution, which in this case could be *Plan B*. The set of all non-dominated feasible solutions is called the *Pareto optimal set*, as solutions in this set aren't dominated by any other feasible solution in the search space. Hence, in a sense, these solutions are optimal for any multi-objective optimization problem.

The trade-off between net revenue and volume is illustrated below, where we seek to maximize these two objectives at once. The continuous line in the figure represents the Pareto optimal front, which displays the trade-off between these objectives. For any solution on the front, an improvement with respect to volume will result in a decrease in net revenue, and vice versa. *Plan A* and *Plan B* on this line represent two of many non-dominated (Pareto optimal) solutions: *Plan A* does a better job on optimizing net revenue, whereas *Plan B* is more effective at maximizing volume.

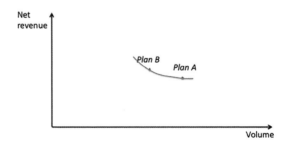

Of course, non-dominated solutions are of interest to us (i.e. solutions that cannot be improved according to one objective without being worse with respect to the remaining objectives). Ideally, an algorithm that deals with multi-objective problems should return several diverse non-dominated solutions, as each of these solutions might be of interest. The concept of diversity is import-ant, because many non-dominated solutions might be very similar to one another. For example, two plans, *Plan R* and *Plan S*, might be non-dominant, but there's little point in returning both of them to a business manager, as they are almost identical (e.g. they differ on a promotional price for one product for one particular week). The simple fact that we make the final decision on which plan to implement has two implications: (1) the number of feasible solutions shouldn't be too large, and (2) the feasible solutions should be diverse. At this stage, human expertise is required to translate different objectives into the same measure (e.g. decide between $1 of net profit versus one unit of volume) or to include some additional, higher-level information, such as the preferred promotional activities of certain retailers. As indicated earlier, we may assert the relative importance of each objective—perhaps by assigning numeric weights or imposing a ranking—and then select subsets of solutions that follow this ordering. Another option would be to select the most import-ant objective and then convert the remaining objectives into constraints, using them as thresholds to be satisfied.

However, when we think about finding a collection of points on a Pareto front, we should recall that population-based methods inherently process a collection of solutions (e.g. evolutionary algorithms, ant systems, etc.). The challenge is designing such a population-based algorithm that will spread out its potential solutions along the Pareto front, rather than seeking out a single "best" solution. And such population-based AI algorithms are the natural choice—instead of requiring multiple runs from different starting solutions in the hopes of landing at diverse points on the curve, such algorithms can be crafted to accomplish this task in a single run. Just an additional mechanism is required for spreading the best solutions in the population!

Understanding the solution (explainability)

In any real-world business environment, one of the key issues related to opti-mization is acceptance of the recommended (optimal) solution by end users. Most business managers—even those that are highly experienced—may not understand the machine-generated recommendation, and as a result, distrust it. This lack of understanding of the recommendations made by AI-based systems is one of the main barriers to their adoption. One way for making such recommendations more acceptable is through "explainability"—in other words, the ability of a decision optimization system to "explain the solution." This is of critical importance, because no value is created if the recommendations of an AI system aren't followed. It's only through the implementation of optimized decisions and recommendations that an organization can realize an improvement in key business metrics, as discussed in Chapter 2.

For these reasons and others, the "explainability" of AI-based systems is a popular topic these days; however, most research efforts are directed towards explaining the outputs of predictive models, rather than explaining the output of optimization algorithms. Indeed, if a solution provides "the best quality measure score," what else is there to explain? As it turns out, quite a bit, and there are several topics related to optimization worth exploring as they can make the solution more "user-friendly." The first is connected with predefined changes that the optimizer may consider when searching for the best solution. The second involves exploration of the boundaries of feasibility that are defined by business rules and constraints. And the third is the maximum number of changes made by the optimization algorithm while searching for the optimum solution. All three topics touch upon the explanatory capabilities of an AI-based system and optimization algorithms in particular, and we'll discuss them in turn.

As mentioned before, all optimization methods work in a similar way. The key step is transforming one solution into another and then repeating this process many, many times. However, due to the enormous size of the search space, random transformations aren't effective, because it would take too long to find an acceptable solution (to say nothing of an optimal solution). Thus, it's worthwhile to provide the optimization algorithm with a "hint"—for example, a table that includes all acceptable changes:

Promotion type	Relevant in retailer	Switch products	Add promos	Subtract promos	Move dates
Non-Promo	Y	Y	Y	Y	Y
Catalog - special	Y	Y	N	N	N
In store	Y	Y	N	N	Y
50% off	Y	N	Y	Y	Y
2 for 1	Y	Y	Y	Y	Y
3 for 2	Y	Y	Y	Y	Y

This table informs the optimization algorithm that if the promotion type is "50% off," then the algorithm shouldn't "switch products" for that promotion (e.g. shouldn't replace one product with another). Also, if the promotion type is "catalogue special," then the optimization algorithm shouldn't add or subtract products but switching products might be allowed. Such meaningful modifications to a promotional plan contribute to the "explainability" of the generated solution, because suggested changes "make sense" and can be easily visualized (as shown in the Trade Promotion Optimization case study in Chapter 9.4).

As explained earlier, all business rules and global constraints divide the search space into feasible and infeasible parts. Going back to our *Plan A*:

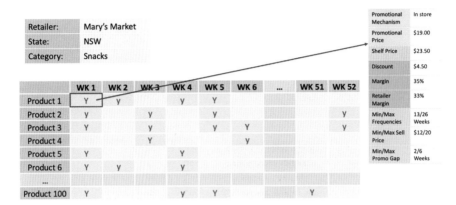

Let's assume this plan is feasible, where all hard and soft constraints are satisfied, and that the optimization algorithm is unable to improve this plan any further. Consequently, *Plan A* is returned by the algorithm as the best solution. However, there might be some ways to communicate with the user. For example, a solution might exist, say *Plan W*, which violates the limit on the minimum and maximum gap between two consecutive promotions, but the quality measure score of *Plan W* is *much higher* than for *Plan A,* which doesn't violate any constraints. Should the algorithm alert us that pushing the boundary of a particular constraint may result in a significantly better business outcome? To some extent, this can be accomplished by setting the appropriate penalty weight for violating such constraints, but too many products participating in such violations may eliminate *Plan W* altogether.

This interaction between the objectives and business rules and constraints is of great importance in many optimization problems, and the more feedback the algorithm can provide the end user, the better. Thus, on the top of the constraint-handling methods we discussed in Section 6.8, the ability to cross over the boundary of feasibility while searching for an optimal solution is always of value.

An additional desirable feature of any optimization algorithm is the ability to return a solution that differs from the starting solution in some limited way. Here we assume that a draft solution exists and the only task of the optimization algorithm is to improve it, which is common in most real-world situations. In promotional planning, an initial plan is often created manually based on experience and we might only want a small number of changes that result in the largest improvement. In particular, it would benefit us to know:

- If only we made a single change to the current plan, what change would result in the greatest benefit (in terms of our objective and quality measure score)?
- If we only made n changes to the current plan, what change would result in the greatest benefit (in terms of our objective and quality measure score)?

It would be easy to extend the business rules and constraints with a "maximum number of changes" allowed to the original plan, as shown in the last row of this table:

Min promos p/ period	Soft	20
Max promos p/ period	Soft	21
Min net revenue	Soft	80
Min retailer margin YOY growth	Soft	80
Gross profit YOY growth	Soft	80
Max number of changes	Hard	1

This additional constraint would result in an additional benefit, namely, that by setting $n = 1$ (as above) we could effectively "ask" the algorithm to show us the single most significant change we could make to the current plan. This new promotional plan (with a single change) could then be loaded as the new original plan, and the same question can be asked again: What's the single most significant change we could make to the current plan? Repeating this process a few times, we can discover the single most significant change (in terms of improving the quality measure score), the second most significant change, the third, and so on. It's also worth pointing out that two separate runs of the optimization algorithm with the parameter n set to 1 may produce a different result than a single run of the algorithm with the parameter n set to 2. The reason for this is that in the former case, the algorithm (in its first and second runs) searches for one change only, whereas in the latter case, it searches for two changes at the same time.

In very advanced decision optimization systems, the optimization algorithm can "explain itself" in natural language. For example, when recommending the best promotional plan, the algorithm can display additional information, such as:

- "Product 76 has a longer gap between promotions that start in Week 6 and Week 13 to avoid significant cannibalization with Product 23"

- "Product 35 has an extended promotional period that starts at Week 30 due to the coming school holidays that will result in a significant uplift in sales"
- "If one additional change is allowed—five changes versus requested four changes—a significant uplift in volume can be realized"

Such statements may also include a variety of other insights, including the consequences of modifying the recommended plans. However, the implementation of such "self explainability" requires significant expertise, both in the scientific aspects of the algorithmic method being used, and also in the business domain of the problem being solved.

6.10 Global Optimization

Although real-world problems are usually comprised of smaller sub-problems (i.e. components) that interact with each other, most organizations realize that these components are related and affect one another, and what's most desirable is a solution for the overall problem that takes into account all components.[18] For example, scheduling production lines (e.g. maximizing the efficiency or minimizing the cost) is directly related to inventory costs, transportation costs, delivery-in-full-on-time (DIFOT) to customers, among other business metrics, and shouldn't be considered in isolation. Moreover, optimizing one component may negatively impact another. For these reasons, organizations can unlock more value through "globally optimized" solutions that consider all the components together, simultaneously, rather than just a single component.

A puzzle example

Let's illustrate the difference between single-component and multi-component problems through a puzzle.[19] A ball has been cut into two parts in a special way, and a cuboid with a hole inside has been also cut into two parts in a special way. The two parts of the ball can be assembled to make a ball, and the two parts of the cuboid can be assembled to form a cuboid, as pictured below:

---

18 There are similar concepts to multi-component problems in other disciplines, such as operations research and management science, with different names such as integrated systems, integrated supply chains, system planning, and hierarchical production planning.

19 The name of this puzzle is the "Cast Marble," created by Hanayama company.

The hole inside the cuboid is slightly larger than the size of the ball, so that if the ball is inside the hole it can spin freely. However, it's not possible to assemble the cuboid and then put the ball inside, as the entry of the hole is smaller than the size of the ball (pictured below, left). Now, the puzzle is to set up the cuboid with the ball inside (pictured below, right), because setting up the ball separately and the cuboid separately is easy but setting up the cuboid while the ball is inside the cuboid is extremely hard.[20]

This puzzle highlights the difference between single-component and multi-component problems. In fact, solving a single component problem (setting up the ball or the cuboid individually) might be easy; however, solving the combination of two simple component problems together is much harder (setting up the ball while it's inside the cuboid).

Terminology

There is also some confusion related to terminology used in academia and industry with respect to "local optimum" (or local solution) versus "global

---

20 The difficulty level of this puzzle was reported as 4 out of 6 by the Hanayama website, which is equal to the difficulty of Rubik's cube.

optimum" (or global solution). We touched upon local vs. global optima in Section 6.3, where we presented the quality measure score landscape for a two-variable function. The graph below displays the quality measure score for every pair of values for the first and second variables, allowing us to visualize the mountain ranges, peaks, and valleys.

As discussed earlier, the highest peak is called global optimum and possibly many lower peaks, which are the local optima:

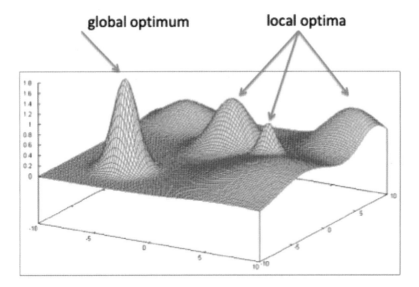

This concept of local vs. global optima in academia is strictly "vertical," where the quality measure score determines the height of a peak, and by having all the measurements we can easily identify all local peaks.

In business, however, the term "global optimum" has a different meaning. Note that when we're dealing with multi-component problems (as the puzzle above showed), we can get a locally optimum solution for each individual component by solving each component separately, and then try to put the components together. Still, the result is likely to be sub-optimal, as the puzzle also showed, where the two pieces didn't fit together and thereby, we really didn't solve the overall problem after all.

Keeping this in mind, modern organizations are interested in "globally optimal solutions"—solutions that allow their organization to reap the greatest possible benefit—rather than solutions to individual components of their operation (which they have to "assemble"). Thus, the concept of local vs. global optima is "horizontal"—the overall quality measure score considers all components of the problem to come up with the best solution to the overall problem.

Business examples

The first example relates to optimizing the mine to port supply chain of a mining company. In such operations, the mining company tries to satisfy customer orders by providing a predefined amount of product (e.g. coal, iron ore, etc.), graded to a specific quality level (such as the percentage of iron within the ore) on a particular due date when the product must be ready for ship loading. The supply chain would have multiple stages—starting with planning the extraction of ore at the mine, all the way to planning when each ship will be berthed and loaded:

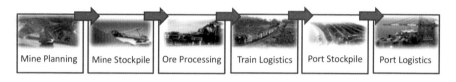

Each of these steps is significantly complex in and of itself—such as coming up with an optimized ore extraction sequence, or the most efficient train schedule (which must consider upcoming maintenance, the availability of various train drivers, the number of junctions in the network where trains can pass)—representing hard optimization problems to solve even on their own.

Apart from the complexity within each component, optimizing a single component in isolation won't lead us to a globally optimized solution for the whole supply chain. As an example, scheduling trains to optimality (moving as much product as possible from the mine to port) might result in too much product at the port or even the wrong product at the wrong time, because we're only worried about maximizing the asset utilization of the train network, rather than what's "good" for the mine or port. Like the puzzle example above, solving each component individually and then assembling the solutions together is unlikely to result in an overall, globally optimum solution; and if all solutions are local, then many business opportunities may be lost. Hence, solving an individual component of the mining supply chain provides us with a local optimum as far as the whole problem is concerned, rather than the global optimum, especially given there are strong dependencies between all components of the supply chain (e.g. some decisions on which blocks to mine directly impact crushing and blending activities, which in turn impact train logistics):

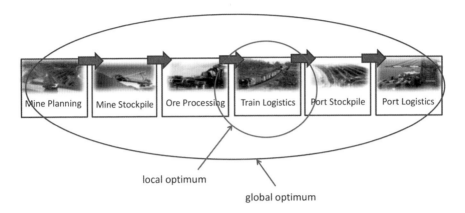

local optimum

global optimum

A global optimization algorithm, on the other hand, would simultaneously address the objectives of each component of the supply chain. However, the implementation of such algorithms for global optimization is scientifically challenging, and there are two general approaches: A centralized approach, where a global "agent" is responsible for "supervising" local agents (that are dealing with the individual components of the overall operation) and tries to synchronize their activities and recommended decisions. Another approach is based on distributed optimization, where local optimization algorithms "talk" to each other in order to synchronize their actions. This can be accomplished in many ways, for example, through co-evolutionary algorithms, which were discussed briefly in Chapter 3.3 in relation to modeling competitor behavior.

Another business example that illustrates the difficulty of global optimization is much smaller in scope and relates to optimizing the transportation of water tanks. In this case, a manufacturer produces water tanks of different shapes and sizes based on customer orders. The total number of customer orders per month is approximately 10,000, which vary in their delivery locations. Each customer orders a water tank with specific characteristics (including size) and expects to receive it within a period of time (usually one month). These water tanks are then transported by a fleet of trucks operated by the water tank company, which have different characteristics and some of them are equipped with trailers. A subset of orders is selected and assigned to a truck and the deliveries are scheduled. Because the tanks are empty and of different sizes, they might be bundled inside each other to maximize the truck's load. A bundled tank must be unbundled at special sites, called "bases," before the tank's delivery to the final destination. There might be several bases close to the various customer locations where the tanks are going to be delivered, and selecting different bases affects the best overall achievable solution. When the tanks are unbundled at a base, only some of them fit back onto the truck, as unbundled tanks require more space. So the truck is loaded with a subset of these unbundled tanks and delivers them to the customer, while the remaining tanks are kept in the base until the truck returns to continue the delivery process.

The goal of the optimization algorithm is to group the customer orders into subsets of tanks that are bundled and loaded onto trucks for delivery (possibly with trailers), and then determine the best base for unbundling the tanks before delivery to each customer, so that the overall delivery cost is minimized. Each of the mentioned procedures in this problem (tank subset selection, base selection, delivery routing, bundling and unbundling) is just one component of the problem and finding a solution for each component in isolation doesn't lead to the optimal solution for the overall problem.

For example, if we just focus on the sub-selection of tanks (based on customer orders), there's no guarantee there exists a feasible bundling solution that would allow this subset to fit on a truck. Also, by selecting tanks without considering the location of customer destinations and location of bases, the best solution might not be a high-quality one, as there might be a customer destination that requires a low-cost tank, but that location is very far from any base, making delivery very expensive. On the other hand, it's impossible to select the best routing for customer destinations before selecting the tanks, because without first selecting the tanks, the best solution (lowest possible cost and distance) is to deliver nothing. Thus, solving each component of the problem in isolation doesn't lead to an optimized overall solution—especially when we add additional considerations, such as the rostering of drivers (who often have different qualifications), fatigue factors and labor laws, traffic patterns on the roads, usability of different trucks for different segments of roads, maintenance schedules, and so on, which further complicate the problem as well as the interaction between its various components.

If we return to promotional planning for a moment, in Section 6.2 we presented the structure of a solution for this problem: a slotting board matrix where rows correspond to products, and columns correspond to weeks of the year. We also noted that this solution might be part of a larger, overall solution, because we might need an overall plan for Mary's Market, whereas this solution only represents one product category (e.g. snacks) in one state (e.g. NSW). However, the problem might be even larger than that. For example, we might need to consider two or more retailers at the same time. So instead of optimizing a promotional plan just for Mary's Market, we may need to globally optimize our promotional planning across many retailers—a sort of cross-retailer optimization. This, of course, would require additional modeling, because in addition to defining the business rules and constraints for each individual retailer, we'd also need to define cross-retailer business rules and constraints. Most likely, the objectives would also change. In other words, we'd suddenly be dealing with a very complex, multi-component problem.

Despite the obvious observation that multi-component problems should be solved as a single problem to come up with the best overall solution, the vast majority of modern organizations don't do this. Why? Because of how overwhelmingly complex the problem becomes when all components are assembled together, requiring not only sophisticated AI algorithms, but significant process of re-engineering and change management within each component of the operation. Hence, for these reasons and others, the design and creation of such algorithms, as well as putting them into software applications that become embedded into business processes and workflows, is beyond the capability of most organizations, despite the significant benefits that global optimization can unlock. We'll return to the topic of global optimization in our discussion of supply chains in Chapter 10.

Also, for more information on some of the issues related to optimization, we encourage you to watch the supplementary video for this chapter at: www.Complexica.com/RiseofAI/Chapter6.

CHAPTER 7

Learning

"An organization's ability to learn, and translate the learning into action rapidly, is the ultimate competitive advantage."

Jack Welch, *former CEO of GE*

In Chapter 2, we combined prediction, optimization, and adaptability to form a Decision Optimization System, which can learn from past decisions in order to improve the quality and accuracy of future recommendations:

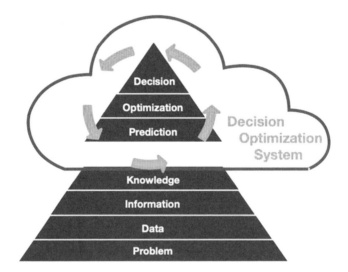

In this chapter, we'll discuss a few ways that Decision Optimization Systems "adapt" to the operating environment as new data becomes available. In the promotional planning example, this would include daily or weekly sales data for all products (whether on promotion or not), along with their prices and promotion type (if any). Note that our prediction model has already predicted these volumes, so now we're getting back the *actual* values. This data will tell us the prediction model's error rate, and whether or not the model requires adjustment. Likewise, our optimization model may require further tuning, because instances of yesterday's promotional planning problem may differ from instances in the future. Given the marketplace is constantly changing and some model adjustments will be required over time, we can do one of two things:

- Repeat the process of manually updating our model at regular intervals. For example, we might update the parameters of the underlying prediction model every three to six months, or we could replace the optimization algorithm by another algorithm that generates better results. However, the process of analyzing new data and updating the model might be expensive in terms of time and effort, so there's an incentive to repeat this process at longer intervals (such as once a year, rather than every quarter). This approach often causes a problematic trade-off, as shorter intervals would be better from an operational standpoint, but worse from a cost standpoint.

- Alternatively, we could implement a learning algorithm that's responsible for automatically updating the model when new data becomes available. By automating this process, new data can be fed into the model quite frequently—even at the end of each business day if necessary—and if a new pattern emerges, it can be captured by the updated model almost immediately.

There are many advantages of using an automated learning algorithm instead of manual updates, including:

- The process is automatic, and therefore inexpensive in terms of time and labor resources
- The frequency of updates can be much higher
- There's a good chance of discovering new, emerging patterns as they occur

If we think about updating a Decision Optimization System in the context of promotional planning, let's first remember that promotional planning has all the inherent characteristics of a complex business problem: The number of possible solutions is beyond astronomical; there are many (soft and hard) business rules and constraints that differ across retailers, geographies, product categories, and individual products; and there are multiple, often competing, objectives and KPIs. Underpinning this complexity is the fundamental challenge of predicting outcomes in the marketplace, as we're making decisions about the future. Let's also recall from Chapter 3 that "all this inherent complexity exists even if we wanted to create just one promotional plan, just once. But of course, promotional planning isn't done just once; it's an ongoing process with decisions being made in a dynamic environment where the unexpected might happen (as COVID-19 recently showed). Amidst all these planning decisions, many variables are also in a state of constant flux. New products are launched and old ones discontinued, retailers open and shut stores, supply chain costs vary, business rules are revised, shelf space and product facings change, brands grow or decline, and consumer preferences and tastes evolve."

Hence, the intelligence of any Decision Optimization System—apart from its prediction and optimization capabilities—lies in its ability *to adapt*. And to achieve this goal, the system must be equipped with some learning algorithms. The same holds true for any Artificial Intelligence system, which must adjust to new environmental conditions through self-correction (self-learning) as new events happen or new data becomes available. In other words, AI-based systems must continuously adapt themselves, or, at the very least, prevent the degradation of their performance in an ever-changing environment. Only such systems deserve being labeled as "intelligent." After all, this is what Artificial Intelligence is all about.[1]

As we delve deeper into this topic, it's worthwhile to distinguish between "adaptability" and "learning," and answer questions such as: Is it possible to "adapt" without learning? How do predictive models learn? How can optimization models learn? To answer these questions, let's start with the topic of "learning" and then build from there.

7.1 Learning

Learning is the process of acquiring new skills or knowledge, which either occurs in a single event (such as being burned by a hot stove) or gradually over time from repeated events and experiences (such as learning how to perform a sophisticated magic trick or understanding a complex theory such as the general theory of relativity). Drawing on these natural processes, Machine Learning[2] is the study of computer algorithms that automatically learn through experience. As a scientific endeavor, Machine Learning grew out of the quest for achieving the original promise of Artificial Intelligence (i.e. General AI), with researchers wanting to teach machines how to automatically learn from data and experience. As discussed in Chapter 1.1, Machine Learning algorithms such as deep learning represent a subset of algorithmic methods within the broader category of Artificial Intelligence algorithms:

---

1   The same applies in nature, as the ability to adapt is a key feature of any living organism and an expression of its intelligence.
2   The term Machine Learning was coined in 1959 by Arthur Samuel, a pioneer in computer gaming and artificial intelligence.

Areas of Artificial Intelligence Research

Robotics	Computer Vision	Natural Language Processing	Cognitive Computing	

Machine Learning Algorithms and Techniques

Deep Learning

Algorithmic Methods

Machine Learning algorithms are used to build models based on "training data," so they can make predictions or decisions without being explicitly programmed to do so. Before proceeding further, let's also clarify that Machine Learning algorithms are first used to create a Machine Learning model, and then used again to update this model:

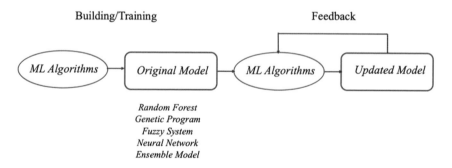

Building/Training   Feedback

ML Algorithms → Original Model → ML Algorithms → Updated Model

Random Forest
Genetic Program
Fuzzy System
Neural Network
Ensemble Model

The feedback loop in the above diagram includes the Machine Learning algorithms for updating the original model, along with any associated update procedures. So "learning" happens twice: once when the model is first built and trained, and then again upon the arrival of new data and/or feedback (which might occur at regular intervals or continuously).

The field of Machine Learning uses many different terms for "learning," such as *supervised learning, unsupervised learning, deep learning, incremental learning, continual learning, feature learning, reinforcement learning, meta learning, robot learning, learning classifier system, association rule learning, federated learning,* among others. Let's clarify and categorize these terms by stating that Machine Learning methods are traditionally divided into three broad categories:

- *Supervised learning*: is where a "teacher" provides the Machine Learning algorithm with labeled examples of inputs and outputs, and the goal of the algorithm is to learn the relationship between these inputs to outputs.
- *Unsupervised learning*: is where the Machine Learning algorithm isn't provided with any labeled inputs or outputs, so it must discover patterns in the data without any "teacher" that can provide examples or guidance. Unsupervised learning can be a goal in itself (e.g. discovering hidden patterns in data) or a means towards an end (e.g. discovering the key variables in a dataset).
- *Reinforcement learning*: is where the model interacts with a dynamic environment in which it must perform a certain task (such as driving a vehicle or playing a video game). As it attempts to perform the given task, the model is provided with feedback in the form of rewards, which it tries to maximize.

Let's discuss each category in turn.

Supervised learning

Supervised learning is best explained through an analogy of teaching a toddler how to say the names of various objects. The process usually involves a teacher (e.g. a parent) showing the toddler an object (e.g. an apple) and saying it out loud ("this is an apple"). The process is repeated for a variety of objects ("this is a pen," "this is a ball," etc.) so the toddler can memorize the names. After a while, the toddler will attempt to name these objects. Sometimes the objects are named correctly and the parent praises the toddler. But at other times, the objects are referred to incorrectly or unintelligibly, in which case the parent corrects the toddler ("It's *not* a ball, this is an apple"). The process of supervising the toddler to learn names (i.e. to map the objects to their respective names) continues until the toddler becomes fluent and there's no further need for corrections.

This example illustrates supervised learning, where the toddler can be thought of as a "model" being trained to correctly name objects. Supervision is conducted by parents as they guide the learning process by providing the actual names of objects (often referred to as "ground truth") and evaluating the names provided by the toddler. Similarly, in Machine Learning, supervision is conducted by presenting the algorithm with sample data (say, an image), asking the algorithm to identify the image, and correcting the answer if necessary ("It's not a school bus, it's a duck"). The key characteristics of supervised learning are:

(1) The learning task is clearly defined (e.g. what kind of objects the model should be able to identify), and

(2) The correct names are known in advance, allowing for knowledge to be built up based on a direct comparison between the names given by the model and the actual ("ground truth") names

Another classic example of application of supervised learning is the problem of identifying fraudulent credit card transactions, with the goal being to build a Machine Learning model for classifying new transactions as either "fraudulent" or "legitimate." Each transaction is described by certain characteristics (variables) such as time, location, merchant, amount, and so on. To build a model in a supervised manner, we would need access to historical data of past transactions. This data should include the ground truth variable for each transaction indicating whether it was fraudulent or legitimate, which is essential to the supervised learning process. Having this historical data, we can then use one of many supervised learning methods (as discussed in Chapter 5), such as random forest, logistic regression, neural network, and genetic programming, among others, to build a Machine Learning model. Also, an example of supervised learning in the context of neural networks was presented in Chapter 5.5.

Let's illustrate the concept of supervised learning with the following diagram:

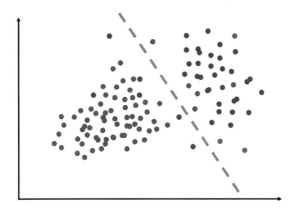

Each dot represents one data point (e.g. one credit card transaction), and the color of the dots represents whether the transaction was fraudulent (red) or legitimate (blue). A model is then created based on the labeled (red or blue) data points, which is represented by the dashed line separating these data points. During the learning process the model tries to position this line so that all blue dots (and only blue dots) are on one side of the line, and all red dots (and only red dots) are on the other. Of course, this might not be possible, and in that case, we may consider using a more complex, non-linear model (or some other structure, like a decision tree). Once the model is built, new data points (grey dots) are classified by the model as either blue or red, depending on which side of the line they fall. And because it's impossible to create a perfect model that can correctly classify all future transactions, we should evaluate its performance by looking at the number of false-positives[3] and false-negatives.[4]

Unsupervised learning

The main distinction between supervised and unsupervised learning is the presence or absence of the ground truth data, which is required for supervised learning and usually not available in unsupervised learning.[5] If this ground truth data is missing, we can't provide explicit feedback to the algorithm on whether the results are correct or not, and this difference implies that unsupervised learning is applicable to a different set of problems than supervised learning. In a nutshell, unsupervised learning methods are presented with a dataset and a general goal of answering the following question: What patterns (similarities and/or dissimilarities) can we identify in the data?

As an example, let's take a situation where ground truth labeling isn't available at all. Let's imagine a broad set of customers visiting an eCommerce site, where after logging in, every customer is characterized by a variety of interactions, such as browsing particular product categories, reading product descriptions, buying certain products, and so on. Using this data, we might want to better understand what "types" of customers are visiting the site, which might allow us to make useful changes to the site (such as the layout of each page, length of each product description, and so on) or create effective promotional campaigns, such as targeted EDM marketing[6] to specific customers. But what are the possible "types" of customers? Do we have a predefined set of possible types? How many of these types might we expect in our dataset? We don't have answers to these questions, and because of that, unsupervised learning methods can be useful in this situation.

---

3   A false positive is an error that occurs when a result incorrectly indicates the presence of a condition. In this case, the model wrongly classified a legitimate transaction as fraudulent.
4   A false negative is an error that occurs when a result incorrectly indicates that a particular condition is absent. In this case, the model wrongly classifies a fraudulent transaction as legitimate.
5   There is also *semi-supervised learning*, which falls in between unsupervised learning and supervised learning. This is applicable in cases where some of the training data are missing ground truth labels. These unlabeled data, when used in conjunction with even a small amount of labeled data, can produce a considerable improvement in learning accuracy.
6   EDM marketing is an acronym for Electronic Direct Mail marketing, which is a form of direct marketing that organizations use to increase sales.

One branch of unsupervised learning is clustering (which we discussed in Chapter 5.1), which tries to find a natural grouping of objects based on similarities and dissimilarities. One of the classic clustering algorithms is k-means, which has one input parameter k and must identify k number of groups in the input dataset. As an output, the algorithm generates a grouping such that objects in the same group (called a cluster) are more similar (in some sense) to each other than to objects in the other groups (clusters). Consider eCommerce example, where some grouping of customers is required. In this case, the algorithm generates a grouping such that customers in the same cluster are more similar to each other than to customers in the other clusters.

Further we might run the k-means algorithm with k = 2 to generate just two distinct groups of customers:[7]

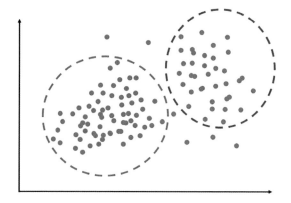

In the above graph, the red and blue circles represent the two groups generated by the k-means clustering algorithm. In comparison to supervised learning, in the unsupervised learning case the input data has no labels (i.e. only grey dots) and the algorithm generated these groups based on the distance measure between them.

After obtaining the result, a detailed analysis of the groups can be conducted to discover their similarities and differences. One of the discoveries, for example, might be that customers in the "red" group made more purchases at regular intervals and their basket size was similar from one transaction to the next. The "blue" group, on the other hand, might consist of customers that made irregular purchases. Based on this analysis, we might conclude that the red group consists of *regular* customers and the blue group of *casual* customers. Having established this grouping, an even more in-depth analysis

---

7   In reality, we often run k-means algorithm multiple times for a range of different values of the input parameter k—for example, between 2 and 10—and then try to determine which result is best.

can be conducted. For example, we might want to understand under what circumstances a casual customer became a regular customer? The answer to such questions might result in further analysis and action, such as targeting casual customers with EDM marketing to encourage more systematic purchases.

Reinforcement learning

The third approach is reinforcement learning, with the most notable and well-known application being AlphaGo—a Machine Learning application that defeated the world champion of Go in 2016 (Lee Sedol). We'll start our discussion of reinforcement learning with this example, as Go is a board game introduced in China more than 2,500 years ago[8] and is played between two players on a grid of 19 horizontal and 19 vertical lines. Each player has 180 stones of the same color (black or white), and puts their stones on the board in turns, starting from black (as opposed to chess, where white has the first move). These stones can be placed on any non-occupied line intersection and can't be removed from this position unless surrounded by stones of the opponent's color. In which case, the stone is "captured" by the opponent. The goal of the game is to occupy the largest territory on the board by surrounding vacant intersections and capturing as many opponent stones as possible.

The rules of Go seem to be much simpler than chess, because unlike chess, we can't move a stone once it's placed on the board, all game pieces are the same (no bishops, knights, pawns, etc.), and there are no special rules, such as pawn promotions or *en passant*. However, with 361 (19 ×19) possible places to place a stone (as opposed to "only" 8 × 8 = 64 places for chess), Go has far more combinatorial complexity than chess. With 361 possible first moves, and 360 possible counter-moves, and 359 possible second moves, and 358 possible counter-moves, the number of possibilities is simply too large to analyze (it would be difficult to even imagine analyzing the game 10 moves ahead, which by comparison, is relatively easy in chess). Because of the large number of possible moves, many scientists believed that no computer program would be capable of beating the best Go player until at least 2025, as the

---

8   According to one of the legends, the game was created as a teaching tool by the Chinese Emperor Yao (2,356 – 2,255 BCE) for his son, Danzhu, to learn discipline, concentration, and balance. Since then, the rules of Go have seen some variation over time and from place to place. The most popular are the ones currently in use in East Asia, but there is a degree of variation even among these.

complexity of chess and Go are very different.[9] This is why the success of Machine Learning over Lee Sedol is considered to be such a significant milestone in the field of Artificial Intelligence.

One of the advantages of reinforcement learning is that it doesn't require a ground truth labeled dataset. We just need to define an *environment* and a Machine Learning model that interacts with that environment. In the case of AlphaGo, the environment is the current board with the location of all white stones, black stones, and empty intersections. The interaction between the Machine Learning model and the board is simply a decision on where to place the next stone. Once a new stone is placed, the environment changes and a *reward* is calculated. If the move eventually leads to the Machine Learning model winning that game, then the reward is high enough so that this move is likely to be repeated in the next game under similar circumstances. Conversely, if the move leads to defeat, then the reward is small (or even negative), thereby discouraging the Machine Learning model from repeating the same move under similar circumstances.

The beauty of reinforcement learning lies in the fact that it doesn't require any supervision or initial knowledge (apart from the rules of the game). The Machine Learning model learns by itself through trial and error. During the initial games, the model makes random moves that typically lead to defeat, but because the model receives reward feedback, it gradually learns which moves to make more often and which to avoid. It should be obvious that such a learning approach requires *many* games, so who should the model play against? In the case of AlphaGo, the Machine Learning model played against itself! For each new game, the model played against another model that was its own replica.

The challenge in reinforcement learning is defining the reward, especially when a single move at the beginning of a game doesn't give much feedback on whether the game would be won or lost. Because of the complexity of assigning rewards, a neural network is usually used to make these assignments. In the case of AlphaGo, an ordinary neural network containing one hidden layer wasn't enough to generate effective reward assignments, so a *deep neural network* was used with multiple hidden layers, representing today's state-of-the-art Machine

---

9   Deep Blue (a chess-playing computer developed by IBM) won its first game against world champion Garry Kasparov in game one of a six-game match on 10 February 1996. However, Kasparov won three and drew two of the following five games, defeating Deep Blue by 4 – 2. Deep Blue was heavily upgraded before playing against Kasparov again in May 1997. Deep Blue won the six-game rematch 3½ – 2½ (winning the sixth game) and becoming the first computer system to defeat a reigning world champion in a match under standard chess tournament time controls.

Learning method. Since AlphaGo uses a deep neural network for reward assignments, it's not pure reinforcement learning, but rather, deep reinforcement learning. The successor of AlphaGo was released in 2017 (AlphaGo Zero) and achieved even better results by relying only on reinforcement learning. Due to its pure reinforcement learning nature, AlphaGo Zero has been successfully applied to many other games, such as chess, shogi (Japanese chess), and even classic arcade-like games from the 1970s and 1980s. This generalization is a step towards the concept of General Artificial Intelligence, as discussed throughout Chapter 1.

Other considerations

Apart from these learning categories (supervised, unsupervised, and reinforcement learning), it's worth covering some additional aspects of learning.

First, is the fact that learning may happen at different stages of a Machine Learning model's lifecycle. In the model for detecting fraudulent credit card transactions, the model learns during the initial training phase. In supervised learning, this means presenting the model with past transactions that the model hasn't seen before, comparing the predicted outcome of the model (fraudulent or legitimate) with actual outcomes, and based upon the result, making any necessary adjustments. After the model has been deployed and becomes operational, it makes predictions (fraudulent or legitimate) for millions of new transactions. Some of these new transactions—especially false positives and false negatives—are later investigated by humans, possibly creating a new dataset that can be used to improve the model (as the model can learn from its mistakes). New data, new feedback.

This process of improving the original model represents another stage at which the model learns. One branch of Machine Learning (called *incremental learning*) deals with situations where the input data is continuously used to improve the model—in other words, to train the model further. Incremental learning is concerned with model learning in an ever-changing environment and represents methods of supervised learning and unsupervised learning that can be applied when new training data becomes available gradually over time. Algorithms that facilitate incremental learning are known as incremental Machine Learning algorithms.

Another branch of Machine Learning (called *continual learning)* emphasizes a model's ability to learn continually from streaming data. This means supporting the model to autonomously learn and adapt as new data streams in. These methods are also known as *auto-adaptive learning* or *continual autoML*. Thus, the learning process doesn't just happen when the model is trained, but also after deployment. Interestingly enough, most scientific research focuses on the first phase of learning (i.e. training of the model), although these other phases are equally important and we'll discuss them further in Section 7.2.

It's also notable that "learning" is mainly associated with prediction models rather than optimization models. This is because the former depends more heavily on past data and inferring (i.e. learning) patterns from such data, whereas optimization models are more dependent on algorithms (i.e. step-by-step instructions). In the case of any prediction model, it's obvious that past patterns can be captured and applied to future cases, and that new data may contain new patterns that can be incorporated together with previously found patterns. At the same time, it's not clear how step-by-step instructions can be changed based on past patterns or new data.

But this doesn't mean that past data (or new data that becomes available at regular intervals after deployment of the optimization model) isn't useful. Although optimization models are used to solve specific problems, there's usually an unlimited number of "instances" of each problem. For example, even though promotional planning is just one problem, there are billions of possible instances of this problem, with each instance having a different number of products, business rules and constraints, objectives, planning horizons, retailers, geographies, and more. Hence, similar to the development and training of a predictive model, past instances of the problem could be used to draw insights on the performance of different optimization algorithms, which could then be used to select the best algorithm along with its various parameters—a topic we'll return to in Section 7.3, where we'll discuss optimization models and learning.

At the beginning of this section, we mentioned that the field of Machine Learning uses many terms for "learning," such as supervised learning, unsupervised learning, reinforcement learning, incremental learning, and continual learning. But in addition to these, there are many other terms like *collaborative learning, feature learning, meta learning, robot learning, learning classifier system, association rule learning,* and *federated learning,* which represent methods that don't neatly fit into the previous categories. To discuss every aspect of every Machine Learning method is beyond the scope of this section, but the following provides a high-level overview:

- *Meta learning:* uses metadata and automatic learning algorithms to improve the performance of existing learning algorithms, and is sometimes referred to as a method for *learning to learn.*
- *Rule-based machine learning:* is a general term for any Machine Learning algorithm that identifies, learns, or evolves "rules" to store, manipulate or apply knowledge. The defining characteristic of rule-based Machine

Learning algorithms is the identification and utilization of a set of relational rules that collectively represent the knowledge captured by the model.

- *Association rule learning:* is a type of rule-based Machine Learning algorithm used for discovering relationships and rules between variables in large databases.

- *Learning classifiers:* are a family of rule-based Machine Learning algorithms that include a discovery component, usually a genetic algorithm that performs supervised learning, reinforcement learning, or unsupervised learning. Learning classifiers seek to identify a set of context-dependent rules that store and apply knowledge in a piecewise manner to make predictions. Several learning algorithms are used to discover better representations of the inputs provided during training. Classic examples include principal components analysis and cluster analysis.

- *Feature learning:* also called *representation learning*, attempts to preserve the information in the input data, but transform it in a way that makes this input data more useful (often as a pre-processing step before generating a classification or prediction). Feature learning can be either supervised or unsupervised. In developmental robotics, *robot learning algorithms* generate their own sequences of learning experiences, also known as a curriculum, to cumulatively acquire new skills through self-guided exploration and social interaction with humans.

- *Federated learning:* also known as *collaborative learning*, trains an algorithm across multiple decentralized devices using only local data. This approach stands in contrast to traditional centralized Machine Learning methods, where all data is uploaded into one location. Federated learning enables multiple algorithms to build a common, robust Machine Learning model without sharing data, thus addressing privacy and data security issues.

- *Deep learning:* deals with algorithms inspired by the structure and function of the brain (thus often associated with artificial neural networks) to aid machines with intelligence without explicit programming.

As mentioned earlier, much can be written about each learning method, but the above provides a sufficient explanation before we move into prediction and learning. Also, to understand the learning process in more detail and see some of these algorithms in "action," we encourage you to watch the supplementary video for this chapter at: www.Complexica.com/RiseofAI/Chapter7.

7.2 Prediction and Learning

There's a strong relationship between training and learning, in that before any prediction model is deployed, it's first trained and tested. And because the model was trained, it "learned." Note that *training* is defined as the provision of information and knowledge through speech, written text, or other methods in a manner that instructs the trainee; whereas *learning* is defined as the process of absorbing information and knowledge in order to increase skills and abilities. So, to predict, the prediction model should first learn.

Also, ensuring that a prediction model is fed with new data is critical, but this only ensures the model doesn't lose relevance. For a model to be truly adaptive, it needs a feedback loop consisting of actual outcomes that can be compared against the predicted outcomes. By examining the prediction bias and identifying particular instances where the prediction bias is consistently high, either positively or negatively, the model can "learn" and begin to weigh its future predictions. Such a model must be capable of identifying problematic instances and then generalizing what the consistently similar variables are.

For example, if a particular product is consistently predicted above actual, then it's relatively straightforward for the model to learn the correct weighting and adjust future predictions accordingly. However, this becomes more difficult if further complexity is added—for example, where a combination of promotions results in cross-category cannibalization or halo effects that weren't expected, or the outcome is impacted by external factors that weren't incorporated into the original model (e.g. a television show that alleges a particular toothpaste to be dangerous, resulting in a drop in consumer demand). Such information couldn't possibly be in any prediction model. Still, a Decision Optimization System should detect any unexpected deviation or increase in prediction error, adjust its own weightings (so that recent data is given more weight when making predictions), and flag such products for further (manual) investigation.

However, to maintain relevance, the prediction model must continuously receive new data and/or feedback on past performance, and learn from this new data or information. This process might take place on a regular but discrete basis, perhaps monthly or quarterly. Furthermore, it may be necessary for the prediction model to "forget" old data that's no longer relevant, and instead, create a moving window of data for making predictions. The main challenge here is for the model to accommodate new data in an incremental way over time, which is different to training the model in static environment to begin with.

When discussing the layers of the problem-to-decision pyramid in Chapter 2, we introduced the concept of adaptability, which leverages a feedback loop to provide the actual outcome of past recommendations *and* new data to make better recommendations in the future:

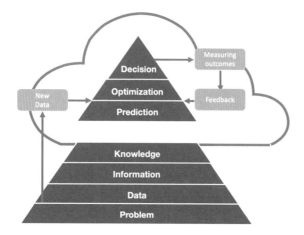

In this diagram, "new data" (left loop) usually arrives at regular intervals, often in a similar format to the historical data used for training; while the "feedback" (right loop) provides a direct evaluation of the predictions or recommendations made by the Decision Optimization System (e.g. prediction error or the appropriateness of the decision recommendation, as evaluated by the end user). Sometimes these two loops are related, as the feedback might be extracted from the new data—for example, new quarterly sales data would allow us to compare predicted volumes against actual volumes for each product and calculate the prediction error. On the other hand, these two loops could be very different, as new sales data might arrive every quarter, but the Decision Optimization System might receive immediate feedback from end users (an example of this is provided in Chapter 8.2).

As previously mentioned, any prediction model based on historical data will gradually lose its relevance over time and might no longer be applicable due to changes in the environment:

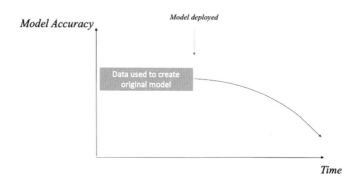

Consequently, if the predictions are inaccurate, then any recommendation made by the optimization model may do more harm than good! For example, the optimization model might recommend changing the promotional plan by lowering prices for several products because the predicted sales volume would be much higher. However, if the actual sales volume turns out to be considerably less than the predicted values, then the total loss might be substantial (selling fewer products at a lower price). Who is to blame? *The prediction model.*

To avoid this situation, we could repeatedly update the Decision Optimization System with new data, which would maintain the model's accuracy at a desirable level as older data might become less and less relevant:

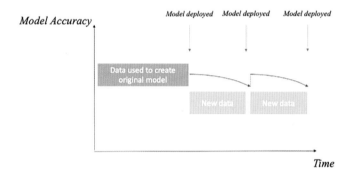

Another way of updating the prediction model is based on direct feedback. This type of updating is much closer to the intuitive meaning of "learning," as the Decision Optimization System receives feedback and learns from its past mistakes and successes (e.g. reducing the probability of repeating similar errors in the future and/or increasing the probability of recommending better decisions). Because of this, there's a real possibility that the Decision Optimization System would actually improve over time, rather than just maintain its accuracy:

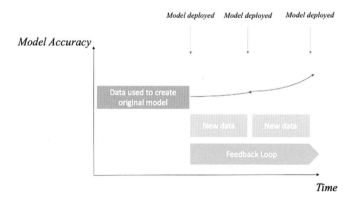

Let's explore the above diagram in more detail through some simple examples. In Chapter 5, we covered exponential smoothing methods, which generalize the *moving average method* (where the mean of past *k* cases is used as a prediction). Note that all exponential smoothing methods assign weights to past cases so that recent cases are given more weight than older cases. They also require at least one parameter *a*, which plays an important role. A prediction for the time interval *t+1* is calculated as:

$$Prediction(t+1) = (a \times Actual(t)) + ((1-a) \times Prediction(t))$$

which simply means that the prediction for the next (future) case is calculated as a total of two values: the last actual case (*Actual(t)*) with weight *a,* and the last prediction (*Prediction(t)*) with weight *1–a.* The prediction model's performance based on this method depends on the selection of the parameter *a,* as the prediction would always be the last actual value if *a = 1.* Because different values of *a* may be required at different times, it might be reasonable to develop a method that would be responsible for adjusting this parameter. In other words, this method would systematically change the values of parameter *a* from interval to interval to allow for changes in the data. The fixed parameter *a* would be replaced by *a(t)* and the above method would assign a new value for *a(t)* at every interval *t*—for example, using a function that's based on the most recent prediction error (i.e. feedback).

In this particular case, developing a method that systematically changes the values of parameter *a* has the following benefits: (1) it's automatic, so the administrative overhead connected with frequent adjustments is reduced, (2) even if the performance is slightly inferior to the "optimal" fixed value of parameter *a,* it reduces the risk of serious errors, and (3) there's no need to specify the initial value of parameter *a*—even if it takes a few intervals for *a(t)* to catch up with the changes in the data, it will eventually do so. However, can we say that by using this method the Decision Optimization System "learns"?

To answer this question, let's consider another example based on a single prediction model, but this time for a classification problem. The feedback loop would observe and record the performance of the prediction model over time, and as cases of misclassification become known, the update procedure within the feedback loop would store these misclassifications (along with the correct classification) for eventual processing. Once a critical mass of misclassifications has been collected, the update procedure would "clone" and modify the original prediction model so it does a better job of handling these incorrectly classified cases. The modified prediction model is then tested offline to ensure it provides an advantage over the original model, and once tested and verified, it replaces the original model. In this case, the learning is more explicit, as the Decision

Optimization System collects the errors and takes corrective measures, so to answer the question above, we can say it "learns from past errors."

In Section 7.1, we introduced the concept of incremental learning, where input data is *continuously* used to extend the model's knowledge through further training. Incremental learning algorithms are frequently applied to streaming data or big data. Many traditional Machine Learning algorithms inherently support incremental learning, and many other algorithms can be easily adapted to facilitate incremental learning. But the goal of incremental learning is for the model to adapt to new data without forgetting its existing knowledge, so there's no need to retrain the model. Some incremental learning algorithms have built-in parameters or assumptions that control the relevancy of old data.[10] We can then use AI-based optimization algorithms to search for the best parameter values that minimize the prediction error. The optimization process is run at regular intervals and results in an updated version of the model (i.e. a model with updated parameters). As for predictions based on a single prediction model, the development of a feedback loop that changes the model's parameters usually works well. Furthermore, it's possible to keep (at least for a while) both the original and updated model, measure the prediction error for both on new data, and then adjust the weights for these two models accordingly.

In the case of a prediction model based on many individual models (as discussed in Chapter 5.6, ensemble models), the possibilities of updating the model are much broader. For example, the feedback loop can be based on:

- Updating some (or all) of the individual prediction models
- Adding or dropping some of the prediction models
- Updating the voting/averaging mechanism
- Updating both the individual models and the voting/averaging mechanism

As discussed, updating each individual prediction model involves updating its parameters while keeping the overall structure unchanged. As an example, a neural network consists of input nodes, hidden nodes, output nodes, and the connection weights in between them. In order to update this model, the update procedure could simply add new data to the training set and then update the weights, without changing the overall structure of the model (i.e. without adding or removing any nodes in the network). If considerable data becomes

---

10 Recall from Chapter 5 that regardless of whether we use exponential smoothing or some other prediction method (like fuzzy system or neural networks), the model includes some parameters.

available within a short interval and the neural network operates in a dynamic environment, then the training data is often like a sliding window where we discard old data, and only use the most recent data to update the model.

Another possibility is to update the voting/averaging mechanism while leaving the individual prediction models unchanged. This is usually straightforward, as it requires rewarding the more accurate models (i.e. increasing their weights) and punishing the less accurate models (i.e. decreasing their weights). In other words, we pay more attention to the models that produce more accurate predictions. If we consider a model based on bagging (discussed in Chapter 5.6), where each individual prediction model is weighted more or less heavily, then these weights can be updated with new data, just like the weights of a neural network.

The third possibility is to update both the individual models and the voting/averaging mechanism, which is usually done in two stages: first when the individual prediction models are updated, and later when the weights assigned to the various prediction models are updated.

It's also worth discussing the pros and cons of a "single prediction model" versus an "ensemble model" from the perspective of updates. On the one hand, ensemble models can increase the useful life of a trained model. For example, a neural network trained to predict sales volume will initially work well, but as time passes, its prediction accuracy is likely to be affected by external factors that weren't considered when the model was initially built and trained. This concept is known as "model drift," with an example being a neural network model struggling to maintain accuracy during a pandemic. However, if we combine this neural network with a time series model, then the time series model can correct the neural network during unprecedented times (and vice versa) resulting in a lengthening of the useful life of the overall prediction model.

On the other hand, the introduction of new components into the prediction model also introduces additional complexity. If any of the models don't function properly, then we need to identify which models are failing. Also, because there are no limits to the number of models in an ensemble, the complexity of the final model may get out of hand and require replacement. Therefore, there's a balance between the complexities and benefits of using ensemble models.

Because of this, there are a number of approaches to improving the maintainability of ensemble models, starting with the regulation of how new models are added to the ensemble. It's useful to analyze the impact of each individual model and compare it to the final result of the ensemble. If the added value doesn't exceed a certain threshold, then the individual model under consideration can be ignored. Although we can update the prediction model within a

Decision Optimization System in a number of different ways, the final decision on what to update, how, and when is always problem-specific and can only be made after evaluating the available data and problem objectives.

For more information on the material covered in this section, including a demonstration of direct feedback within a Decision Optimization System, please watch the supplementary video at: www.Complexica.com/RiseofAI/Chapter7.

7.3 Optimization and Learning

The term "learning" is usually related to prediction models: we train the model, and it learns; we get new data, the model is retrained, and it learns; and we get direct feedback from end users, the model is adjusted, and it learns. The question is whether there is any room for learning within optimization?

In promotional planning, the optimization algorithm runs hundreds of times during each planning period in search of the optimal plan for each retail chain, state, and product category. And if we have 20 retail chains, operating in 6 states, selling products across 10 categories, then we have 1,200 instances to consider (i.e. to optimize) and each of these instances may have different business rules and constraints. As explained in the previous chapter, there are many optimization methods and constraint-handling techniques we could use, begging the question which method will produce the best result for a particular instance of the promotional planning problem (such as Mary's Market in NSW in the snacks category)? Is there any single method that can provide us with the best outcome across all possible instances? Or rather, is one method better for some instances, and another method for other instances?

To answer this question, let's consider twelve instances of the promotional planning problem, with each instance being for a different retail chain (e.g. Mary's Market) within a particular state (e.g. NSW), for a particular product category (e.g. snacks). Although the optimization objective across all these twelve instances is to maximize volume, each of these instances has different planning horizons, business rules and constraints, and tables for defining what changes are acceptable (as discussed in Chapter 6.9, understanding the solution). For example, the business rules for one instance:

	Shelf Price	Min Freq	Max Freq	Max Promo Length	Min Promo Gap	Max Promo Gap	Min Promo Price	Max Promo Price	Price Step	Allow rounding?
Product 1	$23.50	13	26	2	2	6	$12.00	$20.00	$0.50	Y
Product 2	$14.50	13	26	2	2	6	$10.00	$12.50	$0.50	Y
Product 3	$19.00	13	26	2	2	6	$10.00	$15.00	$1.00	N
Product 4	$22.00	20	39	3	2	4	$11.00	$18.00	$1.00	Y
Product 5	$52.00	20	39	3	2	4	$40.00	$48.00	$1.00	Y
Product 6	$48.00	13	39	3	2	4	$40.00	$45.00	$0.50	Y
...										
Product 100	$65.00	13	13	1	2	6	$55.00	$60.00	$1.00	Y

might be very different to the business rules for another instance, along with a different number of products in potentially a different category altogether.

These business rules and constraints are very important, as they define the size of the feasible search space. An instance with a shorter planning horizon (e.g. 6 weeks), lots of business rules, and very tight price ranges (e.g. $2 difference between the min and max promotional price with 50 cent price steps), would have a much smaller feasible search space than an instance with a longer planning horizon, few business rules, and large price ranges (e.g. $8 difference between the min and max promotional price with 10 cent price steps). Similarly, the range between the min and max numbers of products allowed on promotion per week may be quite small (e.g. between 40% and 50%) or large (e.g. between 20% and 80%).

The size of the feasible search space determines the *ratio of feasibility*, which is the percentage of the whole search space that's feasible. This ratio, together with many additional factors (such as the size of the entire search space, the table that defines all acceptable changes to the original promotional plan, the number of soft and hard business rules and constraints, etc.) would influence the selection of the most efficient constraint-handling approach, whether penalties, repair algorithms, or decoders, along with all their details, such as the use of dynamic penalties or the Lamarckian approach (as discussed in Chapter 6.8).

In summary, these twelve instances differ from each other in terms of:

- Number of products (between 25 to 200)
- Planning horizon (between 6 to 52 weeks)
- Number of active business rules and constraints (between 10 to 100)
- Split between hard and soft business rules and constraints (between 10% to 90%)
- Size of min/max ranges for business rules and constraints (from "small" to "large" ranges—for example, a $2 min/max promotional price range versus a $20 min/max range)
- Tables that define acceptable changes to the original promotional plan (from only a few changes to unlimited)

With this understanding, let's return to the question of whether there's any single optimization method that can provide the best result across all possible instances of the problem? Or rather, if one method is better suited for some instances, while another is better suited for other instances?

To find the answer, let's experiment with eight different optimization methods—including evolutionary algorithms, tabu search, simulated annealing, and ant systems, with each method using one of two different constraint-handling approaches along with specialized operators (such as local search). Note also that each method requires the setting of various parameters, like population size, probability of mutation in evolutionary algorithms, evaporation ratio for pheromones in ant systems, cooling rate in simulated annealing, or the size of memory in tabu search. Without getting into the details of each algorithm, let's just label them as Alg_1, Alg_2, ... , Alg_8 and apply them to the twelve promotional planning instances with a single objective of maximizing sales volume.[11] The following table summarizes the results (in volume) of these eight algorithms for these twelve instances:[12]

	Alg_1	Alg_2	Alg_3	Alg_4	Alg_5	Alg_6	Alg_7	Alg_8
Inst #1	32,112	31,445	32,881	32,112	30,905	31,445	32,112	30,905
Inst #2	39,878	38,105	39,878	38,545	40,865	39,878	38,105	38,545
Inst #3	61,348	61,348	60,405	60,405	62,007	61,304	60,980	62,304
Inst #4	57,210	58,912	57,745	58,720	58,105	58,315	58,315	57,210
Inst #5	5,102	5,211	5,435	5,376	5,650	5,650	5,755	5,286
Inst #6	16,952	17,211	17,507	17,507	17,633	17,990	17,009	17,234
Inst #7	11,065	11,453	11,453	12,840	12,220	12,878	12,775	11,234
Inst #8	7,745	7,881	8,701	8,701	8,444	8,723	9,002	7,987
Inst #9	93,144	89,087	89,441	88,992	89,765	88,987	88,775	91,455
Inst #10	29,976	31,108	32,114	31,027	31,333	31,333	30,204	30,988
Inst #11	41,745	42,088	42,555	42,077	42,878	41,348	40,976	42,555
Inst #12	83,012	82,224	80,034	81,773	79,992	79,865	79,348	81,234

The table above illustrates some interesting, but to some degree expected, results. To start with, no single algorithm had the best performance[13] on every instance of the promotional planning problem (so we answered our question), and different algorithms performed better or worse depending on the characteristics of each instance. Secondly, there was always a single "winner" for every instance (i.e. the algorithm that produced the highest predicted volume). There were also cases where two different algorithms found the same promotional plan, like Alg_2 and Alg_6 on Instance #1, although in each instance, there was always a unique overall winner. Third, irrespective of the nature of the algorithm itself (whether tabu search, evolutionary algorithm, simulated annealing, or ant systems), algorithms using repair methods (e.g. Alg_3

---

11 These instances would differ from one another in overall volume, because each instance is dealing with a different number of products in different states for different retailers.
12 The results presented in this table are based on a real experiment conducted by Complexica.
13 The best result (i.e. the highest sales volume) for every considered instance is marked in red.

and Alg_5) performed better than algorithms using penalties (e.g. Alg_1 and Alg_7) on instances with tight business rules and constraints (e.g. instances #4, #7, and #8). Also, instances with long planning horizons (i.e. which have a much larger search space) required different algorithms than those with short planning horizons to return optimal (or close to optimal) solutions. We can make many other observations concerning the relationship between the particular characteristics of each instance of the problem and the performance of these different algorithms and constraint-handling methods.

Which raises a new question, namely, how can we "guess" which algorithm is best suited for a new instance of the problem? Let's explore a few options here. As with ensemble models for prediction, we can run several optimization algorithms in parallel and select the best result. Thus, different optimization algorithms can compete with one another and the best result is implemented. We can also enhance this competition by controlling the run time for each optimization method, so that algorithms making the most progress receives more run time.

Such a hybrid approach for optimization offers some additional possibilities. For example, by introducing mechanisms for exchanging information, these algorithms can cooperate with one another to identify the best solution in the shortest possible time. Hence, the final solution is often much better than running the algorithms in parallel or sequentially.

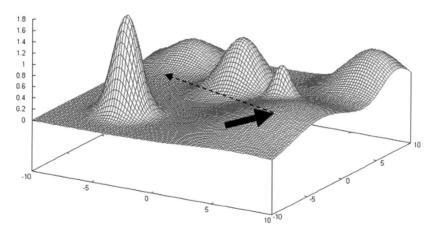

The above graph illustrates a quality measure (fitness) score landscape (as discussed in Chapter 6.3), where two optimization algorithms are cooperating to find the highest peak—in this case, the peak located on the left-hand side of the landscape. The thick arrow shows a simulated annealing algorithm searching up the hill to the right, while a genetic algorithm searches for the highest peak on the left. If the genetic algorithm finds a better solution in some other area of the search space, then this information is passed to the simulated annealing algorithm, which can "jump" to the new area to continue its search.

Of course, such an exchange of information could occur among any number of optimization algorithms, and herein lies the fundamental strength of cooperation: by exchanging information, more optimization algorithms can explore the most promising parts of the search space. Consequently, a hybrid approach for optimization allows for competition and cooperation at the same time: competition, in that each algorithm tries to "win" by finding the best solution; and cooperation, in that all algorithms are cooperating with one another by exchanging search information. Such hybrid models often out-perform any single optimization method—especially when instances of the same problem are diverse—and guarantee that the overall hybrid model never performs worse than any single algorithm. However, running a few optimization algorithms in parallel might be too demanding on computa-tional resources, especially in situations where the evaluation of an individual solution takes a significant amount of time. In such cases, another approach can be used, which contains a significant learning component.

Earlier in this section, we explained that different instances of the same problem may have different characteristics, such as the number of products, planning horizon, active business rules, and so on. Because of this diversity, we could train a prediction model (whether a decision tree, neural network, or fuzzy system) to select the best algorithm for each new instance of the promo-tional planning problem:

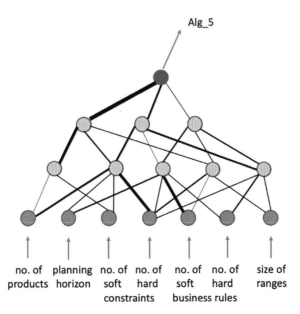

In this diagram, a neural network model is trained on a number of artificially created problem instances, with each instance having different characteristics. Then, for any new instance of the problem, the neural network uses the instance characteristics as inputs (e.g. number of products, planning horizon, etc.) and selects the best optimization algorithm to deliver the final solution! In other words, the neural network learns how to assign the best algorithm for any given instance.

We can go even further with this approach, because each optimization method has several parameters that need tuning, such as:

- In evolutionary algorithms, the population size, probability of operators (crossover and mutation), and selection and replacement pressures
- In simulated annealing, the size of the neighborhood and the rate of cooling
- In ant systems, the number of ants, and the balance between attractiveness and pheromone levels

Typically, when an optimization method is selected for the problem at hand, its parameters are tuned by experimenting with a variety of instances of the same problem. The question is, how can we tune the parameters of one particular algorithm to get the best performance across a wide range of different instances? One option might be to adjust the parameter values during the optimization run (rather than in between runs). For example, at some stages of the optimization process a small population of solutions may perform better, whereas at other stages of the optimization process a larger population may do well. Thus, changing the population size parameter during the run might be beneficial, as a better solution might be found in less time.

So how can we change these values during the optimization process? Well, the methods for changing the parameter values can usually be classified into one of three categories:

- *Deterministic parameter control:* uses deterministic rules to change the value of a strategy parameter (i.e. a parameter that controls how the search is conducted). This rule modifies the strategy parameter deterministically without using any feedback from the search. A time-varying schedule is often used, where a rule is applied after a set number of iterations have elapsed since the last time the rule was active.
- *Adaptive parameter control:* uses feedback from the search to determine the direction and magnitude of change in the strategy parameter. New values are generally assigned based the current state of the search.
- *Self-adaptive parameter control:* uses "evolution" to determine the strategy parameter value by encoding the parameter into the data structure of

individual solutions and letting it undergo variation (mutation and crossover). Better values of this encoded parameter lead to better solutions, which in turn are more likely to survive and produce offspring solutions, and hence propagate these better parameter values.

In summary, there are many ways to adapt the parameters of any single optimization method. In the case of using multiple algorithms, some additional parameters may undergo adaptation during the optimization run, including:

- Parameters for controlling the type of information that's exchanged between the algorithms. For example, should the optimization algorithms exchange complete or partial solutions?
- Parameters for controlling the frequency of information exchange.
- Parameters for controlling the transmission of the information. For example, should information be exchanged "across the board" to all algorithms or provided just to a selected few? In the latter case, which ones?
- Parameters for controlling how information is sent from one algorithm to another. For example, should the existing solution be replaced? or recombined with an existing solution?
- Parameters for controlling the run time for each algorithm.

While we can adapt all these parameters during the optimization run, most hybrid models use both adaptive and static settings: adaptive settings for the algorithm parameters (e.g. population size, probability of operators) and static settings for the hybrid model parameters (e.g. type and frequency of information exchange, the way solutions are exchanged and used by other optimization algorithms, the allocated run time, etc.).

All these topics related to optimization and learning are of great importance, because the size of the search space is often beyond astronomical and any search for an optimal solution would take time. As we pointed out in Chapter 2, for many optimization problems, even if we had a computer capable of evaluating 1,000 solutions per second and began the search billions of years ago (at the beginning of the Universe), we would have evaluated less than 1% of all possible solutions by today. So, for most real-world problems, any effort to make the algorithms smarter through learning will provide a significant benefit in terms of performance.

7.4 Adaptability

The previous two sections discussed various learning aspects of prediction and optimization models, while this section covers the broader topic of *adaptability*—meaning the automatic updating of a model to improve prediction accuracy, run time of the optimization algorithm, and appropriateness of the decision recommendations.

As explained earlier, the need for updating a prediction or optimization model usually corresponds to a situation where new data becomes available, and we realize that the older data used for training the model may no longer be relevant. Such realizations may occur through a steadily growing prediction error or a lengthening of the optimization run time. Traditionally, the process of updating the model takes place every so often (quarterly or annually), because:

1. The process of updating a model is usually manual, so it's time consuming, error prone, and possibly expensive
2. Changes in the operating environment are usually gradual (relatively small), so there isn't much benefit in updating the model every week

Although situations may arise where it's necessary to manually extend or rebuild the model—perhaps to factor in new variables such as additional go-to-market channels or new competitors—such situations are rare. More often than not, we just need to update the model with new data and/or feedback, which usually becomes available on a regular basis (e.g. weekly, monthly, quarterly). It might also be beneficial to immediately take advantage of this new data and/or feedback by having the model update itself automatically, rather than waiting for the next scheduled update. Models and software systems that can update themselves are often referred to as being "adaptive."

When a prediction model is built and trained, all these activities happen in a static environment: we split the initial data into the *training set, validation set,* and *testing set.* The first dataset is used for building the model, the second one for tuning the model's parameters (training), and the third to evaluate the model's performance.

When the prediction model is ready for deployment, it's placed into an operating environment and begins making real predictions based on new data. At this stage, the prediction quality is strongly related to the quality of the model. Then, over time, two types of data are available: new data coming in a similar format to the original training data (e.g. the most recent sales volume data), and direct feedback on the model's performance, which may include various comparisons between predicted values and actuals that show the magnitude of error.

In the timeline below, we marked the point when the model goes live, and when new data (yellow rectangle) and direct feedback (red arrow) become available.[14] As discussed earlier, these two inputs may come at different intervals or together; for example, the model receives new quarterly sales volume data, and then extracts direct feedback by comparing the predicted volumes against actual volumes:

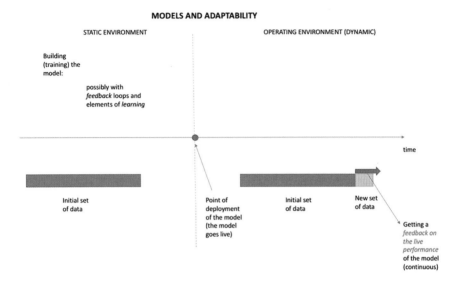

As time passes, we should monitor the number of significant and moderate errors, and if both are low, it means the model is performing as expected! On the other hand, if this isn't the case, then some action should be taken to allow the model to learn and adapt. To accomplish this, we can consider one of three options:[15]

- **Option A** is to *rebuild the model,* meaning that we discard the old model and use all available data (both old and new) to rebuild the model from scratch

---

14 We can illustrate the distinction between "new data" and "direct feedback" through the example of cross-selling. Although we discuss cross-selling at some length in Chapter 8.3, for the purposes of this section it's only sufficient to note that cross-selling is the action of offering an additional product or service to an existing customer. Most cross-selling methods are based on an analysis of items bought by such similar customers, where new data of what those customers have recently purchased influences all future recommendations. However, at the same time, the model is getting direct feedback on whether the recommendation was accepted or not by each individual customer, which represents the second type of input.

15 As indicated in Section 7.2, when a prediction model is based on many individual models (as discussed in Chapter 5.6, ensemble models), there are many possibilities of updating the overall model, as we can update some (or all) of the individual models, add or drop some individual models, etc.

- **Option B** is to *retrain the model*, meaning that we keep the old model and use new data to update the model
- **Option C** is to *adjust the model*, meaning we keep the old model and use cases of significant error to update the model

Let's explain these three options of *rebuilding*, *retraining*, and *adjusting* the model through a simple example. Let's assume that our predictive model is a random forest (discussed in Chapter 5.2), which is a collection of decision trees where each decision tree is trained on a different subset of data with a different subset of available variables for splitting nodes.

Option A then, *rebuilding the model*, refers to a situation where it's necessary to discard the old model (all decision trees are deleted) and build, train, and deploy a new model (which might be a random forest or not). This option might be taken if we realize that some key variables weren't included in the original model, or if a significant shift has occurred in the environment, such as a major recession or pandemic.

Option B, on the other hand, *retraining the model*, refers to situations where one or more additional decision trees are added to the random forest, and these new trees are trained on a combination of old and new data. These new decision trees replace the oldest trees in the model, thus mimicking the sliding time-window effect. This is relevant to situations of natural market growth.

And Option C, *adjusting the model*, refers to situations where additional decision trees are added to the model to address cases that result in significant errors. This usually happens when the model experiences new phenomena, such as a major new competitor entering the market.

The figure below illustrates these three approaches:

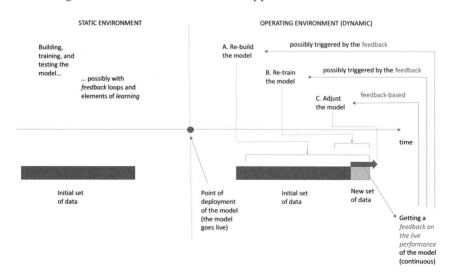

It's also worth looking at these three options from the perspective of learning: When rebuilding a model, the knowledge in the old model is lost and replaced by the knowledge in the new model; thus, it's difficult to label this process as "learning." This is equivalent to replacing one brain with another, which can hardly be classified as a learning experience. But even though we can't claim that the model "improved automatically through experience," the new model did learn during the initial training process, so some learning did occur (as discussed in Section 7.1). For example, before a new neural network model replaces an old neural network model, the new model undergoes training, and thus learning. Hence, elements of learning take place when rebuilding a model, but this learning occurs in a static environment and is only related to training the model, rather than any process of improvement through experience.

On the other hand, retraining and/or adjusting the model may include some other forms of learning. In retraining, we may keep the old model and use new data to train a separate, new model. Both models then work together (bagging, see Chapter 5.6), meaning that the old knowledge was preserved, and new knowledge was generated from new data. Similarly, in the case of adjusting, we may keep the old model and use only cases of significant error to train a separate (new) model. Both models then work together (boosting, see Chapter 5.6), so that again, the overall ensemble model learned by preserving old knowledge and generating new knowledge on the basis of recent errors.

The question now is: Under what circumstances should a Decision Optimization System trigger an update of the underlying models? And once triggered, what type of update should be executed? There are many possibilities here and different operating environments may require different approaches. One approach might be based on monitoring the number of errors and classifying them into two categories: significant errors and moderate errors. If both counts are low (bottom right quadrant below), then "all good":

	High No. of Moderate Errors	Low No. of Moderate Errors
High No. of Significant Errors		
Low No. of Significant Errors		All good

However, in the three other quadrants (when one or both types of errors are high), some action should be taken—the Decision Optimization System should learn and adapt. Again, the question is under what circumstances should we execute Option A (rebuilding), versus Option B (retraining), versus Option C (adjusting)? Based on the count of moderate and significant errors, we might do the following:

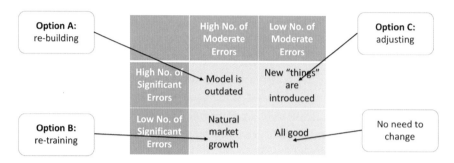

If there's a large number of significant and moderate errors, then the original model is no longer appropriate and requires rebuilding. More often than not, an automatic rebuild of the model is sufficient, as a high number of significant and moderate errors is usually attributable to insufficient training data when the model was originally built, and this can be corrected by combining both old and new data during the rebuild. However, if the model needs to be extended with additional variables, then this will require manual intervention (but as mentioned earlier, such situations are rare).

On the other hand, a low number of significant errors with a high number of moderate errors usually calls for retraining of the model. In such situations we would use new data to train a separate, new model and then use the old and new model together in an ensemble. Conversely, a low number of moderate errors coupled with a high number of significant errors may indicate the emergence of new phenomenon (e.g. a new competitor, a new product, etc.), thus calling for adjustment of the model. Here we would only use the significant error cases to train a separate (new) model, and then use the old and new model together.

Very large changes in the marketplace might require a complete overhaul of the prediction model, as any tuning would probably be insufficient. The need for such massive adaptation often shows up as a sudden and significant increase in prediction error, which can't be reduced by updating the model or by adding additional models to the ensemble. However, this is quite rare, as most changes are gradual. Furthermore, the feedback loop's update procedure would have a threshold that defines what prediction error is acceptable.[16]

---

16 If the prediction error exceeds this threshold, then the update procedure would indicate that an overhaul of the prediction model is required.

Our discussion on adaptability has focused on prediction models, but the same concepts apply to optimization. As discussed in Section 7.3, different instances of the same problem may have different characteristics (e.g. number of products, planning horizons, number of active business rules, and so on). Each new instance that we process (i.e. optimize) provides new data that can be used to improve the optimization model. So the concept of "new data" corresponds to new instances of the problem that come in for optimization. These new instances could update a neural network model that selects the best optimization algorithm for new instances of the problem, as we described in Section 7.3. But because each optimized plan can either be accepted or rejected by the end user, this acceptance or rejection can be fed back into the optimization model as direct feedback (to be used for further tuning of future recommendations).

Note, that all three options of rebuilding, retraining, or adjusting the prediction model, as well as options for making the optimization model smarter (whether by tuning the algorithmic parameters, retraining a neural network or some other machine learning technique on recent instances of the problem, or synchronizing a few optimization algorithms to compete and cooperate) can be referred to as a form of "adaptation" as long as these options are executed in an automated way. So, when we talk about the adaptability of Decision Optimization System, we mean automation of the update process as opposed to manual updates:

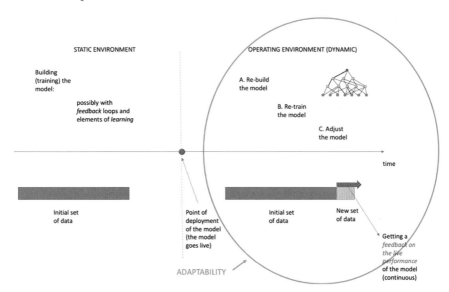

The prediction and optimization methods presented in the past two chapters can also be viewed as "hardware tools," in a sense that each has different strengths and weaknesses, characteristics, and applications. To further stress this point, imagine that we have a hammer—which is a convenient tool for solving problems that involve nails—but we come across a problem that involves screws. What should we do? Well, if the hammer is our only tool, then we might be tempted to "hammer in" the screws. This, of course, is the wrong approach, and we should instead find a better tool (e.g. a screwdriver). The point of this analogy is that the right method (i.e. tool) should be selected for the right problem (or instance of the problem). Unfortunately, many traditional models and software applications are based on a single method (e.g. whether a decision tree for prediction, or evolutionary algorithm for optimization), which is applied to each and every instance of the problem. This is similar to only having a hammer and seeing everything as a nail. This approach is clearly wrong, as the method should be selected for the problem (or instance of the problem), not the other way around!

In general, there are some limitations of using a single prediction method (see Chapter 5.6) or a single optimization method (see Section 7.3). The first and most obvious one is that the selected method may perform superbly on some instances of the problem and abysmally on others. Recall that in promotional planning, we're dealing with different instances of the same problem, as different plans must be created for different retailers, product categories, and states. Although it would be great to claim that a particular method is the "best" for this problem, it's usually impossible to guarantee that this "best" method would provide the best result on every instance. Also, since the problem varies with changes in the environment (e.g. weather, economy, consumer preferences), a single method that performs well in the beginning may deteriorate over time. This is why adaptability is key. For more information on learning and adaptability, we encourage you to watch the supplementary video for this Chapter at: www.Complexica.com/RiseofAI/Chapter7.

This chapter concludes Part II of the book, where we discussed prediction and optimization from the perspective of learning and adaptability. In Part III, we'll return to these topics, but this time from the perspective of real-world applications and case studies.

PART III
Application Areas for Revenue and Margin Growth

Overview

Bridging the gap between "having the right data" and "making the right decision" isn't straightforward—a fact we illustrated with the problem-to-decision pyramid in Part I, where a gap separates the lower levels representing the past, and the higher levels representing the future:

As organizations climb the pyramid, the sophistication of the tools and technologies increases with each layer, with Artificial Intelligence methods having the most applicability and delivering the most value within the higher layers of the pyramid (which is where the greatest complexity resides). For this reason, organizations that climb higher up on the pyramid usually make better decisions than those organizations that get stuck in the lower levels.

In this part of the book, we'll present real-world applications of Artificial Intelligence for various business functions. These application areas are presented as case studies that explore the problem-to-decision pyramid in the context of a real-world problem and business objective, covering both the lower layers of the pyramid focusing on the data and analytical landscape of the organization (i.e. information and knowledge), as well as the upper layers focusing on the enablement of Artificial Intelligence methods for prediction, optimization, and self-learning capabilities.

For ease of comprehension, we've divided Part III into three chapters, with each chapter being dedicated to a specific business function. Also, the case studies are presented through fictitious names, due to confidentially reasons and also because each case study draws upon several similar implementations, allowing us to pull together a variety of interesting observations and lessons for discussion purposes.

With the intent of making this part of the book as relevant and realistic as possible, we have referenced existing and commercially available Artificial Intelligence software for each case study. However, it's important to note there are many AI-based software products and tools available in the marketplace, which can be divided into two broad and general categories:

- *Technical tools*: designed for data analysts, computer scientists, and other technically-oriented people working "hands on" with data and models; for example, tools that allow data scientists to build predictive models based on neural networks.
- *Enterprise applications and platforms*: designed for non-technical, business-oriented people trying to solve a particular problem or challenge; for example, software applications for promotion planning or production scheduling, which embed AI-based functionality to help the business user generate an optimized plan or schedule.

The case studies presented in this part of book reference an enterprise software platform called Decision Cloud®, which is a modularized, cloud-based platform that empowers staff within sales, marketing, and supply chain functions to make better and faster decisions through the use of Artificial Intelligence algorithms. Hence, we've relied upon a business software platform for business users rather than a technical tool for technical users, largely because the application of technical tools requires significant in-house expertise that most organizations lack, and also because this part of the book was written for business managers and executives wanting to understand how Artificial Intelligence applications relate to real-world business problems, rather than more technical readers wanting to understand the application of advanced algorithms and modeling tools (which we discussed at length in Part II).

For more information on Complexica's Decision Cloud®, please visit: https://www.complexica.com/software/decision-cloud

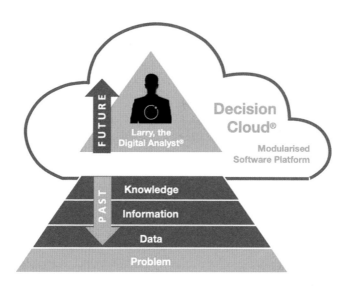

It's also worth noting that Decision Cloud® is powered by an AI engine called Larry, the Digital Analyst®, which does the actual work of converting data into optimized decisions through the use of Artificial Intelligence algorithms.[17] Because of the many AI methods and algorithms within the Larry engine, as well as the fact that each implementation is tuned to the specific constraints, business rules, and available data of each organization, it's highly unlikely for any two implementations to ever be configured in exactly the same way, or operate in an identical manner. In other words, even if the problem is "identical," the objectives or KPIs are still likely to be different, as well as the available training data and myriad business rules and constraints.

For these reasons and others, the case studies presented below could have been addressed through different configurations of a Decision Optimization System, had the constraints, business rules, or available data differed, or if the organization had a different strategy or value proposition it was trying to execute in the marketplace. We should keep this in mind when considering each application area, in the sense that the application of AI in each case study largely depended on a specific organization's situation and immediate objectives, and could have differed (and even been broader) had circumstances been different.

---

17 Larry, the Digital Analyst® was named the 2018 Australian Innovation of the Year and 2019 Australian Software Innovation of the Year, and powers Complexica's Decision Cloud®. For more information on Larry, the Digital Analyst®, please visit: https://www.complexica.com/software/larry-the-digital-analyst.

CHAPTER 8

Sales

*"I have never worked a day in my life without selling.
If I believe in something, I sell it, and I sell it hard."*
Estée Lauder, *Public talk.*

In an attempt to maximize product reach and capture the largest possible share of market, sales structures have grown increasingly complex within most organizations. This complexity has also been driven by the realization that customers aren't equal in size, needs, or expectations, and often require different strategies and channels.

Within the wine industry, for example, a different kind of sales rep is required for calling on Michelin-starred restaurants versus pubs and cafes, because these conversations require different skillsets and knowledge. Add to this the fact that businesses are increasingly competing on the quality of their customer experience, and it becomes obvious why modern sales organizations and service channels continue to grow in complexity. In response to this situation, organizations have begun segmenting their customers by the potential they represent (and consequently, by the level of attention they should receive), as well as introducing metrics such as *share of wallet, customer lifetime value*, and *cost to serve* to measure and optimize the returns from their ongoing investment in maintaining a complex sales structure.

Set against this backdrop of increasing complexity, in this chapter we'll explore a range of case studies to show real-world applications of Artificial Intelligence for revenue and margin growth. These case studies also cover the three possible levels of decision-making that exist within the sales function, which are:

- *Structural (or strategic) decisions*: are high-value, infrequent decisions that can deliver big returns or big losses depending on whether they were right or wrong—for example, radically re-structuring the sales organization only to find that the new structure costs more and delivers less revenue and growth than the original one.

- *Operational (or business-as-usual) decisions*: are lower-value, repetitive decisions that directly impact an organization's KPIs—for example, deciding which customers to visit during each call cycle, or what prospects to target that are likely to provide the best return.
- *Transactional decisions*: are low-value but high-volume decisions made within the context of individual transactions, such as discounting or cross-selling offers presented upon checkout within eCommerce portals (where personalized and optimized offers can amount to a significant bottom-line impact over time), or pricing decisions during the quoting or order taking process, where optimized pricing can lead to improved margin.

Each case study follows the same problem-to-decision pyramid, starting with the problem definition and business objective, then followed by data, reporting, and analytics, before moving into the higher layers of the pyramid representing prediction, optimization, and self-learning.

We'll begin with a case study at the operational level of sales, where a distributor of commoditized goods wants to reduce customer churn and improve share of wallet metrics through the daily optimization of sales activities, before moving down into two case studies at the tactical, transactional level—one based on quoting and ordering through sales reps, the other based on autonomous, digital transactions—before completing the chapter with a few case studies at the strategic, structural level.

For additional information on some application areas presented in this chapter, we encourage you to review material available at: www.complexica. com/software/decision-cloud.

8.1 Call Cycle Optimization for Reduced Customer Churn and Increased Share of Wallet

Unlike marketing and supply chain business functions (discussed in Chapters 9 and 10, respectively), the sales function in most organizations has remained largely untouched by scientific thinking and advances in technology, and as such, remains open for significant improvement. Executives often lament the fact that sales reps have been calling on the same customers in the same way for decades, and there's no "button" they can "push" to dynamically optimize visits, activities, and conversations undertaken by in-field reps.

That said, such "buttons" do exist, and there are many benefits that organizations can realize by optimizing their sales operations, including:

- *Increased utilization and yield on sales resources*: as measured by monthly calls per sales rep, average cost per lead, and average revenue per order, among others

- *Reduced customer churn*: as measured by retention and churn rates, purchase frequency, and net promoter score
- *Greater share of wallet*: as measured by average revenue per customer
- *Improved customer lifetime value*: as measured by average revenue and profitability of each customer over some period of time

Organizations that benefit most from the use of Artificial Intelligence methods for optimizing field sales activities typically have a long tail of customers (in the thousands or tens of thousands) and a large sales force. Their selling style is usually characterized by frequency customer interactions and a high number of weekly and monthly sales calls. From a timing perspective, sales optimization projects are usually prompted by a broader business driver or compelling event, such as a shortfall in revenue, loss of market share, margin pressures, or new competitors entering the marketplace.

In this case study we'll focus on DEF Supplies, a distributor in the hospitality industry that employs hundreds of sales reps calling on tens of thousands of business customers spread geographically across the country. The customers in each sales territory are hospitality venues that vary in size, type (e.g. restaurants, cafes, hotels, etc.), and purchase patterns. In this type of environment, it's not uncommon for customers to spread their purchases of cleaning supplies or disposable tableware across a variety of distributors, while aggregating other products to a smaller set of suppliers. Given this dynamic, if sales reps visit customers on a frequent and regular basis, they might secure their usual order alongside additional, incidental items. However, because of the large volume of customers and transactions, it's difficult for sales reps to stay on top of everything that's happening within each individual customer, and lost sales and customer churn aren't usually detected until it's too late. As an example, a sales rep would be unlikely to notice that a small customer has gradually stopped purchasing some products until the customer has left altogether (in favor of a competitor).

Without increasing the number of sales reps (to improve coverage of the long tail of customers), DEF Supplies wanted to proactively identify and act upon churn risks and customer growth opportunities as they arose. This required DEF Supplies to develop insights and knowledge on what the early signs of risks and opportunities looked like within its customer base, and implement technology that could *automatically* identify these risks and opportunities, as well as *dynamically* "re-focus" individual sales reps on the most important risks and opportunities. In other words, optimizing this level of sales would require sales reps to make optimized decisions on a daily basis for what customers and prospects to visit, what personalized messages to deliver,

and what journey plan to use. This also
meant that DEF Supplies would move
away from traditional, static call cycles,
and embrace dynamic, opportunity-based
call cycles to maximize the utilization of
sales reps. Hence, DEF Supplies defined
their business problem and objective as:

Reduce customer churn and increase customer share of wallet, without increasing headcount within the sales organization

Keeping this business objective in mind, the types of risks and opportunities that sales reps routinely faced in their territories were defined as:

- Churn risk due to competitor activity, or quality or delivery issues (which might have resulted in a customer complaint)
- Cross-selling opportunity
- Up-selling opportunity
- Product replenishment opportunity
- Wallet-share growth opportunity

A manual approach (based on the experience of individual sales reps) for identifying these risks and opportunities within each customer wasn't feasible, both from a knowledge and time perspective. This prompted DEF Supplies to implement a Decision Optimization System capable of automatically identifying churn risks and customer growth opportunities at scale—a task that was ideally suited for Artificial Intelligence algorithms.

Data, information, & knowledge

To fully appreciate the complexity of DEF
Supplies' problem and business objective,
it's worth noting that just executing on the
statement "to gain insights and knowledge
on what the early signs of risks and opportunities looked like within its customer
base" required significant data analysis to understand each risk and opportunity type! Because DEF Supplies wanted to model a varied set of risks and opportunities, this required different sources and types of data to answer questions such as: What does customer churn look like? What are the early signs of churn? What are the contributing factors that lead to churn? Can those factors be found in the historical data? To answer these questions and calculate the risk of churn, DEF Supplies had a substantial amount of internal data, including:

- Product master data
- Core pricing data
- Promotional pricing data
- Historical sales transactional data
- Customer information
- Customer complaints data
- Sales rep activity data
- Sales rep profile data

Despite the extensiveness of the above datasets, they weren't sufficient to generate accurate predictions for several opportunity types, such as cross-sell and up-sell, or wallet share growth. The reason being that the "Customer information" dataset was inaccurate and incomplete, and also lacked a sufficient level of attribute detail to allow for Machine Learning methods to undertake a meaningful segmentation. External data was thus necessary to augment the internal datasets, providing the model with additional customer attributes that were used for *micro-segmentation* purposes.[1]

Although DEF Supplies used its internal data to provide sales reps with a large number of reports and dashboards, limited value was extracted from these informational sources due to the overwhelming number of customers and transactions. As an example, sales reps were provided with reports on customers that were trending down or which hadn't ordered over some period of time, but there was no way for sales reps to know if this behavior was normal or not, or how to act on this information, as shown below:

---

1   In general, segmentation is a process that classifies customers into groups to assist with decision-making, usually within the context of sales and marketing business functions. The main goal of any segmentation is to identify significant differences among customers that will influence their purchase decisions or buying behavior. This segmentation then allows for better decision-making (e.g. developing a value-based offer for one segment of the market, a premium offer for another, and so on). To understand the concept of *micro-segmentation*, it's best to first mention *macro-segmentation*, which classifies customers into broad categories based on a small set of static, structural attributes (e.g. customer size, type, and geographic location). Micro-segmentation, on the other hand, requires additional knowledge and attempts to find *homogenous* groups of customers, classified into a significantly higher number of narrow categories based on a large set of dynamic, behavioral attributes (e.g. purchasing patterns, price sensitivity).

Secondly, in order to properly establish the extent to which sales reps should divert time away from their usual selling activities in order to address items raised in these reports, sales managers needed to understand which risks and opportunities were more important than others across the multiple and highly-varied reports. The sheer number of customers made this difficult, requiring sales reps and managers to become analysts and spend their time evaluating reports rather than being out with customers! Hence, despite the elaborate reporting and data visualization available within DEF Supplies, this information provided limited value in terms of optimizing sales activities and decisions, and couldn't deliver on the business objective of reducing churn or dynamically identifying growth opportunities.

Even the knowledge and insights extracted from DEF Supplies' data and overlaid with other sources proved difficult to "action," as sales reps were interested in knowing what to do next (i.e. what customers they should focus on, with what messages and actions) rather than being provided with more reports and insights that explained the past. For these reasons and others (such as growing competitive pressures in the marketplace requiring DEF Supplies to do something "different" in sales), traditional and manual analytical processes were not sufficient to realize DEF Supplies' business objective.

Decision Optimization System (prediction, optimization, & self-learning)

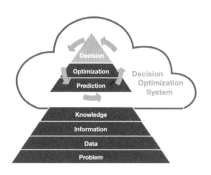

To reduce churn and increase share of wallet within its customer base (without a corresponding increase in headcount), DEF Supplies implemented a Decision Optimization System capable of:

- Automatically identifying and ranking customer risks and opportunities
- Directing sales reps to the highest-value risks or opportunities within their territories
- Creating dynamic call plans
- Optimizing journey plans
- Predicting the opportunity value of each customer
- Predicting the next best conversation or action for each customer

To enable these capabilities, DEF Supplies undertook iterative data analysis on different internal and external datasets to establish and validate correlations. This process was replicated for building and training prediction models for churn, cross-selling and up-selling, and share of wallet estimation. Moving from the knowledge layer of the pyramid towards prediction, churn models were built and trained based upon an understanding of what churn looked like, while the share of wallet model relied on an underlying segmentation model that could dynamically classify customers and prospects into micro-segments based upon their attributes and purchase behaviors.[2] The same segmentation model was used to map each customer's purchase behavior against other customers in the same micro-segment, thereby identifying a list of gap products for each customer (i.e. products they should be buying from DEF Supplies but weren't).

Once the Decision Optimization System was configured and deployed, the new workflow began with sales reps accessing the opening screen below:

> **What would you like to do?**

> ▸ Larry, what customers & prospects need my attention this week?
> ▸ Larry, let's create a quote or order
> ▸ Larry, I need some information about a customer or prospect
> ▸ Larry, let's run some reports

The first option of asking "Larry, what customers & prospect need my attention this week?" provided sales reps with the ability to plan and optimize call cycles through the use of Artificial Intelligence algorithms. This functionality was presented through a "Customer Opportunity Profiler" screen, where sales reps could generate a dynamic call list of prioritized customers within their territory, as shown below:

2　Using both internal and external data, DEF Supplies applied unsupervised Machine Learning techniques to identify patterns that weren't immediately apparent or obvious (e.g. the impact of attributes such as customer complaints, income levels per location, moving holidays and weather in decision-making for purchasing volume and timing, among others) to create a meaningful micro-segmentation of customers. DEF Supplies also created a list of questions to further refine the micro-segments at a later date, once the sales reps completed the required data collection.

By continuously analyzing new data to identify new risks and opportunities across the entire customer base, the Decision Optimization System was able to consider the aggregate set of risks and opportunities for each individual customer and dynamically determine their position on the call list (moving them up or down as needed). The Decision Optimization System would also flag the most important risk or opportunity identified for each customer as the main reason to visit during any given call cycle (shown above as "Reason," in the third column).

Because sales reps could only fit a limited number of visits into each call cycle, the Decision Optimization System prioritized customers in each cycle by considering their revenue, potential, and risks and opportunities. In doing so, the most important customer visits were presented at the top of the list, along with associated *Next Best Conversations*. These suggested conversations (which could also be described as "Next Best Actions" for each customer, or "Messages" that the sales rep should deliver during the visit) were automatically generated by the Decision Optimization System, and were contextual to the customer, sales rep, and point of time (using the latest available data), as shown below:

The application of Artificial Intelligence algorithms within this dynamic sales environment enabled sales reps to not only answer the question of "what customers & prospects need my attention this week?" but also "what should I focus on for each customer?"—in effect providing a double level of automated analysis for each sales rep. Self-explainability of the AI engine (Larry the Digital Analyst®) was also an important consideration for DEF Supplies, so that sales reps could understand the reasoning or logic behind each visit recommendation or Next Best Conversation, as shown below:

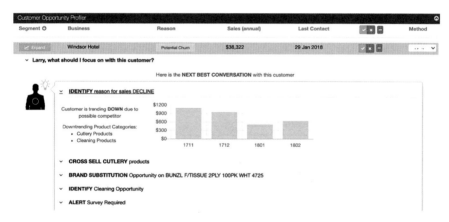

The Decision Optimization System was able to explain each recommendation in natural language, in this case explaining (in red text above) that the customer in question was trending down in core product categories (and overall volume), regardless of seasonality impacts, which was an early indicator of potential churn.

This is a good example of Artificial Intelligence automating the problem-to-decision workflow (which we explored in Part I) to direct sales reps to the highest-value risks or opportunities in each territory. To enable opportunity identification, the underlying prediction models were also trained to estimate wallet share for each customer, down to a product category and individual product level:

These opportunity values and share-of-wallet estimates[3] enabled sales reps to automatically identify and prioritize customers with the greatest potential, allowing them to not only "fish where the fish are," but also lift the sophistication and effectiveness of their customer interactions by using the system-generated Next Best Conversations. In this particular implementation, the Decision Optimization System was able to learn through direct feedback from sales reps on the effectiveness of each recommended visit and Next Best Conversation, as shown below:

This feedback loop was augmented with new data that arrived daily in the form of new transactions,[4] enabling the prediction models to automatically

---

3 Random forest (see Chapter 5.2) was used as a predictive model for wallet share estimation.
4 See Chapter 7 for an extended discussion on the significance of feedback loops and incorporation of new data.

update themselves by evaluating their past performance (in terms of whether past recommendations resulted in new sales).

And finally, to maximize efficiency, the Decision Optimization System was able to geographically cluster the prioritized visits and optimize journey plans for each sales rep by minimizing travel distance and time, as shown below:

By embedding AI-driven recommendations into the call planning process, DEF Supplies was able to increase utilization and yield on sales resources within the first year of deployment. This was largely related to the rapid adoption of the technology by the sales team, which found that the Decision Optimization System made their life easier. Sales reps regularly stated that the Decision Optimization System helped them hit quarterly targets, gave them an understanding of the true potential of each customer, and cut through the clutter to assist in prioritizing daily opportunities and risks in each territory. As an outcome, the Decision Optimization System paid for itself within one year of deployment, customer churn was reduced, and DEF Supplies benefited from increased productivity and improved average revenue and profitability per customer.

8.2 Quoting Optimization for Improved Basket Size and Margin

Having explored the application of Artificial Intelligence to the operational level of sales, we'll now move down into tactical, transactional decisions. Given the high volume of decisions undertaken at this level, incremental gains can amount to a significant bottom-line impact when all these individual transactions and decisions are aggregated. In this case study we'll also continue with DEF Supplies, a distributor in the hospitality industry that employs hundreds of sales reps calling upon tens of thousands of business customers. Recall that DEF Supplies' sales reps visit a long tail of customers spread geographically across the country, where they create quotes and take orders with the goal of securing their usual order, along with any additional, incidental items.

Having addressed the business problem of reducing churn and increasing wallet share within its customer base (as discussed in the previous case study), DEF Supplies turned its attention to the quoting and order taking process with a view to improve transactional metrics, such as "average revenue per order" and "average margin per order." Meeting this business objective would require sales reps to make customer-specific product recommendations that increased basket size (i.e. more line items per order) and overall margin (through brand swapping, product substitution, and up-selling).

However, with the very large number of customers and products (both in the tens of thousands), the only way sales reps could provide margin-optimized recommendations is by becoming analysts and conducting considerable research on each customer prior to each visit (i.e. by researching each customer's particular type of business, geography, industry sector, order history, and so on). On top of this research, sales reps would need to become familiar with DEF Supplies' entire product range (along with associated prices, margins, product substitutions, promotions, and availability across each warehouse) in order to make the right recommendation to the right customer at the right time—an impossible feat for any human being without the use of enabling technology.

Given the large number of in-field sales reps, there was considerable scope for improving the average revenue and margin of orders through the use of technology that could dynamically generate personalized product/basket offers for each individual customer. Hence, the business problem and objective were defined as:

Improve average revenue and margin per order, without increasing the non-selling time of sales reps

In other words, the business objective of increasing margin and average line items per order needed to be achieved without burdening sales reps with any additional research or analysis tasks that might reduce their customer-facing selling time.

Data, information, & knowledge

The data requirements for addressing DEF Supplies' business objective were similar to those described in the previous case-study, such as:

- Product master data
- Core pricing data

- Promotional pricing data
- Historical sales transactional data
- Customer information
- Customer complaints data
- Sales rep activity data
- Sales rep profile data

DEF Supplies used this data to provide sales reps with reports on how individual customers were trending on metrics such as volume and margin:

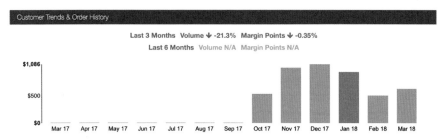

as well as products that were dragging down the overall margin of each customer:

DEF Supplies was also able to leverage the Machine Learning segmentation model from the previous case study, to provide reports on how each individual customer compared with the micro-segment (in terms of revenue and margin performance), as shown below:

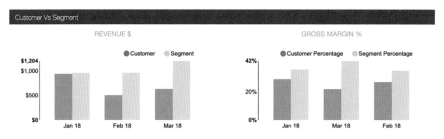

As insightful as these reports were, they were unable to address the business objective of improving the basket size and margin of individual orders, because it was difficult for sales reps to turn these reports into personalized recommendations that could be presented to customers. Part of the problem was that sales reps had limited time to create a quote while standing in front of the customer, and as such, this wasn't the right time to be looking at reports and charts to decide what to recommend!

Even the creation of additional knowledge—such as the analysis of product and transaction data to discover which products shared similar demand profiles, or which products were commonly purchased together, where, when, and at what price—did little to address the problem of generating offers during the quoting and order taking process that were personalized and margin-optimized. Another challenge was that such knowledge was produced by traditional methods and provided to sales reps as "static outputs" which were of limited use. Static outputs are valuable for the creation of promotional bundles in catalogues and in-store retail offers, but using such an approach during a dynamic quoting and order taking process results in generic, one-size-fits-all offers, with the best example being the famous: "Would you like fries with that?"

Decision Optimization System (prediction, optimization, & self-learning)

To improve average basket size and order margin during the quoting and order taking process, sales reps would need to dynamically consider the basket composition of each order as it's being built; the margin, substitutability, and availability of all products relevant to that customer; and the customer's wallet share within each product category, as well as their purchase history and micro-segment information. Given that such dynamic analysis would be impossible to undertake through traditional (non-automated) reporting or analytical approaches, DEF Supplies implemented a Decision Optimization System capable of providing sales reps with:

- Automated research and insights on individual customers and prospects prior to each sales call
- Customer-specific product/basket recommendations and personalized offers that have a high probability of being accepted (based upon that customer's profile, micro-segment, order history, etc.)

- Margin-optimized product substitution recommendations
- Margin-optimized pricing recommendations
- A pre-populated *Predicted Next Order* for each individual customer

As a starting point, the underlying Machine Learning segmentation model was re-used from the previous case study, allowing the Decision Optimization System to dynamically classify customers and prospects into micro-segments based upon their attributes and purchase behaviors. Because this segmentation model was used to map each customer's purchase behavior against other customers in the same micro-segment, it was extended to provided product-specific cross-selling and up-selling recommendations that were contextual and highly relevant to the needs of individual customers.

A fuzzy system (see Chapter 5.4) model was then trained for both product substitution and dynamic pricing. The fuzzy rules for product substitution were based on DEF Supplies' product hierarchies (extended by available package sizes for particular products), cost and margins, as well as character-istics of each micro-segment. Hence, the same product could trigger different substitute offers for different customers. Fuzzy rules were also used to generate margin-optimized prices, which took into account external factors (e.g. com-petitor pricing, supply and demand, geography) as well as the characteristic of each micro-segment (we'll explore the topic of dynamic pricing in more detail in Chapter 9.1).

Once the Decision Optimization System was configured and deployed, sales reps were provided with an order entry screen with which to begin the quoting or order taking process:

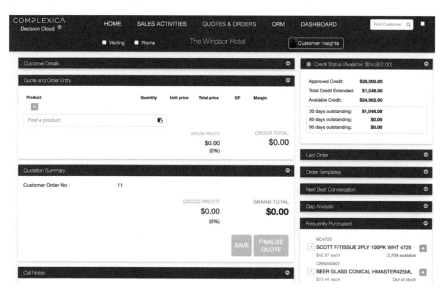

Aside from the usual quoting and order management functionality featured in most order taking applications, the Decision Optimization System deployed by DEF Supplies enabled a variety of AI-driven capabilities. For example, the system would consider the purchase history of an individual customer and their micro-segment, seasonality factors and key selling dates, and use this data to generate a *Predicted Next Order*. Similar to the concept of vendor managed inventory, the system was able to predict when it was time for a customer to replenish specific products and then use these predictions to pre-populate an order to ensure that no items were forgotten during the visit:

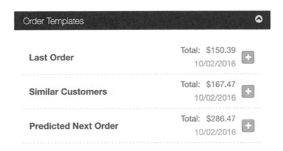

Predicted Next Orders were also used with prospective customers, allowing sales reps to start conversations with a suggested list of items the prospective customer should be interested in buying (and most likely, was already buying from a competitor), thereby increasing the efficiency and effectiveness of each sales call. Even though no historical transactional data existed in such situations (because the prospective customer had yet to purchase anything from DEF Supplies, and hence was still a prospective customer rather than an existing customer), the system was able to place the prospective customer into a micro-segment through the use of publicly-available attribute data, and then use the micro-segment's purchase history to generate the Predicted Next Order.

The Decision Optimization System also generated personalized cross-sell, up-sell, and product substitution offers, which dynamically changed as the basket composition of the order changed:

To avoid lost sales, these product substitution offers were also provided for out-of-stock items:

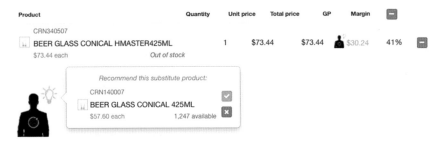

The Decision Optimization system also analyzed product prices, margin, and descriptions in natural language, looking for substitute products that met the same description, and which provided a better margin for DEF Supplies along with a better price for the customer (even if the originally requested product was in stock). In the example below, an in-stock product selling for $42.87 at a 9% margin:

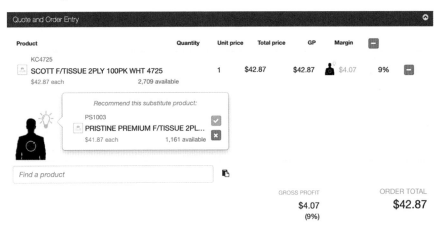

is substituted for another product with the same characteristics (2 ply, 100 pack) that has a lower selling price ($41.87) and higher margin (39%). By accepting this offer, the customer spends less, while DEF Supplies improves its margin, as shown below:

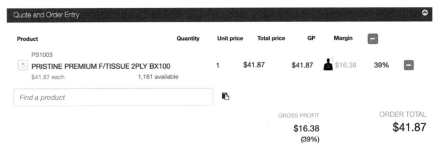

By embedding AI-driven recommendations into the quoting and ordering process, sales reps were able to provide personalized offers within each sales call that were margin-optimized and likely to be accepted by the customer. In terms of business outcomes, the Decision Optimization System resulted in increased number of lines per order, with 11.5% of the orders having additional lines from cross-sold items, while at the same time driving an overall improvement in the average gross margin of orders (with AI-assisted orders achieving a 42.85% average gross margin, compared to 38.95% for non AI-assisted orders).

Also, because the AI-based functionality was embedded within a Decision Optimization System that provided standard quoting, ordering, and order management functionality, DEF Supplies was able to minimize change management efforts when it came to adoption of the system, as well as accelerate speed to competency when new sales reps joined the business.

8.3 Optimizing Digital Sales for Improved Basket Size and Margin

Continuing on at the tactical, transactional level of sales, we'll now explore this business process from a digital and autonomous perspective. One immediate observation worth noting, however, is that deployments of AI-based Decision Optimization Systems into autonomous environments carry a much lower level of change management effort than when the same technology is applied to the other two levels of sales (i.e. operational and strategic). Hence, deployments encounter less resistance when getting into production, and thus, are generally easier to manage.

The opportunities and benefits that organizations can realize from applying Artificial Intelligence within digital environments such as self-service apps or eCommerce portals is varied, but includes:

- Improved "average revenue per order" through dynamic cross-selling & up-selling recommendations that are specific to each customer type (i.e. generating personalized offers)
- Improved "average margin per order" through dynamic and optimized pricing during the online ordering process (i.e. generating personalized prices for each product)
- Reduced churn and increased share of wallet through ongoing, personalized communications and offers
- Improved customer experience through personalization

Although most organizations can benefit from AI-driven recommendations within digital sales channels, the benefits are particularly pronounced for manufacturers and distributors with a large product range, complex pricing structures, and a long tail of business customers, as well as retailers wanting to improve margin and personalization—in other words, organizations where sales are characterized by low-volume, high-frequency transactional purchases that can be made from a variety of suppliers.

In this case-study we'll focus on RST Gourmet, a distributor of food products servicing a variety of hospitality venues such as pubs, restaurants, and cafes. Similar to DEF Supplies, the environment faced by RST Gourmet was characterized by a highly fragmented, long tail of customers, numerous sales channels (in-field sales reps, phone-based sales team, self-service apps, eCommerce portal), along with multiple suppliers providing a large range of products (many of which had a short shelf life or seasonal availability).

RST Gourmet wanted to improve its customer experience within digital channels, as well as the basket size and margin of each order. Hence, the problem and business objective were defined as:

Improve digital channel performance on three key metrics: customer experience, basket size, and margin

The intended approach was the deployment of an AI-based recommendation engine, which could be leveraged across all sales channels to optimize customer interactions and offers.

Data, information, & knowledge

Digital channels are typically more "data rich" than other channels—particularly within business-to-business selling environments—allowing an organization to capture data such as:

- Which customer logged into an app or portal and when?
- How long did they stay logged on?
- What products did they view, in what order?
- How long did they stay on each product page?
- What was their final purchase basket, if any?
- How long did their transaction take?

- After the transaction was completed, did the customer log in again to order forgotten items?

Besides this data from digital channels, RST Gourmet also collected the same internal data as DEF Supplies, including:

- Product master data
- Core pricing data
- Promotional pricing data
- Historical sales transactional data
- Customer information
- Customer complaints data

External datasets were also available, such as census data, bureau of meteorology weather data, major events data (e.g. sporting finals, festivals, etc.), public holidays calendars (including moving holidays), and social media information. RST Gourmet used this data to generate a standard set of reports focused on the transactional performance of the self-service app and eCommerce portal (including comparisons to other channels, product-level performance, and cross-customer comparisons).

In addition to these reports, RST Gourmet developed a significant amount of knowledge through manual data analysis, which enabled the organization to understand which products had similar demand profiles, which products were commonly purchased together, what customer segments purchased what, when, and at what price, and so on. This traditional (and static) approach to knowledge discovery was insightful, but in the context of digital channels it could only be used to create static recommendations that generated low acceptance rates (in effect providing customers with "one-size-fits-all" offers, rather than personalized "just for me" offers—which could improve customer experience, as well as order size and margin metrics). Such personalized offers required dynamic consideration of a customer's segment, purchase history, and current basket composition, and therefore lay beyond the information and knowledge layers of the pyramid.

Decision Optimization System (prediction, optimization, & self-learning)

To improve the customer experience within its digital channels—as well as increase basket size and margin metrics—RST Gourmet turned to

Artificial Intelligence methods that could enable dynamic recommendations for:

- Product bundling and up-selling
- Differential pricing
- Product substitution (against multiple objectives)
- Targeted offers to mitigate churn

Training the prediction models within the Decision Optimization System was a complex endeavour, because each recommendation type required a separate underlying model, which in turn needed to be trained on a different set of data. A few ensemble models (as discussed in Chapter 5.6) were built and trained to provide the required predictions and recommendations. These models also relied upon a segmentation model that classified customers into micro-segments that were sufficiently granular to capture similarities in buying behaviors shared by similar customers. Thus, as the first step of the modeling process, a Machine Learning segmentation model was built based on both internal and external data (in a similar process as that described for DEF Supplies in the previous two case studies) to ensure that recommendations were as contextual and relevant for each customer as possible.

In the diagram below, the first row represents the outputs from one of the ensemble models, such as (for a particular customer) the predicted total spend in different product categories, as well as products the customer should be buying from RST Gourmet but wasn't (i.e. gap products):

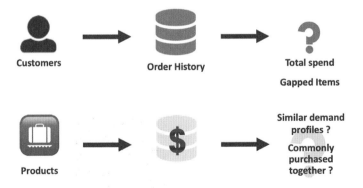

The second row of the diagram represents another ensemble model that was trained on product and transactional data to group products that share similar demand profiles and predict which products are likely to be purchased together. The final output of these models were dynamic recommendations that appeared within the self-service app and eCommerce portal during the customer ordering process, such as the screen below for product substitution:

and cross-selling based upon the customer's micro-segment:

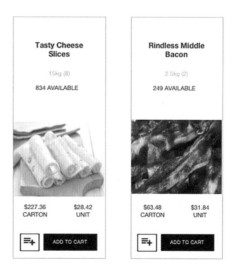

The results achieved by RST Gourmet exceeded those attainable through traditional approaches, such as affinity analysis. Following the deployment of the Decision Optimization System into RST Gourmet's digital environment, surveyed customers reported improved convenience through shorter sessions and reduced instances of forgotten items. The Decision Optimization System also leveraged self-learning algorithms for automated "self-tuning," thereby enabling improvements in the conversion rate over time (the cross-sell acceptance rate rose from 3.5% to 7.1% within 7 months, while the product substitution acceptance rate rose from 3.8% to 11.2% during the same period)—all of which contributed to an improvement of 1.5% additional gross margin from the baseline.

8.4 Sales Structure and Territory Optimization for Improved Market Coverage and Organizational Efficiency

Having discussed the operational and transactional levels of sales in the previous sections of this chapter, we'll now move from voluminous, tactical decisions, all the way up to infrequent, strategic decisions, such as sales structure and territory optimization. As most executives know, optimizing the sales structure of any organization involves a significant level of risk and uncertainty, and is therefore undertaken infrequently. Answering the question of how many sales reps are needed, in what location, calling on what customers, has all the characteristics of a complex business problem, including an almost infinite number of possible solutions, countless business rules and constraints, many conflicting objectives, and a dynamic operating environment (as the customer and competitive landscape is never static).

If the structure we're trying to optimize is small and simple, then manual analysis might produce a reasonable result. But if the structure is large and complex and includes a long tail of customers that require frequent visits or interactions, then the benefits of applying Artificial Intelligence become compelling. These larger, more complex sales structures may feature multiple channels to market—including account managers and sales reps in the field, and inside sales teams on the phone—as well as a sales style characterized by frequent interactions in a highly structured environment (with metrics on the number of sales interactions per day, visibility on the duration of each conversation, profiling of customers and sales staff against a variety of key attributes, and so on). Such organizations may undergo a sales restructure for a number of reasons, including:

- Cost-out initiatives that require a headcount reduction within the sales organization

- Mergers or acquisitions that require a broader product portfolio to be serviced by a combined sales team
- Launch of a digital or omni-channel strategy which requires reallocation of customers to a new primary channel
- Creation of customer growth targets (such as share of wallet) that require customers to be segmented and serviced by opportunity value rather than historical sales

In this case study we'll explore a pharmaceutical manufacturer, ABC Pharma, with approximately 100 sales reps aligned to various drug and therapy areas. The primary task of these sales reps is to drive awareness and education of ABC Pharma's products by visiting thousands of doctors, so these products can be prescribed to patients when required. The total number of doctors in Australia is approximately 70,0000, comprised of approximately 43,000 general practitioners and 26,000 specialist medical practitioners.

The specific problem that ABC Pharma faced was the need to re-organize its sales structure and territories to make way for a new product division, without increasing the overall cost of the sales organization or suffering a decline in sales of existing products. Hence, the problem and business objective were defined as:

Introduce a new product division into the sales organization without increasing costs or decreasing existing product sales

Addressing this complex business problem required the consideration of various possible sales structures and territories, each of which represented a what-if question that needed to be answered, such as:

- Could a number of sales reps be re-assigned from their current product division into the new product division, without any material impact on existing product sales (total number of prescriptions)?
- Instead of re-assigning sales reps to the new product division, could ABC Pharma instead create hybrid specialization roles for staff with transferable skills and knowledge relevant to the new product, allowing them to operate in a dual role without significant productivity impact (i.e. in effect maintaining the same number of sales reps visiting the same number of doctors, but carrying a broader product portfolio)?
- If sales reps were hired for the new product division, would these additional costs be off-set by greater sales from both the new and existing

products? If so, how many reps would be required? How much additional revenue would be generated?

The very large number of these what-if questions led to a very large number of possible scenarios (i.e. solutions) for the new sales structure and territory mapping, with each scenario representing some combination of sales reps, shape and size of territories, and allocation to a product division. Furthermore, the evaluation of each scenario required an accurate prediction of marketplace outcomes for that particular sales structure and territory mapping, as well as consideration of the trade-offs between cost of the overall sales organization, revenue generated from their efforts (i.e. doctor awareness that translates into prescriptions), and the customer experience (in this case, of the doctors) of receiving too many or too few visits from ABC Pharma.

Data, information, & knowledge

Working with data in the healthcare industry brought to light some peculiarities worth mentioning, with Australian Government regulations mandating that prescription data could only be accessed via anonymized and aggregated formats for privacy reasons. The geographical

aggregation levels were carefully calculated to include several pharmacies and doctors together, so that it was impossible to create a one-to-one mapping of individual doctors to individual prescriptions. To understand the challenge this posed, it's worth noting that patients may visit a doctor close to their home but fill their prescription at a pharmacy on the other side of town, which might be close to where they work! As such, this anonymized data could only be used to approximate the prescribing behavior of specific doctors.

Because of these government regulation, ABC Pharma found itself in a situation where it had a significant amount of internal data pertaining to the activity of sales reps (such as their visits and communications), but very poor outcome data, making it difficult to attribute specific sales to specific activities undertaken by sales reps. As an example, it was very difficult to establish that the activities of sales rep "Bob" led to an increase in prescriptions from Dr. "Sam," because there was no way to know how many prescriptions Dr. Sam wrote.

ABC Pharma also had other data challenges, such as their internal data being in different formats with varying levels of completeness and cleanliness. As a case in point, the data produced by account managers from each visit included highly structured and consistent data, such as the date, time, and location of each visit, as well as unstructured and inconsistent data, such as call notes on the visit itself, which varied in length, detail, and structure, or were sometimes missing altogether! The structured data was easy and straightforward to use, whereas the unstructured data was not.

Like most other pharmaceutical manufacturers, ABC Pharma generated an abundance of reports and visualizations that were delivered to sales reps and managers via apps, spreadsheets, and presentation decks. These reports effectively communicated what happened in the past—for example, the fact that certain geographical areas had more product sales (total number of prescriptions) over some time period, compared to the population and number of doctors within the area, as shown below:

Brick Code	Brick Name	Revenue	Doctors	Population
31270	Surrey Hills	$190.58	13	2,940
31260	Canterbury (Vic.)	$1039.50	3	2,500
31030	Balwyn	$57.75	3	1,482
31040	Balwyn North	$33.33	1	790
31020	Kew East	$441.08	3	3,392
31030	Deepdene (Vic.)	$190.58	3	1,294
31230	Hawthorn East	$57.75	3	1,482
31240	Camberwell (Vic.)	$190.58	13	2,940
31250	Burwood (Vic.)	$441.08	3	3,392
31510	Burwood East	$441.08	3	3,392
31280	Box Hill South	$441.08	3	3,392
31301	Blackburn South	$441.08	3	3,392
31260	Box Hill (Vic.)	$190.58	13	2,940
31300	Blackburn	$441.08	3	3,392
31261	Box Hill North	$441.08	3	3,392
31270	Mont Albert	$441.08	3	3,392
31290	Mont Albert North	$441.08	3	3,392

These reports, alongside others that were informational in nature, such as particular postcodes with a higher percentage of multiple products dispensed or sales reps that were more productive with more doctor visits per call cycle, were ultimately insufficient for solving the problem at hand.

Despite the limitations posed by government regulations on data availability, as well as the intermediated nature of the sales process (where sales reps could only influence the doctor, rather than sell anything directly), ABC Pharma developed knowledge and insights by using external data to better understand the opportunity potential of each territory. For example, by augmenting internal data with census data, ABC Pharma developed a demographic profile of each suburb (including age and income), which contributed to explaining why the company achieved various degrees of market success in various geographical areas. This, in turn, enabled ABC Pharma to identify other suburbs with similar characteristics and opportunity potential.

Also, a list of practicing doctors and their speciality area (if any) was publicly available, and ABC Pharma overlaid this data with census data to understand the "catchment area" that each doctor served, providing knowledge on how large or small each doctor's local market of patients might be. ABC Pharma also mapped each doctor to a multi-doctor clinic, hospital, or sole practice operation, so that sales reps could group their visits together when calling on a doctor that was part of a larger clinic. Furthermore, data analysis was conducted on visit call notes to understand each sales rep's sentiment towards each doctor, with the hypothesis being that negative sentiment led to poor engagement and in such cases, the doctor needed be re-assigned to another sales rep.

Although some organizations rely solely on such historical information and knowledge to restructure their sales organization (i.e. to re-map territories, balance the number of customers from each segment across different sales reps, and so on), the complexity of ABC Pharma's problem couldn't be solved at this layer of the problem-to-decision pyramid. In addition to introducing a new product division into the sales organization without increasing costs or decreasing sales, ABC Pharma also wanted to consider the number of visits per call cycle that would create the best engagement and awareness among doctors, as well as the nature, type, and volume of multichannel interactions and how they influenced customer experience metrics—all of which needed to be modeled and taken into account when optimizing the overall sales structure and territories.

Decision Optimization System (prediction, optimization, & self-learning)

To solve the problem of how to introduce a new product division into the sales organization without increasing costs or decreasing existing product sales, ABC Pharma needed to generate many scenarios to answer many what-if questions. This required sophisticated prediction capabilities for evaluating each scenario, which weren't available in the information and knowledge layers of the pyramid.

Knowing that the problem was complex and the decision would carry significant consequences for the organization, ABC Pharma implemented a Decision Optimization System based on Artificial Intelligence methods that could:

- Predict marketplace outcomes (in terms of pre-defined metrics, such as profit, cost to serve, net promoter score, or sales volume) for many complicated scenarios, each one representing a different what-if question

- Optimize any particular scenario (e.g. either the current sales structure and territory mapping or some hypothetical one) based on business rules, constraints, and objectives (which may differ from scenario to scenario)

The prediction model used for ABC Pharma was based on agent-based simulation. As explained in Chapter 4.3, the idea behind agent-based modeling is that each simulation is based on the interaction between agents. In this problem, there were many "natural" agents: sales reps, doctors, and patients, who interacted with one another. The figure below presents a few sales reps (left), doctors (middle) and patients (right):

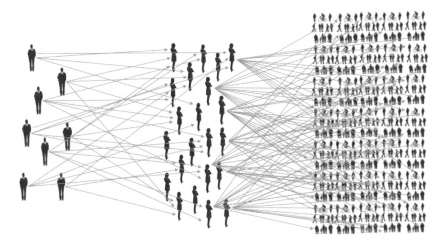

Sales reps and doctors were characterized by a set of associated variables—for example, each sales rep is characterized by age, gender, years of company experience, years of industry experience; while the doctors were characterized by location of their practice, speciality, year and place of their graduation. For both sales reps and doctors, additional variables were inferred from historical data, such as a performance index for sales reps or receptiveness index for doctors, which were included in some probabilistic rules that determined each agent's behavior and actions. Also, external data was used to model the environment for factors such as population density, age and income distributions, and so on. The complex interaction of the agents with the environment and one another resulted in behaviors and business outcomes (such as the number of prescriptions written) that were impossible to foresee or calculate due to all the complexity of their interactions:

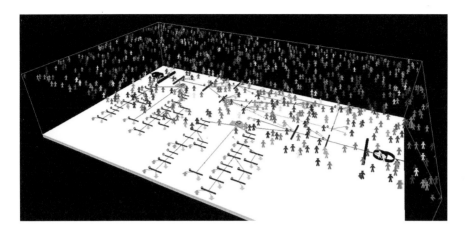

Once the models were trained and the Decision Optimization System deployed, ABC Pharma began testing various what-if questions in an iterative way. The process began with ABC Pharma creating a new scenario for each what-if question, as shown below:

The number and types of variables available for configuring each scenario covered all components of the sales channel, including sales reps, teams, sales divisions, and so on, as shown below:

The core components of each scenario included:

- *Problem variables*: the number, characteristics, and specialization levels of sales reps and doctors, team and territory structures, as well as other key variables that affected the business metrics ABC Pharma wanted to generate predictions against.

- *Business rules and constraints that defined a feasible solution*: which included, among others: (1) only sales reps with the required transferable skills could be re-allocated to a new or hybrid team; (2) a minimum number of doctor visits in each territory; and (3) a maximum distance travelled per day. Many of these business rules and constraints were defined as "soft," in order to explore a wider range of sales structures and territory shapes and sizes.
- *Objectives*: that were scenario-specific, but included minimizing the sales volume impact for existing products when some number of sales reps (or some percentage of their capacity) was re-allocated to the new product division.

When it came to configuring the problem variables, an input screen in the Decision Optimization System allowed ABC Pharma to define a broad set of values that could be changed for different scenarios. The below screen shows the input table covering each sales rep's demographic information, industry expertise, as well as qualitative and quantitative data on performance:

COMPLEXICA   Market Details   Larry Setup ▾   Scenario   Simulation   Optimisation

Sales Representatives

Active	#	Age	Gender	Company Exp.	Industry Exp.	Rep Performance	Call Quality	Activity Level
☑	1	24	Yrs	M	0	Yrs	1	Yrs	0.15	0.68	0.88
☑	2	22	Yrs	M	0	Yrs	0	Yrs	0.53	0.15	0.58
☑	3	31	Yrs	M	0	Yrs	0	Yrs	0.46	0.50	0.36
☑	4	44	Yrs	F	2	Yrs	4	Yrs	0.28	0.93	0.44
☑	5	41	Yrs	M	0	Yrs	6	Yrs	0.78	1.00	0.32
☑	6	34	Yrs	M	4	Yrs	6	Yrs	0.68	0.41	0.63
☑	7	45	Yrs	M	1	Yrs	3	Yrs	0.72	0.62	0.08
☑	8	48	Yrs	F	1	Yrs	6	Yrs	0.67	0.29	0.12

In this configuration screen, ABC Pharma could change the values in specific rows to create a scenario—for example, replacing two experienced sales reps with more than 10 years of industry experience each, with four junior sales reps with less than two years of experience—and then use the prediction model within the Decision Optimization System to evaluate this scenario against different business metrics, as shown below (in this case, predicted revenue by geographic area):

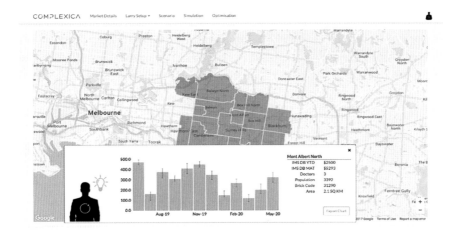

The Decision Optimization System would generate predicted business outcomes against each individual scenario through the use of agent-based simulations. As mentioned earlier, there were many possible configurations of ABC Pharma's sales structure and territories, and these were captured within a variety of different scenarios. ABC Pharma compared the performance of particular scenarios against predicted business outcome metrics, such as profit, cost to serve, net promoter score, and sales volume:

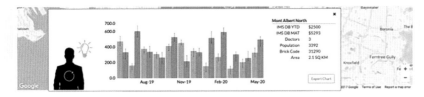

After creating a shortlist of favorable scenarios, ABC Pharma then applied optimization to further improve each scenario. This was done by selecting either "Profit" or "Volume" as the primary optimization objective:

and then adjusting the optimization parameters. For example, by setting the "Sales Force Size" parameter to 90% as shown above, ABC Pharma could further optimize a scenario by evaluating the predicted business impact of reducing the sales rep count by 10%. In this example, the Decision Optimization System would select the specific sales reps to remove, re-draw the territories, and restructure the sales divisions in an attempt to minimize impact on overall business metrics such as revenue and volume. Other parameters that ABC Pharma could influence included the number of doctors to visit, as well as the geographic reach of sales reps, allowing the Decision Optimization System to select the highest value doctors and regions to serve, as shown below:

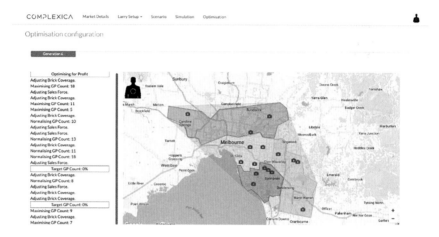

Through this iterative approach of fine-tuning and optimizing each individual scenario, ABC pharma was able to explore and identify specific sales structures that outperformed others. This approach also enabled ABC Pharma to drill-down on the predicted performance of each scenario against multiple KPIs:

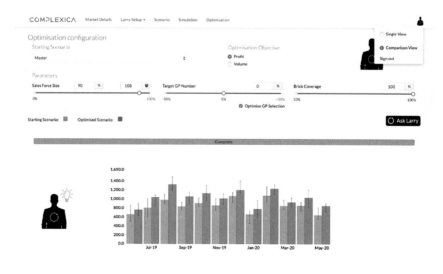

After analysing a large number of scenarios, ABC Pharma restructured its sales reps and territories in such a way that capacity within the existing workforce was re-allocated to the new product division. By leveraging the AI-based prediction and optimization capabilities of the system, ABC Pharma was able to find an optimized sales structure that incorporated the new product division without any increase in headcount costs or predicted decrease in existing product sales—in effect, allowing the organization to implement the highest-value decision.

8.5 Customer Segmentation and Sales Channel Optimization for Improved Growth and Reduced Cost to Serve

Carrying on at the strategic, structural level of sales, we'll take a look at FINE Liquor in this last case study—a manufacturer of alcoholic beverages with approximately 150 field sales reps. The primary task of these sales reps is to visit a universe of approximately 65,000 customers and prospective customers, classified as being either:

- *On-premise venues*: where beverages are sold and consumed on the spot, such as restaurants, music venues, cinemas, hotels, tourist and sporting facilities, and so on
- *Off-premise outlets*: where beverages are sold but then consumed elsewhere, such as retail liquor stores

Even though the dynamics of beverage sales are different for on-premise venues versus off-premise outlets, FINE Liquor wanted to unlock further growth across both customer types while also lowering the overall cost of servicing customers. To accomplish this ambitious goal, FINE Liquor needed to segment its customers by opportunity value and then allocate each customer to either a lower-cost phone-based team, the existing sales team (which would be reduced in number), or a specialized in-field team that would make more frequent visits to a smaller group of higher-value customers and prospective customers. Hence, the problem and business objective were defined as:

Increase overall sales through a lower cost sales structure comprised of segmented customers allocated to different channels

Realising this business objective would require FINE Liquor to analyze various scenarios that balanced the value and

future potential of each individual customer against cost-to-serve metrics, as well as answer numerous what-if questions, such as:

- Would sales decline if lower-value customers were allocated to a phone-based team and no longer received in-person visits? If so, could the drop be mitigated by increasing the frequency of phone calls per month or quarter?
- To what extent would the allocation of very high-value customers to a specialized team (i.e. more experienced sales reps that made more frequent visits) unlock new growth? How many visits should these customers and prospective customers receive per year, quarter, or month? How many specialized sales reps would be needed?
- Could this new specialized sales team be recruited from within the existing field sales team? Based on the required skills and geographical areas, what would be the best selection of sales reps for that team?
- By allocating lower-value customers to a lower-cost phone-based team, could additional capacity be freed up within the existing field sales team to "over-service" a subset of higher-potential customers and unlock growth? If so, which customers should be allocated to the additional "over-service" frequency? How many additional visits would be required per year, quarter, month?

FINE Liquor required a sophisticated technological capability to answer these questions, as each hypothesis would need to be represented by a scenario with predicted business outcomes (measured in total quarterly sales revenue and cost to serve).

Data, information, & knowledge

In comparison to the previous pharmaceutical case study, FINE Liquor's reps were less specialized in terms of skills and qualifications, but the sales process was equally rich in terms of available data for analysis, including:

- Sales team structure
- Territory structure
- Individual rep data (including demographic data, quantitative and qualitative performance data)
- Activity data (call frequency, call duration, call notes, etc.)
- Product master data

- Core pricing data
- Promotional pricing data
- Quotations (at line item level)
- Sales transactional data
- Customer master data
- Outlet format
- Customer complaints data
- Newly created customer segmentation data

A wide range of external datasets were also available (either free of charge or on a commercial basis), which could be used to more accurately profile customers. Some of this external data was particularly useful for estimating the sales potential of individual venues and retail stores, as their revenue wasn't known—in other words, even though there was no way for FINE Liquor to know the revenue and volume of each particular pub, restaurant, or store, these unknown data values could be estimated by analysing a variety of external datasets, such as a venue's digital footprint, social media activity and popularity, social-demographic catchment area, comparison to similar venues, and so on.

FINE Liquor also supported its sales process with various data visualization and reporting tools, which provided aggregated information to sales reps, managers, and executives. These included year-to-date channel and territory reports, sales rep performance dashboards, and customer orders graphs.

In addition to such traditional reporting and data visualization, FINE Liquor also employed in-house data analysts that continuously updated the organization's knowledge base. Some examples of this knowledge included an in-depth understanding of how visit frequency and duration impacted sales performance, or the extent to which field reps were able to influence the commercial outcome across different customer types and segments. This ever-growing knowledge base became an essential stepping-stone for enabling FINE Liquor's predictive capabilities, but in itself wasn't enough for advanced scenario modeling or optimized decision making.

Decision Optimization System (prediction, optimization, & self-learning)

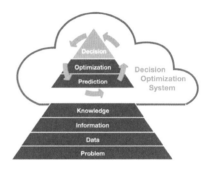

To meet the business objective of "increasing overall sales through a lower cost sales structure comprised of segmented customers allocated to different channels," FINE Liquor implemented a Decision Optimization System that introduced AI-based prediction and optimization capabilities into the organization. Using both internal and external data, unsupervised Machine Learning methods were used to build and train a segmentation model for classifying venues and outlets into micro-segments and predicting their potential and value. This process of micro-segmentation was a key enabler of the Decision Optimization System, and the data available within FINE Liquor's internal systems was extended with additional customer attributes using external datasets.

The prediction model was again based on agent-based simulation, where there were two types of agents: sales reps and customers; the latter divided into on-premise venues and off-premise outlets. Each micro-segment was characterized by a particular "behavioral pattern" that emerged from historical data (such as activity data and transactional data), and a scenario was defined as a particular arrangement of individual sales reps, working on a particular territory, within a particular channel, with a defined activity pattern. Due to the computational expense of evaluating a single scenario, simulated annealing was selected as the optimization method (which we covered in Chapter 6.5) to explore a number of possible scenarios and return the best one in terms of the selected objective (e.g. sales volume) within pre-defined business rules and constraints.

Once the Decision Optimization System was configured and deployed, FINE Liquor searched for a new sales channel structure through an iterative process of formulating and exploring different scenarios (which were based on the Machine Learning customer segmentation model). Each scenario represented a hypothetical configuration of the sales channel structure and allocation of customers to each channel, including all problem variables, business rules and constraints, and business objectives. To capture the necessary level of detail, FINE Liquor created and uploaded a configuration file for each category of variables for each scenario, as shown below:

Create Scenario

Enter scenario name

BDEs	Type to search scenarios	▼	Upload File
Outlets	Type to search scenarios	▼	Upload File
Territory	Type to search scenarios	▼	Upload File

Submit   Close

Because different scenarios represented different what-if questions, the configuration file often differed in the number of sales reps (shown as "BDEs" in the above), their skillset, experience, and channel allocation, as well as the size and shape of sales territories, channels, and customer allocation to those channels and territories. Also, to provide the Machine Learning segmentation model with additional data (and thereby improve accuracy), the characteristics of individual on-premise venues and off-premise outlets could also be directly defined, as shown below:

Once a scenario was successfully configured and loaded, FINE Liquor then used the Decision Optimization System to generate probabilistic predictions for a wide-range of business metrics at the channel, territory, sales team, and individual customer level:

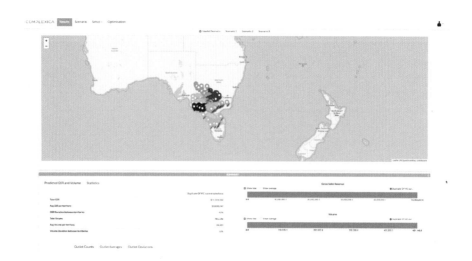

For some scenarios, FINE Liquor created a larger specialized sales team for executing longer and more frequent visits to high-value customers (as identified by the Machine Learning segmentation model); and in other scenarios, FINE Liquor replaced all in-person visits for customers below a certain sales threshold with telephone calls (which varied in frequency and duration, depending on the particular scenario being investigated, as well as the customer segment and historical sales). To fine tune this investigative process, FINE Liquor modified the frequency and duration of calls—for example, one scenario had 25 minutes for "Small Format Retail," 35 minutes for "On Premise," and 45 minutes for "Hybrid Large Format Retail," while other scenarios had alternate values:

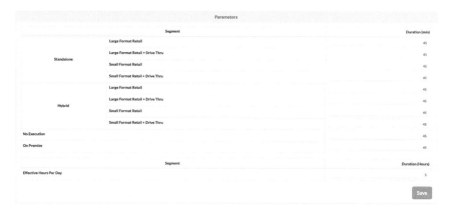

FINE Liquor then selected specific objectives and constraints to further improve individual scenarios, often limiting the amount of change introduced through the optimization process by selecting "Minimal" versus "Complete redraw," as shown below:

OPTIMISATION

Optimisation Objective Function
- Gross Sales Revenue (GSR)   ○ Volume

Existing Territory Modification
- ○ Minimal   ○ Medium   ● Large   ○ Complete redraw
- ○ Include new BDE hire   ○ Specialised

The Decision Optimization System tried to improve each base scenario of the optimization process, attempting to find a higher-performing scenario without violating any hard business rules or constraints:

Predicted GSR and Volume   Statistics

	Original	Optimized
Total GSR	$50,858,015	$52,867,167
Average GSR per territory	$2,825,445	2,937,065
GSR Deviation between territories	47%	40%
Total Volume	476,586	495,158
Average Volume per territory	26,447	27,509
Volume Deviation between territories	41%	32%

Gross Sales Revenue

● Show total ○ Show average   ▓ Original ▓ Optimised

$0 $500,000 $1,000,000 $1,500,000 $2,000,000

Volume

● Show total ○ Show average   ▓ Original ▓ Optimised

0   50,000   100,000   150,000   200,000   250,000   300,000   350,000   400,000   450,000   500,000

Predicted GSR and Volume   Statistics

	Original	Optimized
Number of territories	18	18
Total Travel Time (min)	590,095	589,898
Average time travel per territory (min)	36	35
Travel time deviation between territories	28%	28%
Total BDE call time (min)	736,060	753,930
Average BDE call time per territory (min)	40,670	41,885
BDE call time deviation between territories	13%	13%

As the final output, the scenario that was implemented in the marketplace provided FINE Liquor with the optimal:

- Size of each channel in terms of sales reps
- Allocation of customers into each channel (along with the recommended frequency and duration of visits and telephone calls)

- Mapping of territories
- Journey plan of each field rep

This final scenario lowered the overall cost of the sales structure and its cost-to-serve metrics, while simultaneously opening up new growth opportunities identified through the Machine Learning micro-segmentation model (to be actioned by the existing field force, as well as the newly formed specialist sales team). Finally, to support change management and benefit realization, the Decision Optimization System showed the difference between the original sales channel structure and optimized scenario.

CHAPTER 9

Marketing

"The best marketing doesn't feel like marketing."

Tom Fishburne, *the founder of Marketoonist*

As a discipline, marketing represents a broad field that covers many different types of activities, from strategic planning to tactical execution. These activities are related to communicating offers for potential customers, and depending on an organization's industry and business model, the undertaken marketing activities can vary greatly. Consider, for example, the differences in marketing activities executed by two manufacturers: One in the fast-moving consumer goods industry producing breakfast cereals, and the other in the industrial sector producing factory equipment. The first manufacturer relies on mass-marketing strategies such as TV advertising (which raises brand awareness and reaches a large number of people), while the other directs its marketing at a small and well-defined set of potential buyers through targeted activities (such as newsletters and trade events).

In either case, however, a key element of marketing is understanding potential customers and their needs, so the right product can be promoted at the right time, place, and price. These four broad categories of marketing decisions are often referred to as the "4 **P**s of marketing" since their creation by E. Jerome McCarthy in 1960: namely, **P**roduct, **P**rice, **P**romotion, and **P**lace. Over the years, these categories have evolved and grown to cater for different business models and the ever-growing sophistication of consumers in the information age.

As data on consumer behavior and preferences became widespread, organizations turned to analytical tools to improve the effectiveness of their marketing decisions. A growing number of organizations, however, are expanding beyond traditional analytics and turning to Artificial Intelligence to unlock benefits in a number of marketing areas, including:

- Promotional planning
- Strategic and dynamic pricing
- Content personalization

- Product recommendations
- Predictive customer journeys.

In this chapter, we'll focus on two categories of marketing decisions out of the original four Ps, namely *Price* and *Promotion*. In doing so, we'll present four case studies that follow the same problem-to-decision pyramid used in the previous chapter. We'll begin with a case study on the use of dynamic pricing[1] in business-to-business field sales, before moving into another pricing case-study, but this time for tiered pricing optimization and automated compliance monitoring. We'll then complete the chapter with two case studies focused on the optimization of promotional activities: one from a marketing spend point of view, and the other from the perspective of optimizing the overall promotional calendar.

9.1 Dynamic Pricing for Improved Profitability and Share of Wallet

Few marketing decisions can impact revenue and margin like pricing, and in this section we'll consider the application of Artificial Intelligence to this important area. Because pricing projects can be diverse and varied, organizations can expect to realize different benefits depending on the project scope, with common benefits being:

- Increased "average order value" through larger purchase volumes and additional products that weren't purchased before
- Improved "basket profitability" and "average margin per order" through individualized pricing
- Improved customer "share of wallet" by securing additional product categories that may have previously been purchased from competitors
- Increased market penetration in specific customer segments, driven by strategically priced products.

The typical organization that would benefit the most from AI-driven pricing has a long tail of customers (in the thousands or tens of thousands) and a complex product portfolio. In business-to-business ("B2B") environments, this is often observed in sectors characterized by low-volume, high-frequency transactions, usually involving commoditized products and a highly fragmented customer

---

1   "Dynamic pricing" means the use of technology to individualize product prices based upon a prediction of how much a specific customer is prepared to pay. The goal is to find the optimum price for any given product and customer, leveraging multiple variables such as the price paid by similar customers for the same item.

base.[2] In such circumstances, sales and commercial teams operate within a complex landscape with a high volume of transactions, where it's challenging to track sales performance at the product level and monitor each customer's share of wallet. Hence, it's difficult to estimate the impact of different product price points on sale volumes and their contribution to improving a particular customer's wallet share.

From a timing perspective, pricing projects are usually prompted by a broader business driver or compelling event. For example, these may include cases where a distributor wants to capitalize on recent customer segmentation initiatives, introduce higher-margin private-label products into their portfolio, or react to operational margin pressures.[3]

In this case study we'll focus on OPX Consumables, a wholesaler of non-food consumable items, such as napkins, plates, utensils, and cleaning products used by restaurants, pubs, live music venues, cinemas, accommodation, and tourist and sporting facilities. The products provided by OPX Consumables were either commoditized or highly substitutable, and the customer base was highly fragmented and characterized by ad-hoc procurement and buying. These customers wanted convenience and weren't loyal, which meant they bought day-to-day supplies from whichever vendor showed up at the right time (whenever their stock was running low). In fact, they also gave the lucky rep a significant number of additional product lines, which were added to the order.

To cater for this environment, OPX Consumables' go-to-market strategy featured a range of sales channels. This case-study focuses on the sales channels where staff talks directly to customers in person or over the phone. Between the field and phone-based teams, 60 sales reps looked after a few thousand customers. To drive sales, either from larger orders on frequently purchased items or new purchases from additional product categories, OPX Consumables empowered the sales reps to modify prices when negotiating with customers. Given that

---

2   Fragmentation in a marketplace is observed in industries where a substantial number of businesses operate (usually of small to medium size) and there is no single dominant player.

3   Given that a 1% price increase can deliver a disproportionate increase in margin when compared to a 1% increase in sales or a 1% reduction in procurement costs, it's critical for distributors to optimize their pricing. See the research from McKinsey & Company; https://www.mckinsey.com/business-functions/marketing-and-sales/our-insights/pricing-distributors-most-powerful-value-creation-lever.

OPX Consumables operated in a high-volume, dynamic, and highly competitive sales environment, management believed that the organization could achieve better financial performance through the application of advanced technology. Hence, OPX Consumables defined their business problem and objective as:

Improve overall business profitability and increase customer "share of wallet" metrics through dynamic pricing

Meeting this objective required the introduction of a Decision Optimization System capable of increasing the average order value and profitability of different customer segments through the use of AI-based, dynamic pricing recommendations.

Data, information, & knowledge

Predicting the exact price that a particular customer is prepared to pay for a particular product is a complex endeavour involving significant modeling, near-real time analytics, and predictive components, each serving a different purpose and requiring different datasets.

Regardless of whether an organization is considering embarking upon an AI-driven pricing project or merely trying to make sense of its profit & loss statements, one thing is certain: Transactional data is of critical importance to every organization! That was precisely the case with OPX Consumables, which had plenty of data on which customers purchased what products, where, when, and at what price, including:

- Product master data
- Core pricing data
- Promotional pricing data
- Historical quotations data (at line item level)
- Historical sales transactional data
- Customer data
- Customer complaints data
- Sales rep activity data
- Sales rep profile data.

OPX Consumables used these datasets to generate a plethora of reports and graphs, which focused on the interpretation of transactional data, including reports for:

- Highest selling products, as ranked by contribution margin
- Customer segments that attracted the highest margins (and prices) for top-selling products
- Lowest margin products by category, by region, or customer segment
- Pricing distribution by customer segment.

Although these reports could identify trends, outliers, and patterns, they couldn't provide an understanding of *why* certain price points worked better than others (in terms of propensity to buy)—and developing an understanding of "why" was a prerequisite for generating optimized price points for each customer and product, which in turn required the aggregation, overlay, and analysis of various datasets. OPX Consumables didn't have the necessary in-house expertise to undertake such a data mining exercise, nor did it collect or use external datasets (such as census data, bureau of meteorology weather data, major events data, public holidays calendars) for pricing decisions.

Decision Optimization System (prediction, optimization, & self-learning)

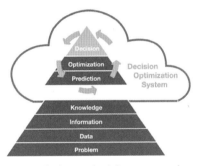

To increase overall business profitability and increase "share of wallet" metrics at the customer level, OPX Consumables wanted to improve the "average order value" and the "average margin per line item per order" in different customer segments and channels. This required a Decision Optimization System capable of:

- Determining the optimal price range for any given product for any given customer (based upon what similar customers were paying for the same product within a specific geography)
- Leveraging this price range to provide sales staff with automated and optimized pricing recommendations in real-time
- Embedding these pricing recommendations (along with automated approval workflows) within the quoting and order taking process.

These capabilities required a variety of algorithmic models to be enabled within the Decision Optimization System. For example, where most organizations rely on traditional approaches to customer segmentation and only have a few, large segments (such as Gold, Silver, and Bronze, or A, B, and C customers), OPX Consumables required a far more granular segmentation that could make meaningful comparisons between similar customers (see Chapter 8.1 for a discussion on micro-segmentation). In the process of configuring the Decision Optimization System for OPX Consumables, AI-driven micro-segmentation was used to allocate customers into microsegments of sufficient granularity to capture the similarities in buying behaviors, alongside other static, structural attributes (e.g. size, geography).

Another core modeling component was price elasticity and cannibalization (across the different products as discussed in Chapter 3.2), to understand how buying behavior for customers in different microsegments was influenced by price changes in specific products. To complicate matters further, the cannibalization effect had to be modeled across key products rather than just considering cannibalization in isolation per product—in other words, modeling how changes in the price of one product would affect substitutable products or other products that complemented the original product. This was a complex data mining exercise that required significant domain knowledge on substitutability across the product portfolio.

As with the Decision Optimization System for DEF Supplies (as discussed in Chapter 8.2), a fuzzy logic model was used for both product substitution and dynamic pricing. As before, the fuzzy rules for product substitution were based on OPX Consumables' product hierarchies and cost and margins— all in the context of each microsegment, so that the same product would trigger different substitute offers for different customers. Optimized prices were then generated by fuzzy rules that included many variables (e.g. competitor pricing, supply and demand, geography) for each customer within each microsegment.

Once these models were trained and deployed within the Decision Optimization System, OPX Consumables could provide sales reps with dynamic, AI-based pricing recommendations. To minimize change management issues (discussed further in Chapter 12.3), these recommendations were embedded in the business-as-usual quoting and order-taking process:

Quote and Order Entry						
Product	Quantity	Unit price	Total price	GP	Margin	
BP72956352G						
160MM SERVING TRAY	22	$1.85	$40.70	$16.72	41%	
$1.85 each	4,501 available					

As shown above, sales reps were presented with optimized pricing recommendations and stock availability for every item added to a customer's order. Of particular importance was the black icon, portraying Larry, the Digital Analyst®, the AI engine that generated the pricing recommendation and associated gross profit ("GP") for each particular item (highlighted in green). By clicking on the Larry icon, sales reps could receive additional pricing information, which was often useful during negotiation:

In addition to providing an optimized price point ($1.85 in the screen above), the Decision Optimization System also provided sales reps with a price range (e.g. $1.31 to $2.00 per unit) that was color-coded to indicate:

- *Red zone*: representing the lowest price that similar customers paid for this particular product
- *Yellow zone*: representing the median price that similar customers paid for this particular product
- *Green zone*: representing the highest price that similar customers paid for this particular product
- This color-coding also matched the color of the gross profit value next to the Larry icon, providing sales reps with a visual representation of where the price sat within the overall range

To create these "dynamic price guides", the Decision Optimization System compared the unit price paid (or requested) by a specific customer for a particular product to what other customers in the same microsegment paid for the same product. The prices paid by similar customers were aggregated and used to create color-coded price guides. The Decision Optimization System would then continuously analyze OPX Consumables' transactional data to update each microsegment and reallocate customers between them, as required, so that the pricing recommendations were as current and relevant as possible.

The sales reps were incentivized to initiate pricing conversations and improve margins whenever possible. To aid with the identification of profit improvement opportunities in each quote, sales reps observed the GP color for each item—especially reds and yellows—as shown below. This facilitated the assessment and selection of specific re-pricing opportunities to be raised with each customer, with the aim of increasing the relevant price into the green zone (or at least out of the red zone):

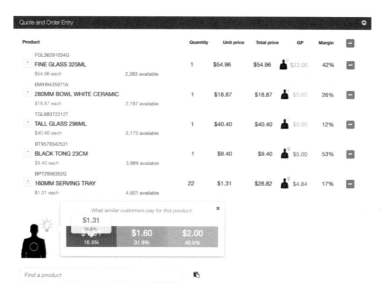

To drive effective pricing conversations, the underlying models also considered volume discounts and each customer's contract terms, as well as the pricing approval process between sales reps and their managers. Sales reps were able to select any price point within the yellow and green range and have it automatically approved, but whenever a price was entered below the yellow range, a pricing approval workflow was triggered that sent an email and sms notification to the sales manager for approval.

And lastly, the green/yellow/red price range for each product was also visible across multiple customer microsegments at once to facilitate comparisons,[4] as shown below:

PRICE REALISED BY SKU, GEOGRAPHY, AND CUSTOMER SEGMENT COMBINATION											
160MM SERVING TRAY	×	NATIONAL	×								
		Min		25%		Median		75%		Max	
Customer Segment	Customer Subsegment	Price	Margin %	Price	Margin %	Price	Margin %	Price	Margin %	Price	Margin %
CLEANERS	HEALTHCARE PRIVATE	$1.17	15%	$1.27	25%	$1.38	35%	$1.52	49%	$1.58	55%
CLEANERS	RESELLER/WHOLESALE/DISTRIBUTOR	$1.13	11%	$1.30	27%	$1.63	40%	$1.48		$1.53	50%
COMMERCIAL	ACCOMMODATION	$1.10	10%	$1.22	22%	$1.36	35%	$1.45	46%	$1.60	60%
COMMERCIAL	AGED CARE	$1.14	12%	$1.22	20%	$1.33	39%	$1.49	40%	$1.64	56%
COMMERCIAL	CATERERS	$1.15	13%	$1.20	18%	$1.01	28%	$1.37	34%	$1.45	42%
COMMERCIAL	CLUBS AND PUBS	$1.25	14%	$1.31	16%	$1.65	31%	$2.00	40%	$6.10	60%
COMMERCIAL	COMMERCIAL	$1.51	16%	$1.69	30%	$1.77	35%	$1.89	40%	$2.00	52%
COMMERCIAL	FUNCTION	$1.36	9%	$1.45	16%	$1.60	20%	$1.63	30%	$1.75	40%
COMMERCIAL	RESTAURANT	$1.32	15%	$1.38	20%	$1.44	29%	$1.61	40%	$1.78	50%
COMMERCIAL	RETAILERS	$1.10	10%	$1.30	30%	$1.33	55%	$1.67	67%	$1.77	77%
GOVERNMENT	FORCES	$1.65	10%	$1.66	20%	$1.96	30%	$2.18	40%	$2.33	55%
GOVERNMENT	HEALTHCARE PUBLIC	$1.04	15%	$1.13	30%	$1.26	40%	$1.36	50%	$1.40	55%

---

4   The algorithmic models within the Decision Optimization System created thousands of microsegments and only an aggregated subset of those is visible in the comparison tables (for practical viewing reasons).

In the table above, the lowest price paid by customers in the "commercial" segment (within a specific geography) is higher than the highest price paid for the same product within the "caterer" segment. Similarly, the median price paid for this product by military customers (classified as "Forces") is higher than the highest price paid by customers in most other segments.

By embedding AI-driven pricing within the quoting process, OPX Consumables was able to lift the overall profitability in underperforming customer segments and product lines, as well as make in-roads in non-core product categories. This, in turn, lifted the average revenue per customer and better positioned the organization against the risk of customer churn. The Decision Optimization System also attracted positive feedback from sales reps, who reported improved confidence in having conversations with customers about price, particularly for underpriced products that carried low margins.

9.2 Tiered-Pricing Optimization and Automated Compliance Monitoring for Improved Margin

In this case study we'll discuss SPARK Electrical, a manufacturer of electrical products servicing the trade, construction, manufacturing, and hardware retailing industries. The customers within any given sales territory were quite diverse in terms of their line of work and size, which invariably determined the type and volume of products they purchased. To service such a wide set of customers, the organization employed more than 100 account managers and provided an extensive product range spread across the security, cabling, lighting, power, and peripherals & accessories categories.

The majority of these products were commoditized, with SPARK Electrical using various pricing mechanisms to maximize sales for each customer, including smaller customers in the long tail. More specifically, account managers utilized a "variable-discount-over-list" pricing model, where each product was assigned a specific base price (i.e. "list price") that was subsequently discounted for each customer. The depth of discount was dependent on the customer's pricing tier: The greater the customer's potential value, the greater the discount.

SPARK Electrical used seventeen different pricing tiers, with each tier unlocking an increasingly deeper discount across the entire product range. Each tier also included an allowance for a number of products to be sold at an even greater discount for a limited period of time. This allowance was referred to as "fixed-term contract pricing" and again, the greater the potential value of a customer, the more products would be brought into fixed-term contract pricing. It's also worth noting that the higher the pricing tier, the greater the depth of discount for products on fixed-term contract pricing.

Over time, a disconnect developed between what management thought customers were willing to pay for certain products and the prices at which those products were offered by account managers. The large number of customers and vast product range, coupled with inadequate controls to enforce appropriate pricing, led to a variety of problems, such as customers inappropriately allocated to higher discount pricing tiers on the promise of higher volumes that never eventuated. Another problem was the excessive number of products on fixed-term contract pricing, which locked in further discounts that were renewed when the fixed term expired. Many of these discounts were continuously renewed and became perpetual in nature, causing SPARK Electrical to define their business problem and objective as:

Improve margin by optimizing the allocation of customers to pricing tiers and monitoring their compliance to volume thresholds

A manual approach for achieving this business objective had already been attempted without success, prompting SPARK Electrical to turn to technology—particularly technology that could assist account managers with decisions regarding tier and contract pricing, and management with automated monitoring of these ongoing customer discounts.

Data, information, & knowledge

Data availability within SPARK Electrical was abundant and diverse, as account managers had been running detailed customer surveys for many years. This meant that the organization had data on each customer's characteristics, line of work, preferences, and the frequency and duration of their projects (for which they needed SPARK Electrical products).

Historical data was also available on the different pricing tiers for individual customers, their compliance with the minimum volume thresholds over time, and the specific products discounted through fixed-term contract pricing. Beyond this, SPARK Electrical also had access to typical transactional, sales activity, and customer-level data, such as:

- Product master data
- Customer information

- Customer complaints data
- Account manager activity data
- Account manager profile data

Since customers spanned the trade, commercial, and industrial sectors, a significant amount of external data was readily available, including building permit approvals, new land releases for residential use, and large redevelopment projects (where previously commercial areas were rezoned for new residential suburbs to be built). On the commercial front, external data included new commercial projects by type, magnitude, and timeframe, whereas on the industrial side, SPARK Electrical subscribed to specialized reports that provided insights and analysis on trends, forecasts, and growth rates for various industry sub-sectors.

Despite having access to high-quality data, the reporting was of limited benefit and focused on the wrong metrics. That is, instead of generating reports to monitor each customer's compliance in meeting the volume thresholds for their pricing tier, the reports were more geared towards disseminating data on the sales performance of each territory. To make matters worse, pricing tier decisions were based on each customer's perceived future value, enabling account managers to allocate customers to more appealing tiers without any compliance reporting to monitor actual volumes.

This skewed approach to reporting—which focused too much on a customer's revenue contribution rather than their compliance to a pricing tier—created a barrier to addressing the margin leakage[5] problem. This meant that the more a customer contributed to sales targets, the less attention was paid to their margin contribution, which paved the way for unnecessarily high discounts across key product lines. A solution to this problem required more than just reporting changes, however, as a technological capability was needed to predict the revenue potential of each customer so they could be initially allocated to the correct pricing tier from the start. Furthermore, each customer's purchasing volume needed to be monitored to detect early signs of not meeting the required sales volumes (rather than just reporting on it after the fact). Simply put, SPARK Electrical faced a significant capability gap between the organization's current technological state and desired future state.

---

5   Margin leakage occurs when an organization increases sales volume and revenue, but makes less margin. There are many possible causes of margin leakage, with the focus of this case study being on the impact of discounts to drive sales—particularly for customers that would have purchased similar volumes regardless of any further discounting.

Decision Optimization System (prediction, optimization, & self-learning)

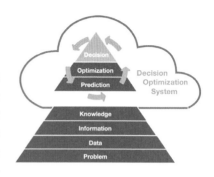

SPARK Electrical knew that its sales culture overemphasized customer revenue contributions to sales targets at the expense of margin contribution to the organization, which was a key driver of margin leakage. Under these circumstances, SPARK Electrical decided to implement a Decision Optimization System capable of:

- Predicting each customer's revenue potential
- Recommending the most appropriate pricing tier based upon that potential
- Continually monitoring each customer's actual performance against volume thresholds for the relevant pricing tier
- Recommending pricing tier downgrades or upgrades where necessary
- Dynamically recommending a discounted price for each product on fixed-term contract pricing

To enable this functionality, a variety of underlying prediction models were trained and deployed using different datasets. For example, a random forest model was used to predict share of wallet and revenue potential using key customer variables like monthly purchases, number of different products purchased each month, average monthly spend during the past 36 months, geography, and so on. This random forest model was used to determine each customer's wallet share given different compositions of the selected variables, with each tree voting and the overall votes aggregated. The training of this model was based on "ground truth data"—meaning customer data where SPARK Electrical knew with a high degree of confidence the true wallet share value (as discussed in Chapter 5.2). Because the Decision Optimization System needed to establish each customer's revenue potential and willingness to purchase different products at different price points, this required both wallet share prediction and micro-segmentation models (as discussed in Chapter 8.1). This also meant that the system not only predicted the propensity of each customer to achieve their minimum required volume, but also each customer's revenue potential and the most appropriate pricing tier.

Another modeling component of the Decision Optimization System was price elasticity and cannibalization (as discussed in Chapter 3.2) across the customer microsegments. By continuously analyzing transactional data, the Decision Optimization System could update the microsegments and real-locate customers between them, thereby changing each customer's predicted sensitivity to price changes. This also enabled a dynamic view of each customer's revenue potential throughout the year, which (when combined with year-to-date, YTD, purchasing volumes), triggered various AI-based prompts (such as pricing tiers to be reviewed for specific customers, or recommendations on the depth of discount for fixed-term contract pricing).

Once the system was configured and deployed, account managers could access a series of recommendations and insights, such as a list of customers that require attention:

In the above screen, the different colors represent varying levels of misalignment between a customer's year-to-date volume and the minimum required volume associated with that pricing tier. Also of critical importance was the "Potential" column (which predicted each customer's revenue potential) and the "Recommended" column (which recommended the correct pricing tier). This combination of insights and recommendations lifted the sophistication of the pricing tier allocation process, and forced account managers to discuss the need for some customers to either increase their volume (to match their pricing tier), or decrease their pricing tier (to match their current volume). These recommended pricing tier changes were held in memory by the system until executed by the account manager, or escalated to management due to inaction.

Once a customer conversation was actioned by an account manager, the change in pricing tier could be executed from the same screen by clicking the "Apply" button. This action presented account managers with a prompt to enter the customer's sales forecast, which was used as an input variable by the Decision Optimization System:

Customer Information			
Business:	**Light & Power Trade**	Current Tier:	**Tier 5**
Code:	**2749105**	Apply Date:	**24 Nov 2020**

Sales Rep Information			
Name:	**Bob Citizen**	State:	**VIC**
Code:	**BCITIZEN**	Mobile:	**0400 000 000**

Change Request			
Target Tier:	Tier 15 ⌄	Sales Forecast:	$
Reason:			

[Apply]   [Cancel]

By automatically analyzing each customer's performance against the required volume thresholds, the Decision Optimization System identified margin leakage and provided relevant recommendations to individual account managers. However, given the sales culture of overemphasizing customer revenue to the detriment of margin, a change management initiative was carried out concurrently to the system deployment. To support this initiative, the Decision Optimization System provided a range of dashboards and reports that visualized whenever a customer's performance might not warrant the depth of discounting provided. For example, SPARK Electrical was able to see a customer's YTD volume, against the pro-rated YTD minimum threshold for their pricing tier, as shown below:

PRICE TIER COMPLIANCE

Spend YTD	Min. $ Threshold YTD	Sales Miss-Match YTD
$1.761.867	$987.521	$774,346

	Jul	Aug	Sep	Oct	Nov	Dec
■ Sales	$60,000.	$88,000.	$140,000	$155,000	$6,800.0	$55,000.
■ Target	$100,000	$108,000	$113,000	$104,000	$78,500.	$150,200

Another critical feature for SPARK Electrical was the system's ability to raise timebound alerts for account managers, indicating that certain customers required urgent action in terms of pricing tier readjustments. These were configured to be triggered when a customers' YTD volume lagged behind the minimum threshold for some period of time (depending on the microsegment). These alerts were also supported by functionality that showed a color-coded list of customers, as shown below:

Tier Change Alert								
Tier-Miss-Match Level	Business	Sales Miss-Match	Current Tier	Should Be	Optimal	Last Change		
High	Electric-ALL	$504,201	Tier 16	Tier 8	Tier 10	17-Mar-20	⌄	⋯
High	Power Up	$306,107	Tier 15	Tier 8	Tier 10	17-Mar-20	⌄	⋯
Medium	Hardware & Supplies	$308,876	Tier 9	Tier 4	Tier 8	19-Nov-19	⌄	⋯
Medium	Retail Power	$222,783	Tier 8	Tier 4	Tier 6	27-Sep-19	⌄	⋯
Medium	Cables and More	$274,499	Tier 9	Tier 6	Tier 8	23-Apr-20	⌄	⋯
Medium	Safety Systems for U	$125,306	Tier 10	Tier 7	Tier 9	19-Feb-19	⌄	⋯
Low	Light & Power Trade	$339,703	Tier 5	Tier 7	Tier 9	23-May-20	⌄	⋯
Low	Security & Electric Specialists	$268,214	Tier 4	Tier 5	Tier 7	05-Aug-19	⌄	⋯

The difference in this screen (versus the previous color-coded list of customers) was that only customers associated with alerts were displayed, along with a different set of insights and recommendations that assisted account managers in diagnosing the situation. This screen used the same colors to indicate the level of misalignment between the pricing tier minimum threshold and actual volume, and also provided two types of recommendations for each customer:

- "Should Be" column recommended a pricing tier for each customer based upon their actual volume during the preceding 12 months.
- "Optimal" column recommended the most beneficial pricing tier for each customer based upon their predicted revenue potential. This tier represented a depth of discount that would entice each customer to increase their volume with SPARK Electrical by potentially bringing over purchases made with other suppliers.

The Decision Optimization System provided these recommendations so that account managers could operate at both a tactical and strategic level. In this approach, "Should be" tiers represented the immediate and straightforward action an account manager should take, whereas "Optimal" tiers represented a possible future state, which required a longer-term, relationship-building effort by the account manager. Although the latter scenario would unlock deeper customer discounts, it required additional volume (usually at the expense of a competing supplier), and that would take time to accomplish. This often resulted in the negotiation and agreement of a pricing tier that sat somewhere in between these two recommendations. Rather than making any sudden change to a pricing tier, these two recommendations encouraged account managers to work towards a gradual repositioning of each customer's tier, and in doing so, gradually unlock a deeper discount level that incentivized further volume (while mitigating the risk of margin leakage in the process).

As mentioned earlier, the pricing model used by SPARK Electrical also relied on fixed-term contract pricing, which meant that a limited number of products could be discounted further (beyond the across-the-board discount provided by the pricing tier). This component of the pricing model represented another source of margin leakage, which the Decision Optimization System aimed to address. Because the reporting undertaken by SPARK Electrical focused on macro-level performance (sales territories, account managers, and customers) and lacked granularity to pick up on product-level issues, it led to a culture where account managers felt comfortable authorizing an excessive number of products to be placed on fixed-term contract pricing, often with high discounts that were regularly extended without any thought or analysis.

To address margin leakage from fixed-term contract pricing, the Decision Optimization System would make recommendations, monitor compliance, then trigger "corrective" actions as required. In the first instance, to help account managers select the appropriate discount for products on fixed-term contract pricing, the system presented account managers with a range of insights and recommendations, including historical price points (as shown below), sales volumes at different price points, as well as current and recommended prices that were inclusive of additional discounts:

| | Sku Code: | ELC73103947Z | |
| | Sku Description: | LED Floodlight BLACK INOX | |

	Current Price: $61.50		Recommended: $61.20
Order Date		**Price**	**Sales**
15/12/2019		$61.50	$13,240
12/12/2019		$61.50	$6,640
03/12/2019		$61.50	$18,240
01/12/2019		$61.50	$22,280
15/10/2019		$61.50	$1,240
20/09/2019		$61.50	$3,240
01/09/2019		$61.50	$4,660
19/07/2019		$63.58	$1,150
02/07/2019		$63.58	$1,020
23/06/2019		$63.58	$840
15/05/2019		$63.58	$710
03/05/2019		$63.58	$560
02/04/2019		$63.58	$542
15/01/2019		$64.80	$334

The inner workings of these price recommendations relied on:

- Dynamic micro-segmentation, which continuously monitored each customer's wallet share potential and updated the associated micro-segment accordingly
- Elasticity and cross-elasticity based on purchase patterns shared by similar customers in the same microsegment
- Business rules that moderated the AI-driven recommendations, and which considered the discount limits for each product and pricing tier

Once fixed-term contract pricing was enabled for any given product, the Decision Optimization System ran automated compliance analytics in the background, monitoring month-to-month volume for each product against the minimum thresholds. To help account managers proactively manage compliance, the system provided a range of insights and recommendations, such as:

PRICE COMPLIANCE

Pricing Tier			Fixed-term contract pricing		
Spend YTD	Min. $ Threshold YTD	Sales Miss-Match YTD	SKU entitlement cap	Contracted SKUs	Non-compliant contracts
$489,500	$798,376	$308,876	30	18	7

CONTRACT PRICING   Sort by: [   ]

Sku Code	Sku Description	Sales Quarter-to-date	Contract Price	Sales Miss-Match Quarter-to-date	Unit	Unit	Valid From	Valid To
TSC52755611	Twin Solar Cable, 4mm 100M Length	$8,000	$160	$7,200	Each	1	01/04/20	30/03/21
LDF88431927	LED Floodlight BLACK INOX	$738	$61.5	$664	Each	1	01/07/20	28/02/21
LTT729688331	Lighting Transformer 220V 240V	$1,250	$25	$1,125	Each	1	01/10/20	30/03/21
WRB74532445	PVC Insulated Building	$7,400	$185	$6,660	Each	1	01/08/20	28/02/21
HWE9336291	2400W Copper Hot Water Element	$1,500	$25	$1,350	Each	1	01/06/20	30/03/21
CBD54271132	PVC Slotted Cable Duct 2M Length	$2,400	$20	$2,160	Each	1	01/09/20	28/02/21
FCC43582719	Fire Cable Control 100M Length	$5,250	$175	$4,725	Each	1	01/10/20	30/03/21

1 2 3 4 5 6 7 8 9 10

In the above screen, the account manager could diagnose the performance of fixed-term contract pricing for a specific customer. This report included the number of products for which the customer was entitled, the number of actual products on fixed-term contract pricing, and the number of non-compliant contracts. Beyond that, the Decision Optimization System allowed for drill-down analysis at the product-level, reporting on the expiry date for each contracted product, the price (inclusive of additional discounts) and current volume against the minimum threshold.

After implementing the Decision Optimization System, account managers reported that it was easier to identify margin leakage in their territory and to have the right conversation with the right customers (at the right time).

The system not only drove better pricing decisions in the first place, but also helped account managers mass-monitor compliance across the long tail of customers. Besides generating an overall improvement in margin, the Decision Optimization System also balanced the sales teams' focus on customer revenue contribution to sales targets, against margin contribution to the organization. In commercial terms, the system achieved payback during its first year of use, solely from the AI-based recommendations executed by account managers.

9.3 Marketing Spend Optimization for Improved Return on Investment

We'll start this section by revisiting the John Wanamaker[6] quote from Chapter 3: *"Half the money I spend on advertising is wasted; the trouble is I don't know which half."* John's statement, which dates back to the nineteenth century, has survived the years and contributed to the evolution of marketing into a discipline, as it neatly describes one of the greatest challenges faced by marketing departments: Simply put, measuring and maximizing the return on investment (ROI) from marketing dollars isn't easy! Marketers have long sought to improve the return from marketing initiatives, as they usually represent a significant portion of an organization's operating budget—most notably among organizations selling directly to consumers.

Not all "marketing dollars" are equal however, and the typical organization that sells to consumers might undertake a variety of different marketing activities throughout the year. These activities could range from promotional events and sponsorships through to search engine optimization and advertising. Given the varied nature of these initiatives, the process of assessing their impact and calculating their respective return isn't uniform. Some initiatives have a straightforward approach for calculating return on investment, with click-through ads being a prime example (where payment is made on the number of clicks a particular advertisement receives), while other initiatives are notoriously difficult to measure (such as sponsorships).

In this case study we'll focus on FUN Splash, a manufacturer and reseller of spa and pool products, accessories, and equipment, which has a national footprint and a go-to-market model based on generating leads through advertising. FUN Splash promoted their products to consumers via advertisements placed on three channels: Facebook, television, and different websites on the internet. To ensure sufficient coverage throughout the year, FUN Splash maintains a well-funded advertising calendar that represented almost 20% of the

---

6   John Wanamaker, a nineteenth century merchant and politician, is considered by some as a "pioneer of marketing".

organization's sales revenues. This advertising budget varied significantly from a channel allocation point of view, in part due to the fact that the cost of TV ads greatly overshadowed the cost of the other channels, and also because Facebook was the top performer for certain types of advertisements during certain times of the year, and thus received a much greater allocation of the marketing budget. Under such circumstances, the budget allocation per channel was inconsistent, return on investment varied greatly throughout the year and across the states, and the business was constantly monitoring performance and trying to improve on where and when to advertise.

In regards to the advertisements themselves, the type and product focus also changed throughout the year and across the states, but the call-to-action remained the same: prospective consumers were to call or email for more information. Each inquiry was processed as a unique lead, with the first step being for a sales agent (based at the organization's head office) to have a qualifying conversation with the prospect. Following this initial interaction, the qualified sales leads—where prospective consumers demonstrated serious interest in FUN Splash products—were sent to the relevant regional office for a subsequent sales conversation, possibly a face-to-face meeting, and eventually, a quote. During this sales process, call centre staff in head office were responsible for tagging each inquiry against the specific advertising campaign and channel that generated the lead. And every month, FUN Splash ran numerous campaigns across the country and (despite the large volume) the call center staff were quite effective at tagging all leads. This made it possible for the organization to attribute leads to their source and measure the success of each campaign (and overall return from each channel).

Overall, FUN Splash ran an effective sales process and generated a sufficient number of leads to meet sales targets. And yet, the advertising model was expensive and hindered FUN Splash from realizing margin and profit objectives. Additionally, management acknowledged the organization struggled to generate balanced returns from advertising initiatives, which meant that:

- Some regional offices received a significantly higher volume of qualified leads than others, creating a situation where some sales teams struggled to process all leads, while others lacked enough leads to meet targets.
- The average cost per lead ("CPL") for some states was significantly higher than others, which meant that return on investment for many marketing initiatives didn't stack up at the state level.
- When certain campaigns performed better than expected and FUN Splash started to experience bottlenecks, the organization couldn't flex its sales capacity. This in turn delayed conversions and face-to-face

meetings, and led to situations where the average cost per lead was appropriate, but profitability suffered.

- The organization wasn't able to effectively determine the point of diminishing returns for different channels, thereby hindering the redistribution of marketing dollars to other initiatives.

Given this situation, FUN Splash defined their business problem and objective as:

Optimize marketing spend across multiple channels and time periods to generate the greatest possible number of leads to regional offices at the lowest possible cost per lead nationally

The inherent complexity and large data volume of this business problem made it ideal for the application of a Decision Optimization System based on Artificial Intelligence methods.

Data, information, & knowledge

To fully appreciate the complexity of the business problem and objective, it's worth noting that just executing on the statement "optimize marketing spend across multiple channels and time periods" involved answering a variety of complex questions, such as:

- What is the optimal timing and spend for each advertising campaign?
- What is the point of diminishing returns for various channels?
- Which campaigns are effective, when, where, and within what channels?
- How should the budget be distributed geographically to optimize for different objectives (number of leads at the regional level and average cost per lead nationally)?

To assist in answering these questions, FUN Splash had collected a substantial amount of internal data, including:

- Historical campaigns by channel and product
- Historical advertising spend by channel
- Historical lead enquires by channel, product, and region
- Historical sales transactions
- Customer information including demographic attributes

FUN Splash had also established that external factors impacted on the success of any given campaign for any given channel, so a lot of effort was placed on collecting relevant data to understand:

- The impact of seasonal weather patterns and their variances throughout the country
- The types of TV advertisements that were most relevant to target consumers
- The time of the day that various advertisements were most impactful

Although FUN Splash amassed a significant amount of internal and external data, the value extracted was disproportionately small compared to the cost and effort of collecting this data. This was noticeable in the limited insights available, which included dashboards and reports such as the one shown below:

Fiscal Week	Facebook	Internet	TV	Total Expenses	Total Leads
W3-Jul/2019	$1,000.00	$1,100.00	$600.00	$2,700.00	6
W4-Jul/2019	$1,000.00	$1,065.16	$600.00	$2,665.16	6
W1-Aug/2019	$800.00	$1,000.00	$600.00	$2,400.00	6
W2-Aug/2019	$800.00	$1,000.00	$600.00	$2,400.00	13
W3-Aug/2019	$800.00	$1,000.00	$600.00	$2,400.00	16
W4-Aug/2019	$800.00	$1,000.00	$600.00	$2,400.00	15
W1-Sep/2019	$800.00	$1,000.00	$600.00	$2,400.00	16
W2-Sep/2019	$500.00	$800.00	$0.00	$1,300.00	15
W3-Sep/2019	$1,000.00	$1,200.00	$600.00	$2,800.00	15
W4-Sep/2019	$1,000.00	$1,200.00	$1,500.00	$3,700.00	15
W5-Sep/2019	$1,000.00	$1,200.00	$1,500.00	$3,700.00	16

The above lead generation report (which focused on expenditure versus the number of leads generated) was effective for general KPI reporting and end-of-month accounting, but otherwise was of limited value because it failed to communicate the reasons why certain campaigns were more successful than others—in other words, this report didn't explain the impact of campaign attributes (such as product, day of the week, time of day, duration of campaign) on the number of leads generated during any given week. Hence, FUN Splash lacked an in-house reporting and analytics capability to identify opportunities to optimize marketing spend that could pave the way for a more profitable business model. Additionally, the lack of more insightful analysis prompted concerns that too much corporate knowledge might exist in the heads of a few staff members, which could put the organization in a difficult situation if they left.

FUN Splash knew it could improve its understanding of what made certain marketing campaigns successful in different states, at different times of the year, and use this knowledge to make better decisions for future campaigns.

Decision Optimization System (prediction, optimization, & self-learning)

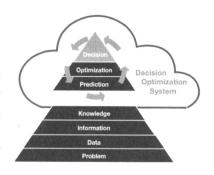

To optimize marketing spend and generate the greatest possible number of leads for each regional office (at the lowest possible cost per lead nationally) required FUN Splash to bridge a significant technology capability gap within the organization. Under these circumstances, a business case was created and endorsed by management to implement a Decision Optimization System capable of:

- Modeling the relationship between various attributes that contributed to the success of different campaigns, including the point of diminishing returns for each channel.
- Predicting the total number of leads, cost per lead, and sales for any given marketing campaign or scenario
- Predicting the total budget required to generate a set number of leads or sales volume.
- Creating and exploring a large number of what-if scenarios, while considering the interplay between various campaign objectives.

This functionality was enabled by a variety of algorithmic models, with the prediction model being the core component of the entire Decision Optimization System. Ultimately, the system's success would hinge on the accuracy of these predictions. Using this prediction model, FUN Splash was able to create many what-if scenarios, compare them, and choose the "best" one (similar to the approach discussed in Chapters 8.4 and 8.5)—with "best" meaning the scenario that generated the greatest possible number of leads to regional offices at the lowest possible cost per lead nationally.

To accurately predict the number of qualified sales leads[7] for a particular marketing campaign, a stacked ensemble model was used (Chapter 5.6 provides more detail on this modeling approach). One of the models within the ensemble generated a baseline forecast (based upon all variables related to the proposed marketing campaign), whereas the other models generated a variety of adjustments (related to demography, correlations among various marketing channels, seasonality, time of the year, and so on). The final model (which combined all constituent models) was trained using both historical

---

7   All inbound leads were qualified and only a portion of them converted into "qualified sales leads," which were then passed down to the regional offices. Hence, the Decision Optimization System needed to predict "qualified sales leads" as opposed to "any" leads.

internal data that FUN Splash had collected over the years, as well as relevant external data.

Once the models were trained and the Decision Optimization System deployed, FUN Splash began testing various what-if scenarios. This involved exploring the interplay of various campaign configurations and objectives in an iterative manner. Each configuration represented a separate scenario in the system, as shown below:

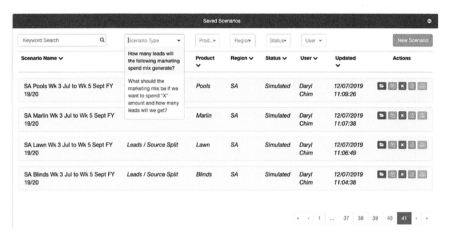

The core components of each scenario included:

- *Problem variables*: such as the product, day of the week, time of day, duration of campaign, marketing mix, full versus partial campaign, and more.
- *Business rules and constraints that defined a feasible solution*: included, among others, maximum cost-per-lead, minimum leads per state, maximum campaign duration, maximum budget per channel per campaign.
- *Optimization objectives*: were scenario-specific, but included maximizing leads for a set budget, minimizing the cost-per-lead for a target number of leads, and maximizing leads while minimizing the cost-per-lead.

To facilitate the process of creating a scenario and selecting the configuration of problem variables, constraints, and objectives, FUN Splash used a "Parameters" screen to input key values, as shown below:

Once the required scenarios were created, the Decision Optimization System then predicted the outcome for different campaigns and budgets. This functionality allowed for comparison between campaign configurations for the same time period, as well as comparison with past performance in preceding periods:

The Decision Optimization System also provided FUN Splash with a campaign prediction, with the performance broken down into key metrics such as expense per channel, total expense, number of qualified sales leads, and cost per lead:

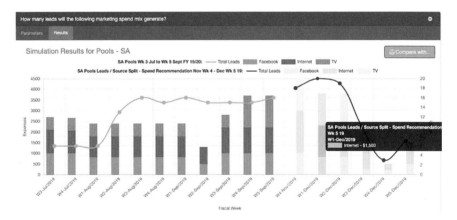

After creating a shortlist of favorable scenarios, FUN Splash then applied optimization to further improve each individual scenario. This was done by selecting (apart from the initial scenario) a primary optimization objective at the national level, such as maximizing qualified sales leads for a set profit target or minimizing cost-per-leads for a target number of qualified sales leads per state.

An optimized marketing plan was generated as an output, which presented FUN Splash with the advertisement recommendations by product, state, and channel. To evaluate the benefits of the recommended plan, FUN Splash accessed an interactive dashboard to drill-down on the predicted performance and explore the impact against multiple KPIs, geographic areas, and timeframes, as shown below:

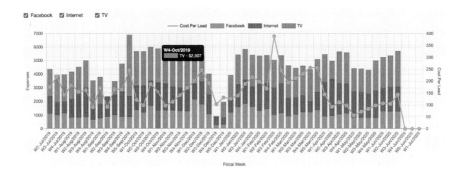

Following the Decision Optimization System's deployment, FUN Splash began experimenting with new campaigns, while maintaining a regular set of "tried and trusted" campaigns. This was part of the change management process to gradually build trust among users, as well as minimize risks. This plan also allowed for an iterative approach over six months, which meant FUN Splash could test and fine-tune state-level plans prior to optimizing at the national level. As an outcome, regional offices that historically lagged behind in qualified sales leads received an increase in volume, which in turn increased their sales performance. FUN Splash observed that at times the cost-per-lead metric increased for some states, but the system compensated through improved performance in other states.

Within two years of deployment, the overall marketing spend had altered significantly. As an outcome, a better return on investment was achieved nationally, and the improved effectiveness allowed for a reduction expenditure from almost 20% of sales revenues to 14.3%. As a direct outcome of the project, the cost-per-lead reduced nationally, regional offices met sales targets, and overall profitability improved.

9.4 Promotional Planning and Trade Spend Optimization

The previous three case studies explored the application of Artificial Intelligence to pricing decisions and marketing spend, whereas this section builds upon those case studies by leveraging both areas to drive revenue and margin growth. We'll also return to the complex problem of promotional planning from Chapter 3, and as such, recommend a review of the "Basic Terminology" section from that chapter as a refresher on core concepts and terminology.

Let's start with the observation that promotional planning and pricing can be considered from two different perspectives: that of a manufacturer wanting to achieve a set of business outcomes through promotional activities, and that of a retailer with their own business objectives. Some of these objectives may align (e.g. growing the overall category) while others may not (e.g. deep discounting to drive foot traffic), and yet both types of activities might be funded primarily by the manufacturer.

For the manufacturer to grow sales within a product category, it can either deploy strategies to increase the overall consumption from that category (while maintaining their market share) or "steal" share from competitors. In either case, a common tactic is to promote products through bundled deals, discounting, and/or high-visibility displays within stores. These promotional activities are primarily funded by the manufacturer, thereby reducing their margin on the units sold. The retailer may also invest in the promotion (reducing their margin as well), in exchange for increased volume and category growth. Such situations create a dynamic of *creative tension*,[8] as the additional revenue gained by the retailer might come at the expense of the manufacturer (and yet both their sales volume might grow on the back of overall category growth).

From the retailer's perspective, the flow-on benefit of running in-store promotions includes increased foot traffic, possible purchases of other products alongside the promoted products, and/or additional purchases made by regular customers (who might buy larger quantities of products, or products they didn't intend on buying—or both). Within this mix of strategies and tactics, the retailer is striving to stretch the manufacturer's promotional dollar in ways that might not be in the best interests of the manufacturer. As an example, if a promotion fails to deliver the incremental sales volume planned by the manufacturer, but enables the retailer to market a well-known consumer brand for a low price, then this promotion might be ineffective for the manufacturer, but effective for the retailer (because of additional foot traffic and/or larger shopping baskets).

From this perspective, the manufacturer has more at stake in the process, and as such, carries a larger burden to ensure their promotions are as effective as possible. This requires the manufacturer to analyze and understand the variables that contribute to a successful promotion, and then negotiate with the retailer so that more promotions feature these variables. The retailer often makes this process more difficult by running many promotions with many manufacturers—either simultaneously or in close proximity—thereby reducing the impact of each promotion. Furthermore, the retailer may impose restrictions and constraints on the manufacturer (in regards to timing and structure of promotions), so that the retailer can maximize the number of promotions across all manufacturers and all categories.

---

8   Creative tension refers to a situation where opposing or conflicting needs have the dual potential of creating opportunity or hindering growth, depending on how they're managed.

In order to maximize the return from promotional activities, manufacturers and retailers have processes in place for allocating promotional dollars to promoted products. In Chapter 3 we discussed these processes from the perspective of a manufacturer, and in this section, we'll explore them from a retailer's perspective. In doing so, let's first clarify some common terminology, as the process of planning promotions, setting prices, and reconciling promotional dollars is usually referred to as *promotional planning* by retailers, while manufacturers refer to these activities as *trade promotion management.*

Furthermore, the process of finding the best promotional plan with respect to predefined objectives and constraints is usually referred to as *promotion optimization* by retailers and *trade promotion optimization* by manufacturers. The term *trade spend optimization* is used interchangeably, as it refers to the budget line item allocated to trade promotions. This process has many commonalities regardless of the perspective, with the most significant being the *slotting board,* which is a visual representation of the promotional calendar where each column corresponds to a week and each row corresponds to a particular product, allowing for promotions to be slotted into each column/row combination (as discussed in Chapter 3).

To understand these similarities and differences (depending on the perspective), let's return to the concept of creative tension between a manufacturer and retailer. Note that from the retailer's perspective, there are creative tensions with multiple competing manufacturers simultaneously:

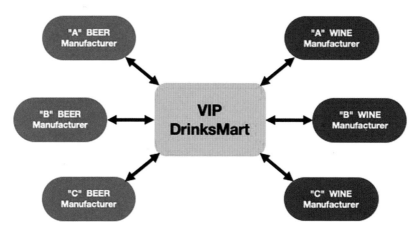

This "multi-faceted" creative tension—between the retailer and multiple competing manufacturers—provides the retailer with a unique opportunity to organize and prioritize promotions according to their preferred pattern and benefit. Even manufacturers with more negotiating power due to ownership of well-known brands may find that their promotional requests clash with other competing products during the same dates, or the retailer imposes constraints that make their requests infeasible.

This planning process requires careful consideration of numerous variables and constraints, with the output being a subset of promotions from the overall mix requested by each manufacturer. Sophisticated manufacturers that leverage Artificial Intelligence capabilities for optimizing promotional activities are better able to meet internal objectives, grow the overall category (driving increased consumption across new consumer segments and/or additional consumption occasions), and satisfy retailer objectives (e.g. margin and growth). These "win-win-win" promotional plans are rare, and retailers are inclined to execute them regardless of whether they are leveraging AI for their side of the workflow.

Sophisticated retailers, on the other hand, are more likely to create "win-win" promotional plans that optimize for their own objectives and deliver category growth. In such scenarios, while the category grows and manufacturers as a whole benefit from the added consumption, some manufacturers are bound to benefit more at the expense of others, depending on the strength of their brands and sophistication of their planning process. In either case, organizations with more sophisticated tools for promotional planning have the opportunity to:

- Reduce time and labor required for creating promotional plans
- Improve promotional and trade spend effectiveness
- Optimize plans to drive category growth and increased market share
- Create a stronger manufacturer-retailer relationship

In this case study we'll build upon the concepts introduced in Chapter 3, but from the perspective of VIP DrinksMart, a retailer with more than 2,000 physical stores (some company-owned, and some independently-owned operating under a common brand banner) and several e-commerce sites. Of particular relevance to this case study are decisions related to which products VIP DrinksMart should sell (from a range of more than 20,000), where (specific product ranging for each store type and geography), at what price (base price). These decisions are handled by a merchandising team of approximately 40 people, who are also responsible for promotional activities to drive various business outcomes.

The challenge this team faced on a day-to-day basis was the creation of promotions that could drive sustained profitable growth, increased foot traffic, and improved return on trade spend dollars. To deliver such promotions, the merchandising team relied on internal investment along with co-funding from a range of manufacturers. VIP DrinksMart wanted to realize a return on their internal investment, as well as minimize any negative flow-on effect on inventory levels, as some promotional decisions might lead to either too much stock (which could negatively impact shelf availability and working capital) or not enough stock (which could hinder the success of a promotion, with store shelves running empty and consumers leaving disappointed).

Despite the strategic importance of promotional planning and pricing, VIP DrinksMart relied on a maze of interdependent spreadsheets, most of which were offline and standalone. This highly inefficient process required staff to spend a significant amount of time on transactional aspects of the workflow, such as:

- Collating manufacturer requests for promoted products
- Slotting all requests on a promotional slotting board
- Analysing the performance of similar past promotions
- Setting the variable values for each promotion, such as promotion type, duration, and support
- Deciding on the depth of discount and constructing the pricing inputs and subsidies
- Repeating the above sequence of steps across each category

Besides being error-prone, this labor-intensive process consumed time that could have been used for higher-value tasks (such as analysing why certain promotions were more successful than others, and then using this knowledge to create more effective promotions in collaboration with manufacturers). This situation led to VIP DrinksMart replicating last year's promotional plan, making a few tweaks or changes, then executing this plan without identifying any significant opportunities for improvement.

Over time, VIP DrinksMart began to fall behind other retailers in the industry, especially those with more sophisticated tools and processes for promotional planning and pricing. This led to fears that competing retailers would command a larger share of each manufacturer's trade spend budget, leaving VIP DrinksMart with reduced investment in their own promotions, reduced foot traffic, and reduced growth. Although VIP DrinksMart had a talented merchandising team, their performance was hamstrung by dated technology and inefficient processes. Every time management requested a promotional plan to drive specific business outcomes, the process of generating such a plan took many days. VIP DrinksMart knew that the right technology and tools

could enable a more efficient process and more effective promotions, which in turn could enable sustained growth—to say nothing of other benefits, such as improved forecasting accuracy for inventory planning purposes (Chapter 10.1 provides a more detailed discussion on demand planning and inventory optimization). With this in mind, VIP DrinksMart defined their problem and business objective as:

> *Drive sustained profitable growth, increased foot traffic, and improved return on trade spend dollars through optimized promotional planning and pricing*

Because the implementation of a Decision Optimization System for promotional planning and pricing would represent a significant departure from VIP DrinksMart's current processes and ways of working, management acknowledged that the project would also require a coordinated change management effort.

Data, information, & knowledge

To fully appreciate the complexity of VIP DrinksMart's problem and business objective, it's worth noting that promotional planning optimization requires both strategic industry knowledge as well as tactical knowledge on the ongoing effectiveness of each promotion. Developing this knowledge required the analysis of different types of data to answer questions such as:

- What drove consumption within each category (e.g. consumer preferences, consumption habits and occasions, key selling dates, price sensitivity, etc.)?
- How did consumers make purchasing decisions (e.g. complementary and substitute products, price-package architecture, etc.)? And how did these decisions differ by segment?
- What were the characteristics of successful/unsuccessful promotions (e.g. promoted products, promotional price, depth of discount, promotion type, promotional period, retail chain, geographical region, etc.)?
- Which factors affected the success of various promotions (e.g. other promotions within the same or other categories during the same period, use of high-visibility promotional displays, out-of-stocks, etc.)?

To answer these questions, VIP DrinksMart leveraged a substantial amount of internal and external data, including:

- Product range per store
- Historical promotional plans
- Core pricing
- Promotional pricing
- Loyalty card information
- Consumer behavior
- Historical sales
- Store network data (including store characteristics and catchment area demographics)
- Promotional support information (including catalogues, promotional displays, off-locations[9] in-store)
- Inventory levels and replenishment data (including out-of-stock instances and duration)
- Shipments from VIP DrinksMart warehouses to each store

Because VIP DrinksMart operated both company-owned stores as well as banner stores that were independently owned and operated under a common brand, the merchandising team faced several challenges in terms of data availability. In particular, these banner stores used different point of sale systems with different data standards, which meant that transactional data varied from one store to the next. Another related (and more serious) challenge was that VIP DrinksMart couldn't force banner stores to share their point of sale data with the merchandising team, so some geographies had missing data (Chapters 3.2, 4.1, and 4.2 discuss various approaches for dealing with missing data). These data issues made it more difficult to run effective promotions: after all, it's only through the analysis of point of sale data that a retailer or manufacturer can establish which products were sold, when, where, in what quantity, and with what other products.

Overall, when it came to assessing the state of VIP DrinksMart's data, management established that (despite some gaps) it had a sufficient volume of quality data to proceed with an AI-based project (Chapters 11.1, 11.2, and 12.1 provide more information on the topic of data quality). However, the processes that VIP DrinksMart had in place for collecting, storing, and analyzing data created other challenges. Some data were duplicated across multiple databases causing version control headaches, while data on the characteristics of historical promotions (including details on the promotion type and

---

9   Off-location displays are an important part of many promotions, with manufacturers and retailers negotiating on whether the promoted product warrants a separate display location within each store. An example of an "off location" within supermarkets is the highly-visible shelves at the end of each aisle.

promotional price) was held by the merchandising team in standalone, offline spreadsheets that weren't accessible by others in the organization. These inefficiencies hadn't been a problem in the past (as these datasets were only used by the merchandising team), but became self-evident as other business units and departments attempted to access this data for analysis.

Despite the required time and effort, VIP DrinksMart used these disparate datasets to produce regular reports and data visualizations, such as the ranging report shows below (for a specific product):

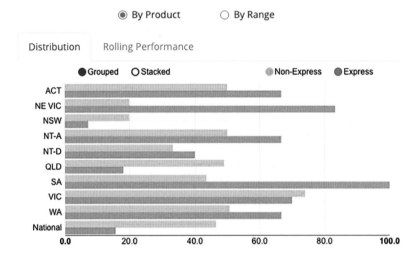

In addition to informational reporting, VIP DrinksMart also evolved its analytics capability over the years to understand the impact of price, geography, promotion type, time of the year, and other variables on the outcome of any given promotion. However, to create optimized promotional plans that could drive sustained profitable growth, increased foot traffic, and improved return on trade spend dollars, VIP DrinksMart needed to gain a deeper level of understanding of how promotional effectiveness differed between various customer segments (grouped by similarities in their purchase behavior), and synthesize this knowledge into predictive models that could accurately predict the uplift of various promotions.

Decision Optimization System (prediction, optimization, & self-learning)

To realize its business objective, VIP DrinksMart deployed a Decision Optimization System that could:

- Replace home-grown spreadsheets and automate labor-intensive processes for data loading and handling, promotional planning, and budgeting
- Automate the assessment of promotional effectiveness
- Automate the creation and exploration of a significant number of slotting boards
- Optimize promotional plans against multiple objectives while considering the interplay of various constraints and objectives (e.g. volume growth and customer margin objectives, trade spend efficiency metrics, price floor and ceiling constraints, and so on)
- Predict the outcome of hypothetical plans, thereby enabling the merchandising team to test various what-if scenarios and provide management with a range of options

The overall project was divided into two phases: The first being digitalization, and the second being the enablement of AI-based capabilities. During the project's first phase, VIP DrinksMart replaced its spreadsheets with a centralized slotting board and digitally integrated workflow (Chapter 12.2 provides more information on the topic of digitalization). This phase of the project digitally connected the interdependent steps of the planning process, providing all stakeholders with one version of the truth, as well as online accessibility to the slotting board, each manufacturer's promotional requests, and the performance of past promotions. Not only did the digitalization phase free up time in the merchandising team so they could move to higher-value tasks (such as creating and testing various scenarios), but it also set the stage for the second phase of the project, which was the deployment of AI-based functionality for prediction, optimization, and learning.

Once complete, the digitalization phase streamlined the end-to-end promotional planning process, allowing manufacturers to execute many steps of the process themselves. This was due to each manufacturer having access to the Decision Optimization System through a self-service portal (as shown below), thereby eliminating long email trails for processes that were now digitalized:

One of these time-consuming steps (previously handled via email and multiple standalone excel files), was the process of gathering each manufacturer's promotional requests. In the new digitalized process, manufacturers could submit requests through a submissions page, as shown below:

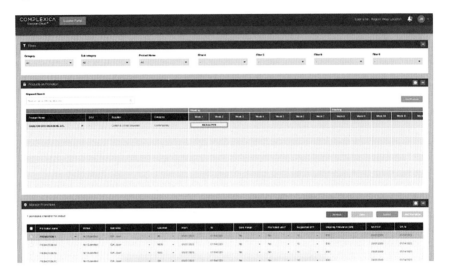

Once all submissions had been received, VIP DrinksMart was able to visualize these requests within a digital slotting board:

This was an important advancement for VIP DrinksMart, as it removed the onus on the merchandising team to follow up with manufacturers and manually enter their requests into a slotting board. It also allowed the team to assess—from a "big picture" perspective—whether the breadth and depth of promotions were aligned with VIP DrinksMart strategic intent for that time period. And lastly, this step marked the beginning of analytical work, during which the merchandising team would analyze the effectiveness of past promotions, as shown below:

The above screen illustrates how VIP DrinksMart could use the Decision Optimization System to access information on sales revenue and volume of each past promotion (broken down by promotion type and baseline vs. promoted uplift), return on trade spend dollars, and predicted vs. actual results (each of which could be clicked-through for more granular detail).

Following this step, the merchandising team began assembling the first version of the promotional plan, which involved shortlisting the relevant set of requested promotions as well as creating other promotions that would be negotiated with the relevant manufacturers. Once this promotional plan was assembled with the right mix of promoted products, the merchandising team would then turn their attention to the pricing and funding mechanisms of each product (e.g. trading terms, promotional allowances, ongoing allowances, off-invoice, scan rebates, unallocated contributions to cooperative campaigns), as shown below:

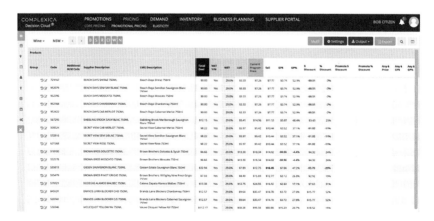

Once all these promotions had been finalized for the different promotional periods, VIP DrinksMart collated the final plan, which included the ability to overlay all promotions across multiple categories into a single view through a cross-category slotting board. At this point, VIP DrinksMart could use the AI-based functionality in the Decision Optimization System to further optimize the plan, which represented the second phase of the project after digitalization.

The success of this next phase depended on the quality of the underlying prediction model, which faced several data issues. On the one hand, company-owned stores used the same point of sale system and provided clean, normalized data on transactions within each store. On the other hand, banner stores used different point of sale systems, with different data structures, and only 62% of these stores shared this data with VIP DrinksMart. This situation required the use of two separate ensemble models (as discussed in Chapter 5.6): The first model was for stores with point of sale data, and the second model was for stores without this data. The first ensemble combined a statistical, time-series model with a few AI-based methods (as discussed in the later sections of Chapter 5), while the second model added an agent-based simulation.

The second ensemble model also used an alternative dataset as a proxy for the missing data. This proxy data wasn't as accurate and granular as the missing point of sale data, which affected the accuracy of the model as it was difficult to attribute sales to specific promotions. However, this deficiency was overcome through agent-based simulations (as discussed in Chapter 4.3), as well as training the prediction model at the national level. This hybrid approach enabled the Decision Optimization System to predict the outcome of promotional plans across the entire store network with an acceptable level of accuracy—"acceptable" meaning more accurate than any result the merchandising team could generate internally, and of sufficient accuracy for the optimization result to be meaningful.

Besides point of sale data, several other datasets were used by the Decision Optimization System to make predictions at the consumer segment level (grouped by similarities in purchase behavior). To achieve this, the prediction model needed to consider not only drivers of consumption within each category, but also variables related to how consumers make purchasing decisions, the extent to which these variables contributed to actual results, the interactions between these variables, and how this "variable contribution" changed across consumer segments (based on historical data for promotions targeted at those specific consumer segments). To train and tune the prediction model, the Decision Optimization System relied upon consumer preference data (such as loyalty-card data and purchased, third-party consumer behavior data). In addition to these datasets, VIP DrinksMart also had access to plenty of historical data, including historical promotional plans, promotional support information (e.g. catalogues, promotional displays, off-location displays in-store), and inventory and replenishment data (e.g. out-of-stock instances and their durations). As a result, the prediction models were able to consider:

- Customer segments based on purchasing behavior similarities
- Price elasticity
- Cannibalization, including pack-size cannibalization, subcategory cannibalization, cross-category cannibalization, and delayed cannibalization
- Promotion characteristics, including products, promotional price, depth of discount, promotion type, promotional period, and geography
- Other factors, such as seasonality, key-selling dates, in-store displays, and inventory levels

Because the prediction models considered the performance of a promotion at the shopper segment-level—including switching between brands and pack sizes, as well as understanding longer-term performance drivers—this allowed the optimization model to target shopper segments with tailored (i.e. optimized) promotions.

Although the prediction models had been trained and tuned using significant amounts of historical data, the models could be thrown off by any major event that affected the usual market dynamics and consumption patterns (e.g. the COVID-19 pandemic). During such circumstances (when the prediction error increased to a level beyond the capability of the self-learning mechanism to address), VIP DrinksMart could "re-baseline" the models by changing the historical data time window used by the Decision Optimization System, as shown below:

Once consumption patterns returned to normal, VIP DrinksMart could change the time window again so that the Decision Optimization System ignored the period containing aberrations. Related functionality also allowed VIP DrinksMart to flag "significant" events within the historical data (i.e. outliers and other patterns not commonly observed), as shown below:

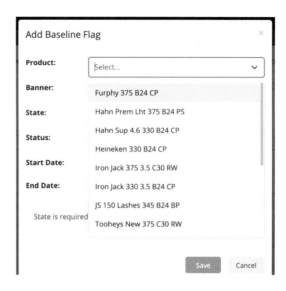

Examples of significant events included new product introductions, or unique, one-off promotions, such as turning each store into Santa's workshop for the two weeks leading up to Christmas. In such cases, VIP DrinksMart flagged these time periods as abnormal, allowing the prediction model to exclude this data from the self-learning process.

With all of these inputs, the Decision Optimization System could predict the performance of any plan across a variety of metrics, such as "net sales revenue" or "volume growth." This, in turn, enabled the system to predict the profit & loss performance of each plan, along with other financial metrics:

Regardless of the business objective for any particular promotional period, creating an optimized plan was a complex process that required both "art" and "science." As a case in point, the merchandising staff had a sound understanding of what drove consumption within each category, how consumers made purchasing decisions, and the characteristics of successful and unsuccessful promotions. The team applied this expertise (the "art") to assemble the best possible plan, which they could then test and iterate by using the prediction model within the Decision Optimization System (the "science").

This represented a process of "manual" optimization (of creating a plan, checking its predicted outcome, making changes to the plan, checking the predicted outcome of these changes, and so on). To explore a significant number of changes for any particular plan would have taken an inordinate amount of time (which is one of the defining characteristics of a complex business problem) and hence, wasn't feasible or practical. The Decision Optimization System could "automate" this process by using AI-based optimization algorithms to automatically iterate and improve the plan (as discussed in Chapter 6). However, this optimization process (also part of the "science") relied on business rules and constraints that defined the feasibility of a plan, as well as objectives that defined the overall goal, making the "art" vital to the whole process (as these business rules, constraints, and objectives were set by the merchandising team).

There were many types of business rules and constraints, some of which were related to promotion types, as shown below:

EXECUTION CODE	EXECUTION NAME	RELEVANT IN BANNER	SWITCH PRODUCTS	ADD PROMOTIONS	SUBTRACT PROMOTIONS	MOVE DATES	PRECEDING EXECUTIONS	FOLLOWED BY EXECUTIONS
	(Blank)	Yes	Yes	Yes	Yes	Yes		
10P	10 Pack	No	No	No	No	No		
26P	2 x 6 Pack	No	No	No	No	No		
2CO	2 x Carton Offer	No	No	No	No	No		
3O	3 for Offer	No	No	No	No	No		
6P	6 Pack	No	No	No	No	No		
AdL	Adv - Lead	No	No	No	No	No		
Adv	Advertised	Yes	Yes	No	No	Yes	Adv, BW	(Blank), Adv
BO	Buy Op	No	No	No	No	No		
BOG	Buy Op - OG	No	No	No	No	No		
BW	Buy Week	Yes	Yes	No	No	Yes	(Blank)	Adv, IS

Product Override For ...   Add Product Override

PRODUCT CODE	PRODUCT NAME	RELEVANT IN BANNER	SWITCH PRODUCTS	ADD PROMOTIONS	SUBTRACT PROMOTIONS	MOVE DATES	PRECEDING EXECUTIONS	FOLLOWED BY EXECUTIONS

Some business rules and constraints (like those shown above) could only be activated or deactivated, while others could be defined by a range of values, such as the min/max duration of promotions, the min/max gap between promotions, the min/max number of products on promotion, and so on. The merchandising team could also define the optimization objective, which was used as the evaluation function to compare various plans against one another. For example, the "maximize volume" objective could be selected by clicking "Volume Outcome (measured in liters)," as shown below:

After testing several optimization methods, simulated annealing was found to be the most appropriate in terms of runtime and overall result. As discussed in Chapter 6.5, simulated annealing relies on gradually improving a single solution (i.e. a single promotional plan) through many iterations. A few constraint-handling techniques were also used, which included a combination of repair algorithms and penalty functions (as discussed in Chapter 6.8).

This optimization functionality also enabled the creation and exploration of what-if scenarios, with each scenario being a promotional plan with slightly different business rules, constraints, and objectives. As an example, VIP DrinksMart could create a scenario in which several well-known brands are placed on promotion every fortnight. The outcome of this scenario would be that VIP DrinksMart could regularly buy well-known brands at a reduced price (instead of ever paying the published price for these high-volume, frequently promoted items). On the flip side, without varying the price from one promotion to the next, the value equation might be reset in the consumers' mind—that is, promoting product "X" too frequently at price "Y" might lead to a situation where consumers are no longer willing to pay full price ("I never pay more than Y for that product," they might start thinking, "I'll just wait for the next promotion").

The Decision Optimization System allowed for many such what-if scenarios to be created, optimized, and evaluated, so that the merchandising team could compare alternate ways of achieving a desired outcome. These what-if scenarios could also be used to explore promotional plans that targeted specific consumer segments, as VIP DrinksMart had different objectives for different segments (which required a different mix of promoted products). Each scenario within the Decision Optimization System could be named and saved, as shown below:

ID	Name	Created Date	Created By	Notes	Edit	Select
20	Test Plan XYZ	2019-07-12	arantes	Improved ROI in all Promo Periods	✏	○
19	Alternate Frequency	2019-06-05	bloggsm	Looking at decreased activity impact on NSR	✏	○
18	Summer Push	2019-09-08	jonesa	Upping frequency and advertising on summer promos	✏	○
17	Footy Final Promo	2019-12-13	brownc	Distributed ads and deals across VIC	✏	○

Whether used for generating what-if scenarios or optimized plans, the Decision Optimization System always provided a "before" and "after" view, as shown below (from a profit & loss perspective):

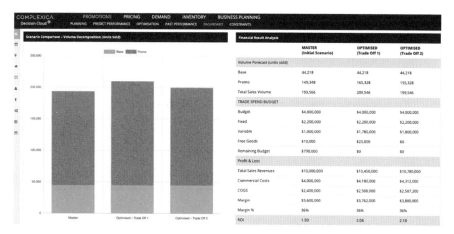

To understand how the promotional plan had changed, the merchandising team could select the "slotting board view" to see the changes, down to the specific promotional period and product, as shown below (where the Decision Optimization System removed products from periods 15 and 16, and added new promotions to periods 19 and 20):

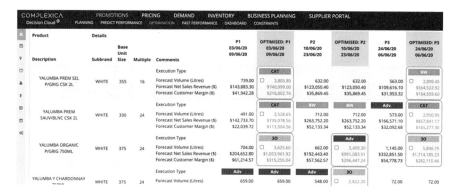

When creating an optimized plan, the merchandising team could also limit the number of changes made by the Decision Optimization System (as discussed in Chapter 6.9). When multiple changes were made, a summary screen visualized the incremental benefit of each change, as shown below:

No Of Changes (Up to)	○ 1	○ 2	○ 3	○ 4	○ 5	○ 6	○ 7	○ 8	○ 9	○ 10
Total Volume (Litres)	3,805.85	6,334.50	9,960.10	13,369.40	16,191.60	19,631.80	23,700.30	26,599.75	29,550.70	35,447.45
Total NSR ($)	$740,999.00	$1,476,077.56	$2,530,039.48	$3,521,122.99	$4,341,536.53	$5,371,544.77	$5,968,393.72	$6,532,916.64	$7,390,757.81	$9,104,943.04
Total Customer Margin ($)	$216,002.74	$329,507.30	$644,762.34	$941,209.58	$1,034,734.76	$1,200,501.88	$1,353,238.21	$1,517,797.81	$1,683,075.11	$1,965,185.57

This functionality enabled the merchandising team to deliver the largest possible benefit for the smallest effort (as measured in number of changes, as each change required negotiation with the manufacturer funding the promotion). To assist with negotiation, the "science" enabled VIP DrinksMart to defend the plan's predicted outcome, showing that both parties would benefit from the proposed change.

These optimization results could also be visualized through a range of charts and reports, as shown below:

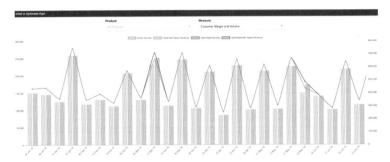

Because the maximization of one objective (e.g. "volume") was likely to come at the expense of another objective (e.g. "net sales revenue"), the concept of a single "best" plan was rarely relevant and a different optimization algorithm[10] was used for multi-objective optimization (as discussed in Chapter 6.9). Instead of a single plan, the Decision Optimization System generated several optimized plans, each being the "best" in the sense there was no better plan on all objectives (Chapter 6.9 provides a more in-depth discussion on non-dominated solutions and Pareto optimal sets), as shown below:

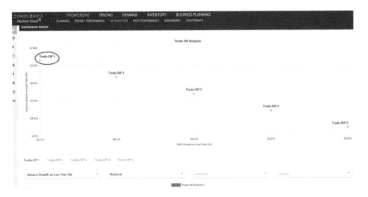

---

10 Evolutionary algorithms (as discussed in Chapter 6.7) were used instead of simulated annealing, as multi-objective optimization requires the parallel processing of many solutions rather than just one (which is the case in simulated annealing). By processing a population of solutions simultaneously, the optimization algorithm can generate many solutions in a single run, facilitating the trade-off analysis.

The merchandising team could then explore the pros and cons of each plan, thus analyzing the trade-offs. Also worth noting is that these trade-offs required consideration from both from a strategic and tactical point of view. This was due to the fact that at any given point in time, two or more "optimal" plans might be better or worse depending on whether they were viewed from a strategic or tactical perspective.

For example, a manufacturer might want to steal share from a competitor during a specific time of year, and might favor volume growth at the expense of profit. If VIP DrinksMart found this promotion appealing (which is critical, as the manufacturer needs to secure VIP DrinksMart's commitment to execute the plan), then the merchandising team could "insert" this promotion into the overall plan. However, the manufacturer's tactical addition to the overall plan might tie up valuable floor space within stores, introducing a bottleneck for competing manufacturers as VIP DrinksMart wouldn't be in a hurry to buy more stock if their inventory for this category and subcategory was already too high (making this promotion attractive from a tactical point of view, but less so from a longer-term perspective).

Because of such dynamics, AI-driven promotional planning and pricing (or trade promotion optimization) is often viewed as sitting at the cross-roads of "art" and "science," (as tying up floor space might have been deliberate on the part of the manufacturer[11] in the previous example). After several years of using the Decision Optimization System, VIP DrinksMart grew to become best in class for promotion planning and pricing, with improved margins and relationships with key manufacturers. More effective promotions led to improved (and sustained) profitable growth, increased foot traffic across the store network, more banner stores contributing point of sale data, and improved return on each manufacturer's trade spend investment. The process improvements and efficiencies gained through digitalization also allowed VIP DrinksMart to reduce headcount within certain teams, allowing staff to be redeployed to undertake more value-adding tasks.

---

11 Under such circumstances, manufacturers lacking well-known brands or advanced capabilities for promotional planning might have their submissions and promotional dollars gridlocked by the retailer's constraints (and be disadvantaged in their ability to negotiate their desired position).

CHAPTER 10

Supply Chain

"Leaders win through logistics. Vision, sure. Strategy, yes.
But when you go to war, you need to have both toilet paper and
bullets at the right place at the right time. In other words,
you must win through superior logistics."

Tom Peters, *business author and speaker*

Supply chains are all about "supplying" the items we buy and consume each day through complex "chains," which move and process raw materials to make the final products we see on store shelves. Depending on the industry, a supply chain could be as straightforward as a few retail shops, warehouses, and trucks, or as complex as a sprawling network of mine sites and processing plants connected by rail and sea transport. From a higher perspective, however, all these supply chains are linked in one way or another through an intricate web of interactions. If we think about any common product—such as a bottle of wine or family car—we can trace the individual components of those products back through their respective supply chains, back through the trucks and ships that brought those products to the liquor store or dealership, back to the factories where those products were assembled, back through the transport network that brought those components to the factory, and so on.

In the case of wine, the grapes have to be grown first, before they begin their multi-month journey through harvesters, trucks, weighbridges, crushers, and other processing facilities to become the colorful accompaniment at our dinner table. And when it comes to cars, some components originate at mine sites, where the iron ore that will eventually become the car's frame and doors and hood is extracted from a pit. Each of these steps in the supply chain has its own challenges and complexities; for example, planning a mine site requires consideration of what grade of ore is required at what point in time, coupled with truck and digger availability, workforce rosters, maintenance schedules, and more—which is just the first step in the process—followed by the scheduling of trains that will transport the ore from various mine sites to the port, where the coordination of stackers and reclaimers happens to ensure that each ship is loaded on time. And after that, there is more, much more, as the ore arrives in another country and is heated by coal (which arrived at the furnace through a similarly complex supply chain) to become steel, which in turn is molded into the car's frame and doors and hood—components that represent just a handful of the 30,000 parts that make up the average car, each of which has their own supply chain from raw materials to finished part. On top of this, the automaker needs to predict consumer demand for its cars across different countries—a difficult problem in itself—all while the cars are being assembled and placed on ships for transport to those markets.

At every step, there is complexity, and the more steps we consider together, the more complex the problem becomes. This inherent complexity makes supply chain problems particularly well suited for the application of Artificial Intelligence and Decision Optimization Systems. Given the multi-component nature of supply chains (as discussed in Chapter 6.10), we'll present each part separately: First demand planning and inventory in Section 10.1, then production planning and scheduling in Section 10.2, followed by logistics and distribution in Section 10.3. At the highest level, these are the core components of a supply chain operation: predicting demand, planning and scheduling production (whether it be the production of iron ore from a mine or the assembly of cars in a factory), and then organizing logistics and distribution. Each component represents a complex business problem in itself, and together, an almost impossible challenge.

10.1 Demand Forecasting and Inventory Optimization

The holy grail of supply chain optimization is predicting what will be sold, in what quantity, where, and when—with 100% accuracy. An organization capable of doing that could run the leanest possible supply chain—with minimal inventory levels—while always satisfying customer demand and never stocking out. Unfortunately, such prediction accuracy is impossible to achieve, and for that reason, all organizations must carry some level of inventory to buffer against unexpected changes in demand.[1]

Despite the fact that organizations of all shapes and sizes have gorged themselves on demand planning software over the past few decades, demand forecasting still remains an unsolved problem within most organizations— "unsolved" in that the forecast error is still high, leading to stockouts, as well as excessive inventory levels and obsolescence. "We still carry a lot of

---

1   Another reason that organizations carry inventory is because the lead times on raw materials can also vary, so without a buffer of these inputs the manufacturing process can come to a halt. Such inventory is often classified as *raw materials inventory*—which could be the car frame, door, or hood—versus finished goods inventory, which is the finished (i.e. completely assembled) car itself.

inventory," these organizations will confess, "not as much as before, but still a lot. *Unfortunately, it's in the wrong place, at the wrong time.*" Hence, not only are working capital and obsolescence costs still high, these organizations routinely suffer lost sales due to stockouts.

MAX Hardware found itself in precisely this situation. After evaluating many software systems for demand planning, the company purchased and implemented a system that best fit their operation from a features, functionality, and workflow perspective. Being a manufacturer and distributor of hardware products—such as nuts and bolts, screws, and various types of tools—the company sold its products through both retail chains and directly to the trade (i.e. electricians, plumbers, carpenters, etc.). Given MAX Hardware's extensive products range—in the thousands—along with significant lead times for certain raw materials and components, the company's flexibility was limited in ramping up production when inventory ran low. For this reason, improving demand forecast accuracy was of paramount importance—especially because trade customers couldn't wait for backordered products, so whatever MAX Hardware stocked out, these customers bought from competitors (sometimes leading to a permanent change in loyalty).

After the new demand planning system went live, however, it became apparent that forecasting accuracy was no better for a large number of products. Using these system-generated forecasts, MAX Hardware was still producing too much or too little of different product lines, creating excess inventory of some lines and shortages of others. As for products produced in the correct aggregate quantity (countrywide), they were frequently sent to the wrong distribution center and required expedited shipping to another distribution center to fulfill demand in that part of the country. Hence, the forecast for these products was accurate at the aggregate level, but highly inaccurate at the granular level of individual states or customer segments.

Because of these forecast accuracy issues, manual overrides became the norm at MAX Hardware as inventory managers and key account managers overrode the forecast in an attempt to improve its accuracy. In many cases, this made the situation worse, and a large amount of time was spent "playing around with the numbers," as management put it. Eventually, the forecasting function was removed from the system altogether, as MAX Hardware reverted to spreadsheets for generating a manual forecast—which meant that, in effect, the company had returned to the same state as before the demand planning system

was implemented (i.e. using spreadsheets, gut feel, and manual processes to create the forecast). This situation continued for some time until management decided to take action and began searching for a system that could provide superior forecasting accuracy. As such, MAX Hardware defined their business problem and objective as:

Reduce inventory levels and stockouts through more accurate demand forecasting

By going through the process of implementing the failed demand planning system, MAX Hardware realized that the vast majority of such systems were based on a standard set of statistical models that were configured in the same way: Namely, by taking the historical sales data for each product line—like the one shown below:

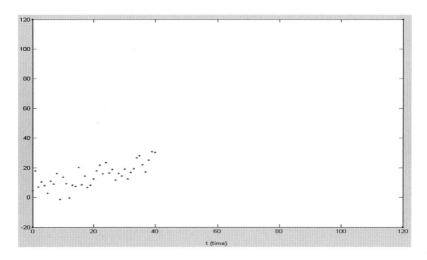

and finding the statistical model that best "fit" this data:

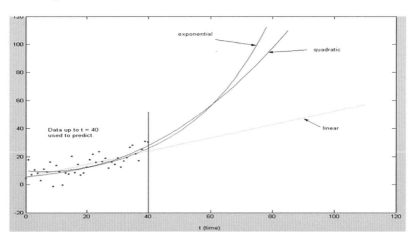

In the graph above, each curve is a statistical model, and each curve fits the data better than any other curve of its specific type. And so for the same historical data, we have three different predictions generated by three different models. But by only using internal data and standard statistical models, what inevitably happens is that the future turns out very differently to what these models predicted, thereby creating a forecast error of varying magnitude for different products:

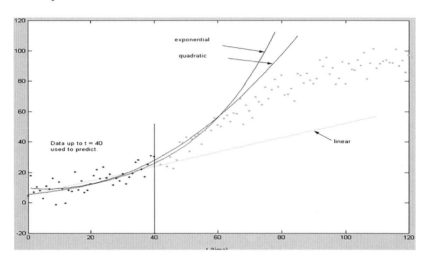

Through this experience, MAX Hardware realized that demand forecasting was a scientific problem of selecting the most appropriate prediction method for the problem at hand and building a model (as discussed in Chapter 5), rather than a software problem of selecting the application with the most features and functionality. The real difficulty lay in predicting the future, which had to be addressed algorithmically within the selected demand planning system.

Data, information, & knowledge

Like many other manufacturers in the building materials sector, MAX Hardware had a substantial amount of internal data, including:

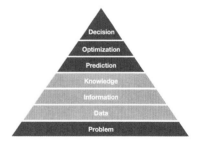

- Historical sales by product by customer
- Historical pricing data by product by customer
- Historical inventory levels by product by week
- Historical forecasts created by inventory managers and by key account managers for retail chains

This data was used to provide inventory managers with a variety of reports and visualizations, including inventory levels against actual sales by product and time period:

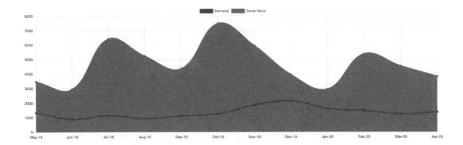

as well as historical performance on KPIs such as stockouts, customer fill rates, and inventory days cover:

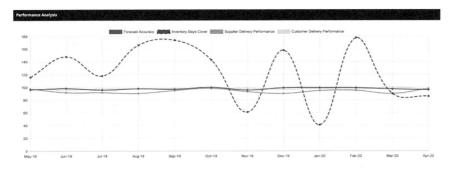

Although these informational reports were plentiful, they didn't provide any predictive capabilities—only a rear-view mirror look at what happened in the past. To gain a better feel for future demand, MAX Hardware began experimenting with external data (such as building approvals and customer forecast data) in search of patterns that might repeat in the future. But such efforts were ad hoc, sporadic, and driven entirely by the analytical capabilities of the staff that undertook such analysis. For these reasons and others, the business case for a Decision Optimization System—one that would allow the company to hold the right inventory, at the right location, at the right time through improved demand forecasting accuracy—was created and endorsed by MAX Hardware.

Decision Optimization System (prediction, optimization, & self-learning)

As is the case with many other complex business problems, MAX Hardware's business objective to simultaneously reduce inventory levels and stockouts was dependent on the accuracy of its prediction model—in this case, the accuracy of predicting future demand. Although it was still possible to improve inventory levels by using an inaccurate forecast (by holding a larger amount of safety stock for the most variable product lines and dynamically changing these safety stock levels throughout the year to account

for seasonality and other demand effects), the largest benefit would accrue through improved forecast accuracy.

Knowing now that the problem of forecast accuracy was algorithmic in nature, the new Decision Optimization System had two fundamental differences from the first demand planning system: First, it used an ensemble model that combined statistical models with AI-based methods such as neural networks and fuzzy system (as discussed in Chapter 5); and second, each model was fed with both internal and external data to improve accuracy:

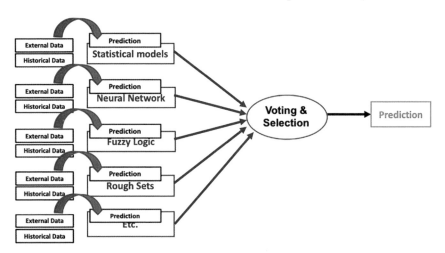

The ensemble model achieved a substantial increase in forecast accuracy over both the manual, spreadsheet methods, as well as the statistical models used by the failed demand planning system. This ensemble model became the prediction component of the Decision Optimization System, providing MAX Hardware with the most probable view of future demand by product, by distribution center, by time period, and in many cases, by customer:

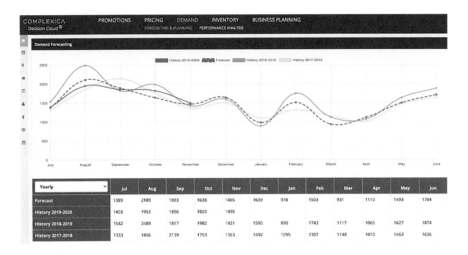

Once the Decision Optimization System was configured and implemented, it considered historical sales, customer forecasts, relevant external data, as well as promotional and pricing information for each product:

Although the demand forecast was generated by an ensemble model that used internal and external data, MAX Hardware still had the capability to override these forecasts:

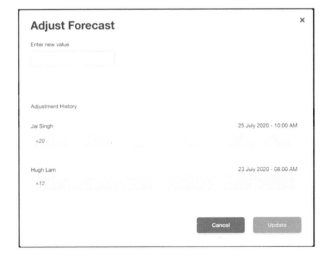

These overrides were captured in an audit log, and then analyzed by the Decision Optimization System to provide feedback on the effectiveness of each manual intervention (which in most cases were inferior to the system-generated forecasts). As for new products that lacked sales data, the Decision Optimization System used the historical sales data of similar products to estimate future demand:

The Decision Optimization System also allowed MAX Hardware to understand the trade-off between working capital levels and customer fill rates. This was done by defining working capital and fill rate targets, which could be set for all products and customers in aggregate, or broken down into individual targets for individual products, customers, distribution centers, and time periods, as shown below:

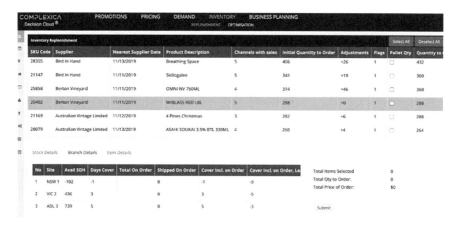

Once these targets were set, the Decision Optimization System would attempt to find a Pareto curve of optimized solutions (as discussed in Chapters 2.3 and 6.9) that illustrated the trade-off between inventory levels and customer fill rates:

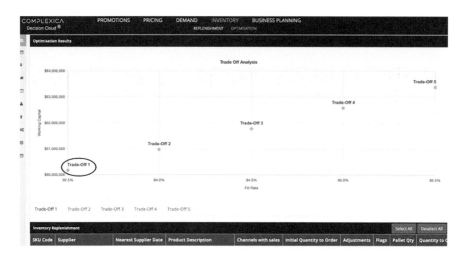

This advanced capability for multi-objective optimization allowed MAX Hardware to implement different inventory policies for different product lines, customers, distribution centers, and time periods, and was based on evolutionary algorithms (discussed in Chapter 6.7).

The reason that evolutionary algorithms were selected for optimization, is because this AI-based algorithmic method could simultaneously produce many potential solutions (i.e. a population of solutions)—hence it could generate a few final solutions at the end of a *single run*. Of course, these final solutions had to be substantially different from one another, because if the five solutions were quite similar (with just minimal differences), the usefulness of the trade-off results would be modest at the very best. To address this issue, the evaluation function of the evolutionary algorithm took into account the "uniqueness" of solutions: "similar" solutions were penalized, so they become less attractive as candidates for the next generation of solutions. Furthermore, the algorithm placed a premium on non-dominated solutions[2]—in other words, solutions where there was no single solution in the population of solutions better on all objectives (e.g. working capital and customer fill rates). Because of this evaluation function, the evolutionary algorithm improved the Pareto curve of solutions from one generation to the next, and the final result (consisting of several "best" solutions) was presented as a diverse set of possibilities that illustrated the trade-off between working capital and customer fill rates (as shown above).

And lastly, based upon the system-generated demand forecast and optimized inventory policies, the Decision Optimization System provided MAX Hardware with product replenishment recommendations that could be

---

2   See Chapter 6.9 for a full discussion on this topic.

reviewed/modified/accepted before being converted into production orders for finished goods, or purchase orders for raw materials:

The Decision Optimization System implemented by MAX Hardware provided a number of tangible benefits, including:

- Improved forecast accuracy, which was particularly important for hard-to-forecast product lines. On average, forecasting accuracy increased from approximately 64% to 89%, with many products exceeding 95%
- A 18% reduction in finished goods inventory
- A 43% reduction in stockouts, leading to a corresponding increase in customer fill rates (as measured by Delivery In Full, On Time metrics, "DIFOT")
- Less time and effort for inventory planning and replenishment, with some tasks being reduced from a few days to a few hours

MAX Hardware also realized additional benefits in metrics such as the cash-to-cash cycle time, stock turns, and customer loyalty, all of which contributed to the company's overall profitability and competitiveness.

10.2 Scheduling Optimization for Improved Asset Utilization, Throughput, and DIFOT

Every factory needs to plan and schedule its production, regardless of whether it's assembling cars, bottling wine, producing cardboard boxes, or extracting iron ore from a mine. These factories can be thought of as "nodes" within a supply chain, where raw materials and components go in one end and finished products emerge from the other. Many of these nodes are interconnected, where the output from one node is an input into the next. For example, the output from a mine could be iron ore or coal, which represents the raw material input into a steel-making factory. And in that steel-making factory, the finished sheets and slabs of steel become the raw material into the next manufacturing node, where the steel is formed into car components—and so on, with each node having its own demand forecasting, production planning, and scheduling process.

In this supply chain context, the words "planning" and "scheduling" are often used interchangeably, despite meaning very different things: Planning refers to *what* an organization will do, whereas scheduling refers to *when* an organization will do it. For this reason, planning is more macro and "higher-level" (i.e. deciding what products to produce each week or month, depending on the forecasted demand), while scheduling is more granular and exact. As a simple example, an airline might plan to provide 100 return flights between two cities for the month of May (based upon the forecast demand for travel)— which represents *what* the airline will do. This planning process is simpler than scheduling *when* these 100 round trips should occur: the exact time, crew, planes, maintenance, and so on. Hence, planning problems are usually easier to solve and optimize than scheduling problems.[3]

When it comes to production planning and scheduling within a factory, the same concepts apply. An automaker would first create a production plan for building a particular mix and volume of cars (again, based upon the forecast demand, production capacities, inventory levels at dealerships, as well as other considerations), and then use this plan to schedule the assembly of these cars (the exact components, production lines, and timing). Hence, the planning process is done at a higher, more macro level, whereas scheduling is granular and involves many complex details, such as the availability of input components and raw materials, labor constraints, production line availability, changeover times, maintenance schedules, and more.

In this case study we'll discuss CAST Metals, an organization with eight foundries spread across different locations, with each foundry operating several furnaces and casting machines. In a foundry operation, products are produced by melting metal inside a furnace and then pouring this heated liquid into a mold. Once the metal has cooled and solidified, the mold is removed to produce the final product (which could be a metal component for a railway network, automobile engine, pipe, or any number of other products). Creating and modifying a quarterly production plan of what products to produce was

---

3   The predominant difficulty in planning problems lies in accurately predicting what will happen in the future; once this has been addressed, the planning process is usually straightforward. The difficulty in scheduling problems, on the other hand, is finding the schedule that maximizes or minimizes certain objectives—such as asset utilization or cost—from an almost infinite number of possible schedules.

part of CAST Metals' sales and operations planning ("S&OP")[4] process, in which confirmed and forecast demand was synchronized with manufacturing capacity and inventory levels. During this regular planning cycle, CAST Metals balanced production across its foundries by considering manufacturing capabilities and capacities, the location of its customers, transportation costs, as well as the overall production load and inventory across the network. Hence, demand forecasting, inventory management, and the global optimization of production across CAST Metals' eight foundries were addressed at the planning level, and were not a consideration for the scheduling process.

When the monthly production plan was converted into a weekly schedule—going down to hourly time buckets at the individual machine level—the objective was to meet customer due dates while simultaneously maximizing asset utilization and factory throughput. However, converting the higher-level plan into a detailed schedule was a complicated and difficult undertaking, requiring CAST Metals to consider many constraints and business rules for each individual foundry. Some of these constraints represented physical limitations (such as melting times and furnace capacities), while others represented operational business rules related to:

- Manufacturing some products during day shifts or night shifts
- Not manufacturing some products at the same time because of their similarity (making these products difficult to sort at the end of a production run)
- Operating certain casting machines on particular days (e.g. from Monday morning to Thursday evening)
- Using certain casting machines for particular products because of efficiency and tooling reasons

In addition to these business rules and constraints, the production schedule had to coordinate many independent processes, such as the preparation of cores and molds, pouring of molds, and the finishing of castings. There were also many relationships between various metal grades to consider, as well as the transition time for changing from one metal grade to another. Because of all these complexities, the result was substandard performance on metrics such as Delivery

---

4   Sales and Operations Planning is an integrated planning process for aligning and synchronizing various business functions of an organization.

In Full, On Time ("DIFOT") and Overall Equipment Efficiency ("OEE"), as well as excessive overtime labor due to last-minute schedule changes. Hence, CAST Metals defined their business problem and objective as:

Simultaneously increase asset utilization, factory throughput, and customer service levels through optimized production scheduling

To achieve this objective, CAST Metals decided to replace the manual, spreadsheet-based approach for production scheduling with a Decision Optimization System capable of:

- Converting production plans into detailed schedules that were optimized for asset utilization, labor costs, and customer service levels
- Dynamically "re-optimizing" the production schedule whenever circumstances changed (customer orders, machine failures, etc.)

Given the complexity of this business problem (i.e. an astronomical number of possible solutions, a dynamic environment with frequent changes, and many problem-specific constraints), Artificial Intelligence algorithms were the natural choice for optimization.

Data, information, & knowledge

When creating the monthly production plan or weekly schedule, CAST Metals had access to a variety of datasets, including:

- Forecast orders by product
- Confirmed orders by customer
- Historical sales by product by customer
- Bill of materials for each product
- Product routing for each product for each foundry
- Historical customer service levels in terms of DIFOT metrics
- Historical factory performance levels in terms of asset utilization, maintenance schedules, labor costs (especially overtime), and breakdowns
- Historical inventory levels by product by week
- Current inventory levels
- Historical demand forecasts and their accuracy

This data was used to produce daily and weekly reports, such as inventory levels against actual sales by product, machine downtime, and overdue work orders, as shown below:

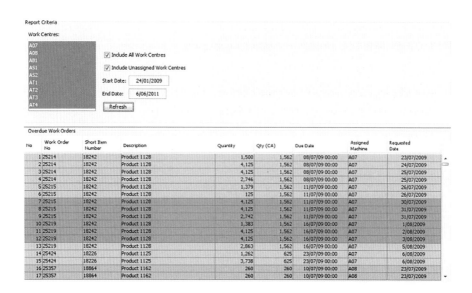

CAST Metals used these reports to balance urgent and overdue orders against run lengths and changeover times, with the output of this spreadsheet-based process being a day-by-day, line-by-line production schedule:

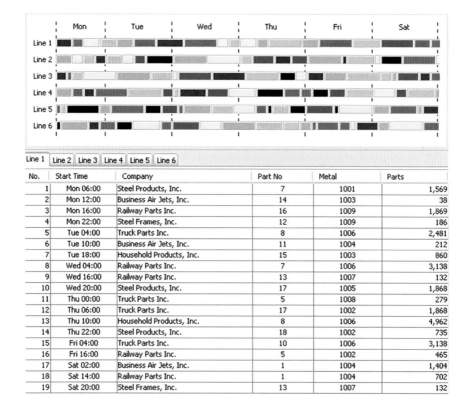

No.	Start Time	Company	Part No	Metal	Parts
1	Mon 06:00	Steel Products, Inc.	7	1001	1,569
2	Mon 12:00	Business Air Jets, Inc.	14	1003	38
3	Mon 16:00	Railway Parts Inc.	16	1009	1,869
4	Mon 22:00	Steel Frames, Inc.	12	1009	186
5	Tue 04:00	Truck Parts Inc.	8	1006	2,481
6	Tue 10:00	Business Air Jets, Inc.	11	1004	212
7	Tue 18:00	Household Products, Inc.	15	1003	860
8	Wed 04:00	Railway Parts Inc.	7	1006	3,138
9	Wed 16:00	Railway Parts Inc.	13	1007	132
10	Wed 20:00	Steel Products, Inc.	17	1005	1,868
11	Thu 00:00	Truck Parts Inc.	5	1008	279
12	Thu 06:00	Truck Parts Inc.	17	1002	1,868
13	Thu 10:00	Household Products, Inc.	8	1006	4,962
14	Thu 22:00	Steel Products, Inc.	18	1002	735
15	Fri 04:00	Truck Parts Inc.	10	1006	3,138
16	Fri 16:00	Railway Parts Inc.	5	1002	465
17	Sat 02:00	Business Air Jets, Inc.	1	1004	1,404
18	Sat 14:00	Railway Parts Inc.	1	1004	702
19	Sat 20:00	Steel Frames, Inc.	13	1007	132

Besides being labor intensive and time consuming, the scheduling process was inefficient for many other reasons, including:

- The final schedule didn't consider many business rules and constraints, largely because these rules were in people's heads. Consequently, the production schedule was usually "un-executable," in that it omitted or abstracted certain variables such as maintenance, changeover times, differences in the defect rate between various machines, and variations in production run times. The production schedule was therefore seldom achieved, and schedule adherence was low.
- The production sequence was suboptimal, as no team of human experts could consider all possible scheduling combinations, which in turn led to factory performance issues.
- The schedule was static and disconnected from the factory floor, as there was no data feed from each machine to understand production from a "scheduled" vs. "actual" point of view. This meant that re-scheduling was a slow and painful process of first realizing that something had happened—such as machine failure, high defective rate, or some other event—followed by updating the spreadsheet-based schedule, before finally printing a new version and pushing it down to the factory floor (by which time it was again out of sync and not reflective of what was actually happening).

The business case for configuring and deploying a Decision Optimization System was based on achieving higher production volumes through each foundry (leading to greater revenue per site) and fewer late orders (leading to fewer financial penalties, greater customer satisfaction, and greater customer loyalty). Consideration was also given to potential future phases, where the Decision Optimization System could be extended to production planning and demand forecasting, thereby allowing CAST Metals to improve demand forecast accuracy and reduce inventory (as discussed in the previous case study), as well as globally optimize across all eight foundries to realize further efficiency gains.

Decision Optimization System (prediction, optimization, & self-learning)

To enable optimized scheduling across its eight foundries, CAST Metals implemented a Decision Optimization System based on Artificial Intelligence methods for optimization. When converting the production plan into an executable schedule for each foundry, the Decision Optimization System considered the current inventory level of each product (as measured in days cover), as well as the designation of each order—namely, whether it was "make to stock" for replenishing inventory or "make to order" for a specific customer:

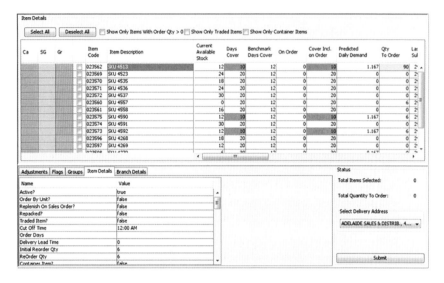

These inventory levels and designations impacted the prioritization of orders, with the Decision Optimization System placing more emphasis on orders where the product was being produced for a specific customer and no inventory existed for buffering the due date.

Another important consideration was the interplay between furnaces and casting machines, which represented the core scheduling issue. The primary objective was to optimize the distribution of production orders over some period of time in a way that maximized furnace utilization and machine throughput. Because the furnaces and machines worked together in the production process (first melting, then casting), the maximization of furnace utilization and production-line throughput had to be considered jointly. Secondary objectives included the maximization of DIFOT metrics and minimization of labor costs when optimizing the production schedule.

To generate a detailed schedule that was optimized (as well as realistic and executable), the Decision Optimization System held a variety of foundry- and machine-specific data that was referenced by the optimization model, such as changeover times between various metal grades:

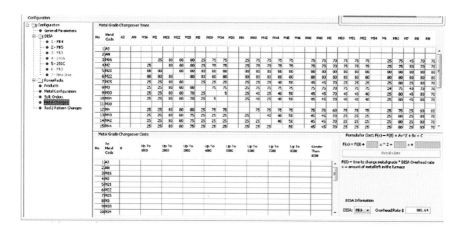

Within the optimization model itself, the approach for handling constraints was based on decoders, which separated between objectives and constraints (as discussed in Chapter 6.8). Using this approach, the optimization model used the constraints to "guide" the optimization process toward feasible schedules of higher quality. This constraint-handling approach also allowed for easy modification of business rules related to labor availability:

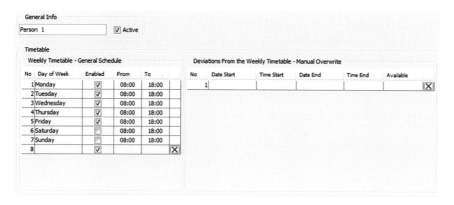

and machine availability:

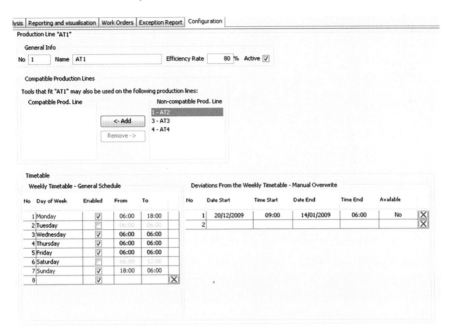

Both labor and machine capacities were treated as soft constraints, allowing the Decision Optimization System to flex production up and down as required. By modifying these constraints (or changing the capacity of the foundry in a more fundamental manner—for example, by adding another casting machine within the Decision Optimization System), CAST Metals could ask "what-if" questions and create alternate schedules, as shown below:

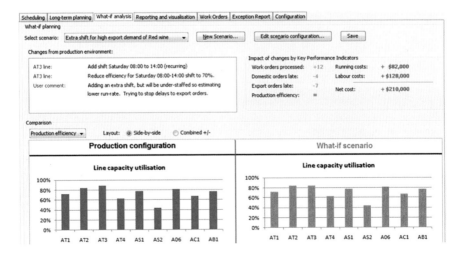

CAST Metals used this functionality to analyze a variety of "what-if" scenarios, including:

- Examining the effect of moving a production order forward or back, or from one machine to another
- Splitting large production orders into smaller work orders
- Examining the effect of constraining certain orders so they couldn't run in parallel
- Examining the effect of changes to the production calendar, furnaces, and production lines.

The most important output of the Decision Optimization System, however, was the production schedule itself. To generate feasible schedules right from the start of the optimization run, the system used a combination of evolutionary algorithms and simulated annealing (plus a decoder responsible for generating near-feasible solutions). Although the quality of the system-generated schedule improved as the optimization run progressed, CAST Metals could stop the run at any point and use the best available schedule rather than waiting to the end. This allowed for flexible usage and provided CAST Metals with ultimate control over the optimization process. The screen below shows an optimized schedule for a particular week, where each bar represents a production order:

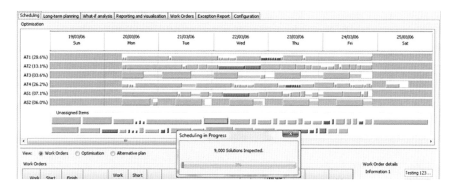

The details of each order could also be viewed by clicking on any bar, or by selecting the order from the items list:

Order Items

No.	Order No.	Customer	Part Code	Part	Metal	Base Grade	Schedule	Ordered	Ex-works Date	Lead Time	Earliest start date	Production Date & Time	Delay hours	Dies	Moulds	Cav-ings	Pattern Set	MPH Eff-ts.	Time (h:mm)
1	COO7393	Pandrol Australia	2427	Adaptor Plate	M4	Ductile	Day/Night	4,000	22/12/2008	7		12/01/09 06:00	666	M4	168	3,360	605	100.0%	00:38
2	COO7386	OneSteel Trak-Lok	2199	Lock in Shoulder	M4	Ductile	Day/Night	60,000	22/12/2008	0		12/01/09 06:38	506	M4	2,128	51,072	379	100.0%	08:11
3	COO7005	PANDROL UK LIMITED	2387	Fastclip	M4	Ductile	Day/Night	32,768	1/01/2009	0		12/01/09 14:49	276	M4	2,712	32,544	565	100.0%	10:02
4	COO7446	Pandrol Inc (NJ)	2647	Shoulder Safelok III	M4	Ductile	Day/Night	18,600	12/01/2009	0		13/01/09 01:08	15	250C	849	15,282	1,159	100.0%	02:39
5	COO7394	Pandrol Australia	2079	Shoulder	M4	Ductile	Day/Night	26,880	22/12/2008			13/01/09 10:39	891	250C	1,715	27,440	271	100.0%	06:21
6	COO7390	Pandrol Australia	2079	Shoulder	M4	Ductile	Day/Night	26,880	14/01/2009	7		13/01/09 10:00	101	250C	1,715	27,440	271	100.0%	06:21
7	COO7395	Pandrol Australia	2079	Shoulder	M4	Ductile	Day/Night	26,880	22/12/2008	7		13/01/04 16:21	706	250C	1,715	27,440	271	100.0%	06:21
8	COO7440	PANDROL UK LIMITED	2890	Fastclip	M4	Ductile	Day/Night	246,888	22/12/2008	0		13/01/09 22:42	554	250C	4,630	74,080	1,232	100.0%	16:13
9	COO7381	Pandrol Australia	2119	Spacer	M3	Ductile	Day/Night	13,550	22/12/2008	0		13/01/09 06:00	523	M4	301	14,448	304	100.0%	01:12
10	COO7440	PANDROL UK LIMITED	2890	Fastclip	M4	Ductile	Day/Night	246,888	22/12/2008	0		13/01/09 00:51	521	M5	1,466	17,592	1,256	100.0%	05:08
11	COO7440	PANDROL UK LIMITED	2890	Fastclip	M4	Ductile	Day/Night	246,888	22/12/2008	0		13/01/09 07:12	10,600		10,600	127,200	1,256	100.0%	37:08
12	COO7433	PANDROL UK LIMITED	2331	Fastclip	M4	Ductile	Day/Night	99,240	26/12/2008	0		14/01/09 24:55	481	250C	6,395	102,320	1,101	99.0%	22:39
13	COO7390	Pandrol Australia	2023	Shoulder	M4	Ductile	Day/Night	26,992	21/01/2009	7		14/01/09 20:20	14	M5	1,367	21,872	221	100.0%	05:03
14	COO7433	PANDROL UK LIMITED	2331	Fastclip	M4	Ductile	Day/Night	82,710	19/12/2008	8		15/01/09 13:35	666	250C	4,862	77,792	1,101	100.0%	17:03
15	COO7390	Pandrol Australia	2023	Shoulder	M4	Ductile	Day/Night	20,992	9/01/2009	7		15/01/09 01:24	306	M5	1,275	20,400	221	100.0%	04:43
16	COO7426	Pandrol Australia	2569	Fastclip	M4	Ductile	Day/Night	12,000	11/01/2009	0		15/01/09 06:07	933	M5	764	9,168	839	100.0%	02:56
17	COO7440	PANDROL UK LIMITED	2890	Fastclip	M4	Ductile	Day/Night	240,000	14/01/2009	0		15/01/09 09:03	62	M5	11,971	143,652	1,256	100.0%	41:56
18	COO7440	PANDROL UK LIMITED	2890	Fastclip	M4	Ductile	Day/Night	240,000	14/01/2009	0		16/01/09 06:39	64	250C	6,328	101,248	1,232	100.0%	22:10
19	COO7150	PANDROL UK LIMITED	2524	Fastclip	M4	Ductile	Day/Night	65,536	22/12/2008	0		17/01/09 04:49	619	250C	487	7,792	1,110	100.2%	01:46
20	COO7150	PANDROL UK LIMITED	2524	Fastclip	M4	Ductile	Day/Night	65,536	22/12/2008	0		19/01/09 11:00	677	250C	1,716	27,456	1,110	100.0%	06:15
21	COO5573	BOSCH CHASSIS SYSTE...	2695	Abutment Bracket	M4	Ductile	Day/Night	15,000	12/2008	0		19/01/09 17:18	488	250C	973	11,676	1,233	100.0%	03:28

Comments

Active Tools for Order Item: 2079-"Shoulder"

Dies	Pattern Set	Tool Name	Tool Date	CPH	Adjusted MPH	QtY/hr	Moulds for Job	Time for Job (h)	Metal for Job (kg)	Dies Tool	Product/CPH per Pattern	Compatible Dies's Defined per Dies		Quantities	
													Ordered	26,880
													Shipped	0
													Inventory	0
250C	271	Shoulder 75197 (Patt Set 2...	8/07/2008	16	270	20	1,715	06:21	34,300 M4	2079/16			Required	26,880
													Reject %	2
													To Produce	27,429

The Decision Optimization System also received a live data feed from each machine, allowing it to compare actual production against scheduled production in real time. Whenever the situation reached a point where the schedule was no longer feasible, the Decision Optimization System would flag that re-optimization was required based upon the current state of production in the foundry—in other words, that the current production schedule would no longer be met, and that re-optimization was required to re-align future production with the current reality on the factory floor.

This streaming machine data allowed for a real-time view into each foundry, providing CAST Metals with not only a "scheduled" versus "actual" perspective, but also an ability to dynamically re-optimize production whenever the unexpected occurred (which unfortunately was often). During this re-optimization process, if the Decision Optimization System ran out of capacity in the foundry to process all orders with hard due dates, it would flag these orders as "unassigned items," as shown below:

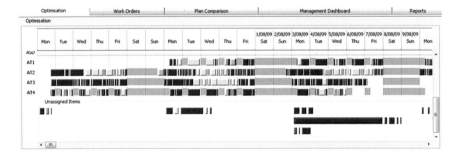

The Decision Optimization System also provided additional reporting on various utilization ratios, throughputs, and other KPIs in both graphical and numerical form. Such reporting was also displayed within the scheduling system itself, as shown below, allowing CAST Metals to evaluate the performance of each production schedule:

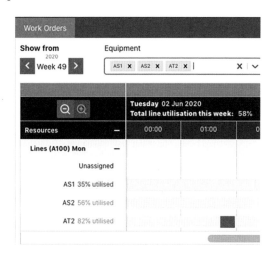

From the very start of the project, CAST Metals had a clear view of the KPIs it wanted to improve, which provided a baseline and benchmark for validating the performance of the Decision Optimization System. Also, given the scale of CAST Metals' manufacturing footprint, management knew that any improvement in these metrics would translate to a direct and significant improvement in financial performance of the entire business. The realization of these benefits, however, was dependent on CAST Metals successfully navigating two change management challenges:

- First, the spreadsheets that CAST Metals had built up over the years had to be replaced by the Decision Optimization System, which was a challenge in itself within each foundry. "But I've been using that spreadsheet for years," the production schedulers would complain. "No system can capture everything I've put into that spreadsheet!"
- And secondly, during the user acceptance testing ("UAT") phase of the project, CAST Metals encountered further resistance from end users because the Decision Optimization System was recommending schedules that "didn't look right."

On this second point, CAST Metal realized that "optimization projects" were very different from "automation projects," and thus required more significant change management. If CAST Metals had configured the Decision Optimization System to exactly replicate what end users did and generate schedules that "looked right" to everyone, then the only value of such a system would have been the time saved in generating these schedules. This would have become an automation project, because CAST Metals would be automating the scheduling process with the end result being exactly the same (only faster). This wasn't the outcome CAST Metals was seeking, so the Decision Optimization System wasn't configured to replicate what end users did, but rather, to generate optimized schedules that could improve various KPIs (and so by definition, these schedules had to be different to those being generated by end users up to that point). The only way the Decision Optimization System could create value was by recommending a *different* decision that led to a *different* result—in the case of CAST Metals, a different production schedule that led to improved asset utilization, throughput, and customer service levels). And because the system was recommending something different to what had been typically done in the past, change management was more challenging—"Hey! That doesn't look right to me," the production schedulers would say. "I would have done it differently."

The first change management challenge was addressed through extensive user training on the new Decision Optimization System, whereas the second challenge was addressed by educating end users on why the system was making certain recommendations. In addition to this education, the Decision Optimization System provided an explanation in natural language as to why a certain schedule or scheduling decision was optimal (i.e. "explainable AI," as discussed in Chapter 6.9). CAST Metals was able to successfully navigate these change management issues in large part because of strong executive sponsorship and leadership (which Chapter 11.3 explores in greater detail). Once the system was fully adopted, CAST Metals began executing the new schedules and realized an immediate improvement in manufacturing performance. Each foundry experienced a jump in DIFOT and asset utilization metrics, as well as reduced overtime labor requirements. The improvement within each foundry varied according to the capability of the production scheduler and the sophistication of their spreadsheets. In other words, the Decision Optimization System outperformed very capable staff with very sophisticated spreadsheets, but the outperformance was modest; as for average staff with basic spreadsheets, the improvement was pronounced.

And lastly, not only did CAST Metals know what was happening within each foundry in real time, but the company could now dynamically re-optimize and re-align the forward schedule with the production realities at each site, thereby running a continuously optimal manufacturing process.

10.3 Logistics and Distribution Optimization

Logistics is the "connector" of a supply "chain," involving modes of transport (such as trucks, trains, and ships), as well as storage locations. Many logistics and distribution problems are "multi-nodal," in the sense that one truck needs to make many deliveries or stops (like the traveling salesperson problem in Chapter 2.1, which is representative of typical routing or journey planning problems). In multi-nodal problems, the optimization objective is to find the route that minimizes travel time and other cost metrics, while satisfying a number of hard and soft constraints (such a delivery times or slots). From that perspective, such problems are "one dimensional" and rarely encompass any prediction component other than a demand forecast used for load planning purposes.

Rather than concentrating on a standard logistics operation, this case study will explore a node-to-node distribution problem where the complexity arises not from the optimization challenge of finding the best route, but from the number of factors that impact the distribution plan (and which need to be considered during the optimization process, such as price changes, inventory levels, seasonality, and more), and the significant prediction problem that underpins the entire optimization result.

With this in mind, the case-study presented in this section is about GMAC, a car financing organization in the United States that leases around one million cars each year to consumers, organizations, and rental agencies.[5] When a car lease agreement expires—which could be from one to five years—the car is either returned to GMAC or purchased by the leasee (in either case, these cars are called *off-lease cars*). GMAC doesn't need to worry about the purchased off-lease cars, but it needs to sell the returned off-lease cars at one of many auction sites located across the United States. Each of these returned cars is different in its make, model, body style, trim, color, year, mileage, and damage level, and the overall number of cars leased each year translates into approximately 5,000 returned off-lease cars each day. The following figure illustrates a particular day, where green circles represent the returned off-lease cars and yellow circles represent the 50 auction sites at which GMAC sells its cars:

---

5   This case study is also covered in the article by Michalewicz, Z., Schmidt, M., Michalewicz, M., and Chiriac, C., called *A Decision-Support System based on Computational Intelligence: A Case Study*, IEEE Intelligent Systems, Vol. 20, No. 4, July–August 2005, pp. 44–9, which can be downloaded from: https://www.complexica.com/hubfs/case%20 studies/Case_Study_An_Intelligent_Decision_Support_System.pdf.

The larger the green circle, the more cars were returned at that particular location, with the sizes and locations of these circles varying from one day to the next (as different people and organizations return their cars at different locations). The yellow circles, on the other hand, represent the designated 50 auction sites where the returned off-lease cars are sold. The locations of these auction sites are fixed.[6]

GMAC's task was to distribute the daily intake of approximately 5,000 cars to the 50 designated auction sites; in other words, to assign an auction site to each particular off-lease car. For example, if the first car is located at a dealership in Northern California, GMAC would consult some reports[7] on what the average sale price for that particular car is at each auction site (after adjusting for mileage, trim, damage level, etc.), and then ship the car to the auction site with the highest average sale price. Of course, GMAC also needed to estimate the transportation cost to each auction site (the longer the distance, the higher the cost, and longer transportation times resulted in higher depreciation costs and risks). Using this method, GMAC's decision for the first car could be visualized in the following way:

---

6   Although the locations of the 50 auction sites are fixed, GMAC may, from time to time, change the auctions it does business with by dropping some sites and adding new ones (thereby changing the location of the 50 yellow circles). This may happen if cars are routinely damaged at some sites, auction fees go up, or some other reason. However, these decisions raise several additional questions, such as: *How do we evaluate the monetary impact of dropping some sites and adding others?* and *Can we increase profits by replacing some auction sites with others?* We will address these important questions later in this section.

7   Many reports are available for estimating the auction price of cars, including *Black Book*, *Kelley Blue Book*, the *Manheim Market Report*, and others.

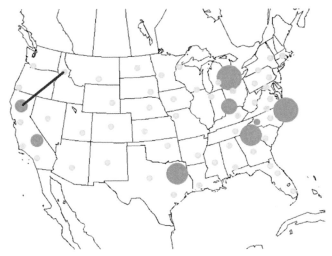

with the blue line representing the decision to ship the car from Northern California (green circle) to an auction site in Idaho (yellow circle). GMAC would then repeat this process for each car. Although straightforward, this approach for distributing off-lease cars didn't work very well, and led to a situation when

GMAC didn't capture the full value of each off-lease car. Because the entire process was based on manual analysis and individual, car-by-car decisions, any small mistake that resulted in a net reduction of "only" $50 per car, would cost GMAC $250,000 in a single day!

As such, GMAC defined their business problem and objective as:

Maximize the aggregate resale value of all returned off-lease cars by optimizing the logistics and distribution to individual auction sites

This was a difficult business problem to solve and objective to realize, because of the following reasons:[8]

1. *Number of possible solutions.* There were 50 possible solutions for each individual car, as GMAC can ship a car to any of the 50 auction sites; for two cars, there were 2,500 possible solutions (50 × 50); for three cars, 125,000 possible solutions (50 × 50 × 50), and so on. For 5,000

---

8   Recall our overview of complex business problems in Chapter 2, where we discussed the astronomical number of possible solutions, dynamic environments, and problem-specific constraints—all of which are present in this problem.

cars, however, there were approximately 50⁵⁰⁰⁰ possible solutions (50 multiplied by itself 5,000 times)! This was an overwhelming number (1 followed by 8,494 zeros) and no supercomputer could evaluate all these combinations in a billion human lifetimes. Nevertheless, GMAC had to make daily decisions for these cars, irrespective of how complex the problem was or the number of possible solutions.

2. *Transportation costs.* When GMAC shipped an entire truckload of cars from one location to another, it would realize a better price per car than when it shipped only one car (or a few cars), thereby lowering the overall logistics cost. This occurred because the cost of transport was primarily tied to individual trucks and drivers, with the number of cars on each truck being of secondary importance. Hence, the relationship between transportation cost and number of transported cars looked similar to the model presented towards the end of Chapter 6.1. Given this model, the cost for sending a single car from one location to another was $250, but the cost of sending two cars was $300 (reducing the cost per individual car to $150), with each additional car being $50. If a truck could hold 10 cars, then the transportation cost of a fully loaded truck was $700, or just $70 per car. But if GMAC needed to transport 11 cars, then a "jump" occurred in cost with $700 for the 10 cars on the first truck, and $250 for the single car on the second truck (for a total of $950).

3. *Volume effect.* Although GMAC wanted to send each car to the auction site where the highest price could be realized, sending too many cars of same color, make, and mileage to the same auction site would trigger the volume effect. For example, if GMAC sent 45 white Chevrolet Camaros to the same auction site (which might have all been returned from a rental agency on the same day), then these cars were likely to sell for the minimum opening price, because with 45 identical cars for sale, there wouldn't be enough buyers to bid the price up on each car (meaning there was a limit to how much supply could be absorbed by each site). On the other hand, if GMAC sent only five Chevrolet Camaros to the same auction site, then these five cars would fetch a higher price because the same number of buyers would be bidding on a smaller number of cars. To illustrate this point, the volume effect for a particular car at a particular auction site might be:

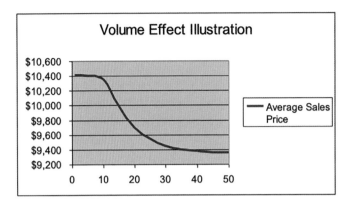

This graph illustrates the volume effect phenomenon, where GMAC could realize more money per car by selling *fewer similar cars*. In this example, the current average sale price for a particular car at a particular auction site might be approximately $10,400, and GMAC could realize this price by shipping up to seven cars to this location. However, if GMAC shipped 30 similar cars, then the average sale price per car would drop to $9,450. Note that the term "similar" could mean more than just the same make, model, or color. For example, many white compact cars of different makes and models often competed for the same buyers, thereby reducing the average sale price per car. Consequently, due to the volume effect, it wasn't effective for GMAC to consider one car at a time.

4. *Price depreciation and inventory holding costs.* To further complicate matters, every auction site had a set day for selling cars (e.g. every second Friday at 10 am). Because of this, if GMAC shipped 100 cars to an auction site and the delivery arrived one or two days *after* the auction day, then these cars would sit until the next sale day, incurring depreciation and holding costs. Because of this, GMAC needed to check the exact sale day and inventory levels across all 50 auction sites before making any new distribution decisions.

5. *Price changes.* Used car prices change over time, and these changes may be slow and subtle (over many years as consumer preferences change), sudden and dramatic (as was the case in March 2020 when the COVID-19 panic set in), or region specific (e.g. convertible cars become unpopular in northern states during the winter months, and consequently, they fetch a lower price—which is part of the "seasonality effect"). GMAC also had to deal with next year's models entering the market during August and September, causing older models to drop sharply in price (also part of the seasonality effect). During this time of year it was better to ship cars nearby and sell them quickly, rather

than shipping them longer distances to more lucrative auction sites. Additionally, new body style models are introduced every few years, causing an even bigger drop in price for the older body style.

Coming up with the daily decision of where to send the returned off-lease cars wasn't easy, as the decision needed to consider the above factors.

Furthermore, the process of transporting a car to a specific auction site could take up to two weeks, as the truck would have to drive to the pick-up location, load the car, pick up some additional cars (possibly somewhere close by), and then finally deliver the cars to the designated auction. Because of this, GMAC had to consider the sale price for each car a couple of weeks ahead of time. For example, for a car located in Jacksonville, Florida, GMAC might consider sending this car to an auction site in Georgia, Pennsylvania, or California. The price prediction for these three auction sites would be different, because GMAC would be predicting the sale price five days into the future for the Georgia auction site, ten days into the future for the Pennsylvania auction site, and fifteen days into the future for the California auction site. The differences in time were due to the transportation distance. However, to predict these prices, GMAC needed to consider the seasonality effect, price depreciation, volume effect, and inventory levels. In making the decision of Georgia vs. Pennsylvania vs. California, GMAC would also need to weigh the possibility of a better price in California against the higher transportation cost, higher depreciation, and higher overall risk.

These challenges were ideally suited for AI-based algorithms and the implementation of a Decision Optimization System, which would rely on advanced prediction, optimization, and self-learning capabilities to improve GMAC distribution decisions.

Data, information, & knowledge

GMAC maintained a historical collection of transactional sales data that could be visualized as a two-dimensional table representing off-lease cars sold at auction. One dimension of the table represented the number of records (cars), and the other dimension represented the charac-

teristics of each car (e.g. VIN,[9] make, model, mileage, etc.):

---

9   VIN is an acronym for "Vehicle Identification Number," which is a string of 17 digits and letters that contains considerable information about a specific vehicle, (including country of origin, manufacturer, and model year).

VIN	Type	Make	Model	Miles	Year	Color	Transmission	Body/Doors	Damage
2G1FP22P1P2100001	Rental	Chevy	S-10	34,983	2002	Silver	Manual	2D	$0
WB3PF43X8X9000331	Lease	Chevy	Cavalier	59,402	2001	Red	Automatic	2D Coupe	$0
4BBG38FJF04JDK000	Lease	Chrysler	Sebring	74,039	2000	Gray	Automatic	2D Coupe	$500
DJOW03FFU990SJ206	Lease	Ford	Escape	37,984	2001	Green	Manual	4D Sport	$250
JD8320DJ2094GK2X3	Rental	Ford	Focus	30,842	2001	Green	Manual	4D Sedan	$0
2JE9F0284JD0213M3	Lease	Isuzu	Rodeo	59,044	1999	White	Automatic	4D Sport	$250
4380JDDD9W02MD001	Rental	Jeep	Cherokee	48,954	2000	Black	Automatic	4D Sport	$500
490DK20285JF0209D	Rental	Mazda	626	38,943	2000	White	Automatic	4D Sedan	$0
10D92JD920KD00002	Lease	Nissan	Altima	39,488	2000	Black	Automatic	4D Sedan	$0
D920DKJ0284JJ9990	Rental	Nissan	Altima	23,584	1999	White	Manual	4D Sedan	$0
JD88D92JJD02K3361	Rental	Saturn	L	21,048	2001	White	Automatic	4D Sedan	$750
10DS0JJ20DXI00093	Lease	Suzuki	Vitara	15,849	2003	Yellow	Automatic	2D Sport	$0
21KD02KD0DJ920M27	Lease	BMW	Z3	49,858	2000	Blue	Manual	2.3 RSTR	$250
389DJ2DD298JWQ082	Lease	Ford	Explorer	42,893	2002	Green	Automatic	XLT 4WD	$0
108DJ2048FJJ20043	Rental	Ford	Mustang	20,384	2002	Red	Manual	GT	$0
DJC82002009DD2J04	Rental	Mercury	Frontier	27,849	2001	Silver	Automatic	SE-V6 Crew	$500
830DMM3029XMW0092	Lease	Honda	Accord	26,849	2002	Yellow	Automatic	EX V6	$0
CNEU200220CCI2202	Rental	Toyota	4Runner	33,483	2000	Silver	Automatic	SR5	$0
CNDJ2940JD88D2JD0	Lease	VW	Beetle	5,459	2003	Blue	Manual	GLS 1.8T	$0
1VC0CMEJ200V9EJJ1	Lease	Toyota	4Runner	81,837	2001	Silver	Automatic	LMTD 4WD	$250

This historical sales data contained the VIN, postal code of the auction site (ZIP), transaction date, and the sale price of each car:[10]

VIN	ZIP	Date	Price
39WWK93309KJ33012	28262	2.11.2004	$12,035
UDJ2293M99DL0K220	30334	2.11.2004	$15,600
4D09WJD92JE93H990	30334	2.11.2004	$10,590
KD37D92JF83NF8822	90012	3.11.2004	$9,265
NKI2389DD974F2235	28262	3.11.2004	$13,450
K29DH38FHW02HD923	48243	3.11.2004	$13,955
MDK293HFDWH299305	90012	4.11.2004	$12,495
28DN39FNDJW2N0024	90012	4.11.2004	$11,925
29H93NFI3HJF93F04	48243	4.11.2004	$11,396
ND920ENF1NAD02834	48243	5.11.2004	$9,835
D39DJ39EHQ8HH9335	28262	5.11.2004	$8,965
02UFIMF03JF9SH935	90012	5.11.2004	$13,960
D932NF93HG9057362	48243	5.11.2004	$8,830
00F8EB3IDNB293758	48243	8.11.2004	$7,920
IE038THJ203TH0234	28262	8.11.2004	$19,250
39FH324MV092HGM39	48243	8.11.2004	$22,640
F92N9F389FH120458	90012	8.11.2004	$13,580
F9485JG03H25495J5	30334	9.11.2004	$16,970
08GN94HJH03J49327	30334	9.11.2004	$14,320
F04JH402KG4509G45	48243	9.11.2004	$9,110

GMAC also possessed data for individual auction sites (e.g. the average number of participating auctioneers during different times of the year) and external data such as historical weather conditions at different auction sites during different sale days, historical petrol prices, color preferences in different areas of the United States, and so on. GMAC used this data to generate a variety of reports for the price difference between auction sites for the same off-lease car on attributes such as color, volume of cars sold at each auction site, number of auctioneers at different seasons at different auction sites, and so on. The following graph illustrates one such report: The price difference between auction sites for one particular make (Pontiac) and model (Grand Prix) with an odometer reading between 20,000 and 40,000 miles:

---

10 We could easily obtain the characteristics of each car by merging it with the previous table.

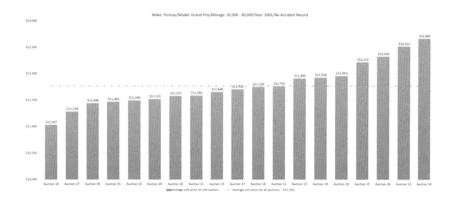

GMAC also studied the volume effect at different auction sites for various types of cars, and this knowledge was presented via graphs and other reports, like the one shown below (which shows the sale price of a Pontiac Grand Prix sold with a number of similar cars at the same auction site, with the different colors representing different odometer ranges—in this example, yellow circles correspond to the lowest odometer range of 0 to 10,000 miles:

Nonetheless, all this data, information, and knowledge were of limited assistance in helping GMAC make the best daily distribution decision, because even if GMAC had "perfect knowledge" and could accurately predict the price of *any* car at *any* auction site for *any* day, they still wouldn't know how to optimally distribute 5,000 cars because of all the complexities of this problem, such as logistics, price depreciation, inventory levels, volume effect, and so on. The number of possible distributions was simply too large to be evaluated in any reasonable amount of time, which drove the business case for an AI-based Decision Optimization System capable of increasing the aggregate resale value of returned off-lease cars.

Decision Optimization System (prediction, optimization, & self-learning)

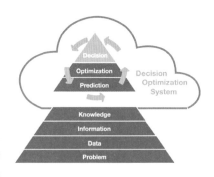

For this particular logistics and distribution problem, the Decision Optimization System had to consider the characteristics of each car, characteristics of each auction site, transportation costs, volume effects, countrywide inventory (as well as cars in transit to various auction sites), price depreciation curves, and market-driven changes in price.

Before the predictive model was built, the data went through a data preparation process that included variable transformation and variable composition, data reduction and normalization, and the generation of missing values (as discussed in Chapter 4.2; also, for more information on this process, please watch the supplementary video at: www.Complexica.com/RiseofAI/Chapter4). GMAC also augmented its internal data with Black Book data (which provided regional sale prices, with each region containing several states and more than a dozen auction sites) and the Manheim Market Report (which reported on the average sale price of all cars sold at auctions owned by Manheim). The resulting prediction model was an ensemble based on decision trees (as discussed in Chapter 5.1) that generated sale price predictions in the following sequence of steps:

1. *Base price.* A predicted "base price" was generated based on the car's make, model, body style, and year.
2. *ZIP-based make/model adjustment.* Because some makes/models sold for a premium or discount in certain regions, the prediction model adjusted the base price for these specific makes/models in certain regions (e.g. Chevrolet Corvettes might sell for a $300 premium in Florida and California, and a $600 discount in Montana and Idaho).
3. *Car group/color adjustment.* Because some car groups/colors sold for a premium or discount irrespective of the region, the prediction model adjusted the base price for these specific car groups and colors (e.g. yellow Chevrolet Corvettes might sell for a $500 premium, while green ones for a $1,000 discount).
4. *Mileage adjustment.* The prediction model adjusted the base price for mileage and *model-year-age*, which was the age of a car according to its model year (i.e. when the 2005 Chevrolet Corvette became available in August 2004—which underwent a complete body style change—the

model-year-age of the 2004 Chevrolet Corvette became 1, as that model year was only one year old).

5. *Depreciation adjustment.* The prediction model adjusted the base price for daily depreciation, as calculated from the car's return date to its predicted sale day. Because the daily depreciation rate was higher in the summer months (preceding the introduction of new models), the depreciation rate increased from June onwards, reached its highest value in August, and then decreased to lower than average values for October, November, and December.

6. *Seasonality adjustment.* Because some makes/models sold for a premium or discount in certain regions at different times of the year, the prediction model adjusted the base price for these specific makes/models during certain seasons (e.g. convertible Chevrolet Corvettes may sell for a $1,800 discount in the northern states during the winter months).

7. *UVC adjustment.* The Universal Vehicle Code (UVC) component provided a more detailed car specification than the VIN, and in cases where the UVC was available, the prediction model adjusted the base price for additional options (e.g. the UVC might reveal that a specific Chevrolet Camaro is equipped with an upgraded suspension package).

For an average daily intake of off-lease cars, the ensemble model would predict each car's final auction price. However, if GMAC received a large number of similar cars on a particular day, then the predicted auction prices for these cars were adjusted further to account for the volume effect. For more information on the predictive model used by GMAC for this particular distribution problem, please watch the supplementary video at: www.Complexica.com/RiseofAI/Chapter5.

The Decision Optimization System also provided GMAC with the ability to add, modify, or delete various constraints and business rules. Constraints that were applied to all auction sites were regarded as global constraints, and an example of this was the "maximal transportation distance" constraint which limited the transportation distance of all cars—as shown in the screen below:

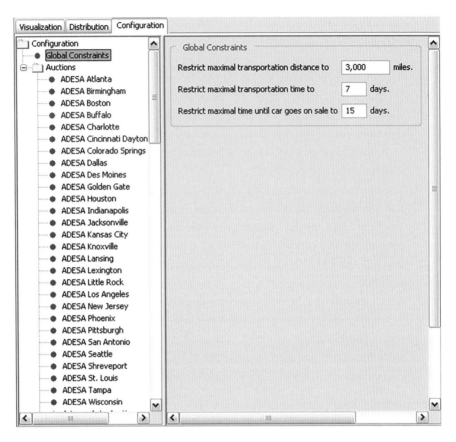

GMAC could also implement a large variety of local, auction-specific constraints within the Decision Optimization System, such as:

- *Mileage constraints*: which defined the upper and lower mileage of cars that could be shipped to a specific auction site. An example of this constraint would be "only ship cars that have between 30,000 and 70,000 miles to the ADESA Atlanta auction site."
- *Model year constraints*: which specified a range of model years that could be sent to a specific auction site. For example, GMAC could specify that a particular auction site could only accept model years between 2002 and 2004.
- *Make/model exclusion constraints*: which specified certain makes/models that were to be excluded from specific auction sites.
- *Color exclusion constraints*: which specified certain colors that were to be excluded from specific auction sites.
- *Inventory constraints*: which specified the desired inventory level at each auction site. For example, GMAC could specify an inventory level between 600 and 800 cars for an auction site at any particular time.

The screen below shows the local constraints set for the "ADESA Boston" auction site:

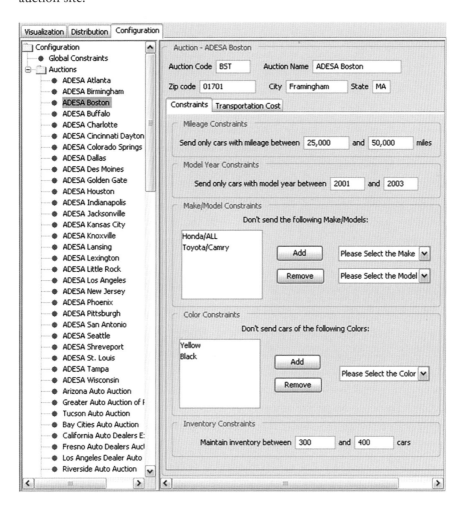

Each auction site could have different constraint settings, which represented the business rules that GMAC wanted to operate under for that particular site. For the ADESA Boston auction site, the constraints represented the following business rules (as shown above):

- "Send only cars with 25,000 to 50,000 miles"
- "Send only 2001, 2002, or 2003-year models"
- "Do not send any Honda or Toyota Camry cars"
- "Do not send any yellow or black cars"
- "Keep the inventory between 300 and 400 cars"

Except for the inventory constraint, all these constraints were defined as hard constraints. If the Decision Optimization System had to break a hard constraint, it would mark this recommendation with the notation "constraint violation." Inventory constraints, on the other hand, were defined as "soft" constraints and a penalty was assigned to solutions that violated these constraints. The penalty for violating a soft constraint would grow exponentially, and so instances where this constraint was violated in a significant way were rare. However, if the Decision Optimization System had to process a very large number of cars on a single day, then the inventory constraint might have been violated at almost every auction site. In such cases, the exponential penalty function would make these violations uniform. For example, in a case where all auction sites have a maximum inventory constraint of 300 cars but the current number of cars to be distributed would increase this inventory level to an average of 400 cars per auction, then the penalty for violating this soft constraint would be evenly distributed across all sites (so that they have the same degree of violation).

These constraints allowed GMAC to set various business rules (e.g. "do not send any red cars to Florida") within the Decision Optimization System, and so the configuration screen served as a link between GMAC and the system. GMAC could also use this configuration screen to investigate various "what-if" scenarios, such as "what would be the distribution of cars if we set the maximum transportation limit to 500 miles?" Because 300 auction sites were configured in the system and only 50 of them were "active," GMAC could activate or deactivate any auction site, and then re-run the optimization process to test a specific what-if scenario, such as "what would happen to the aggregate resale value of all cars if we used 60 auction sites instead of 50?"

GMAC could also use different what-if scenarios to investigate different transportation cost options available from different suppliers. The Decision Optimization System calculated the transportation cost from any distribution center to any auction site for any number of cars, and takes into account two factors that influenced this cost: (1) the distance between a distribution center and an auction site, and (2) the number of cars being shipped. The screen below shows the transportation costs for the ADESA Boston auction:

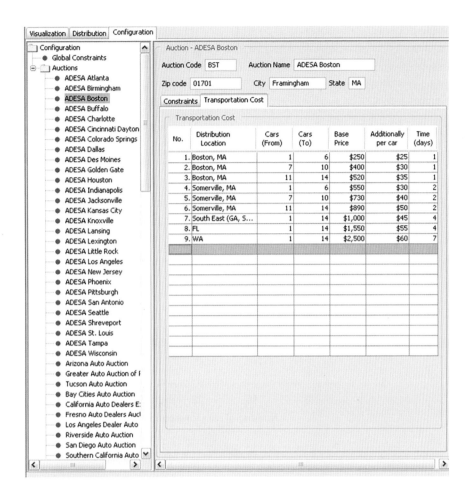

In this screen, the transportation cost is defined for cars sent to the ADESA Boston auction from five different locations.[11] The first two locations are defined by the cities Boston, MA and Somerville, MA; the third location is defined by a region containing the states Georgia, South Carolina, and North Carolina; while the fourth and fifth locations are defined by the states Florida and Washington, respectively. According to the transportation prices above, it would cost $250 to send a truck to Boston, MA, plus an additional $25 for each additional car. If GMAC wanted to ship six cars, then the transportation cost would be $400 ($250 + $25 × 6 = $400).[12] Also, row "No. 9"

---

11 The transportation cost was defined in terms of how much it would cost to ship a car (or group of cars) from a particular ZIP code, city, state, or region to the auction site.

12 If GMAC wanted to ship more than six cars, then the cost would be $400 for the first six cars ($250 plus $150 for six cars), plus $30 for each additional car. Hence, to ship 8 cars, the cost would be $400 for the first six cars, plus $60 for two additional cars, for a total of $460. Another price break occurred at the eleventh car, reducing the incremental cost per car to $35.

above defines the transportation cost between the ADESA Boston auction and the state of Washington. Because of the long distance (approximately 3,000 miles), it would cost $2,500 to ship a car to Washington, plus an additional $60 for each car on the same truck. Although the cost of shipping one car would be $2,560, the cost of shipping fourteen cars would be $3,340 ($2,500 + $60 × 14 = $3,340), or about $239 per car (which is ten times less!). As the following graph illustrates, the more cars transported from the same location, the smaller the transportation cost per car (in this particular case, cars that are transported from Boston, MA to the ADESA Boston auction):

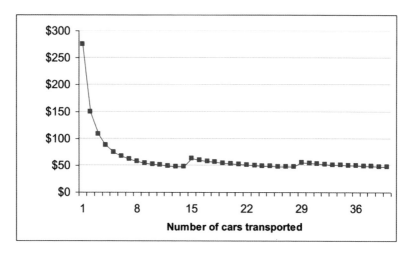

In this graph, the average transportation cost per car decreases from $275 for one car to just $47 for fourteen cars. The graph also illustrates that the average transportation cost increases to about $62 when we need to transport fifteen cars (because an additional truck is needed for the extra car). After the fifteenth car, the average transportation cost goes down again, with smaller spikes when additional trucks are needed.

Besides these transportation costs, the Decision Optimization System also used inventory levels for each auction site to calculate several important parameters for the optimization process. One of these parameters was the volume effect, which was based on how many similar makes/models (or cars of the same color) were present at a specific auction site. Another important parameter was the anticipated sale date. If GMAC had 1,200 cars at a particular auction site (or in transit) and approximately 500 were sold during each auction session, then GMAC could assume that a car shipped today would be sold in the third auction session. Therefore, the Decision Optimization System needed to consider the additional depreciation and seasonality effect during this additional time. Once the system-generated distribution plan was approved, the auction inventory was updated with the new cars assigned to each auction. And lastly, the cars that had been recently sold at these auction sites were removed from inventory.[13]

The optimization model generated a variety of possible distribution plans that served as input to the prediction model. This input provided a destination assignment (i.e. auction site) for each off-lease car, which the prediction model used to generate a predicted sale price. The optimization model then summed all these predicted prices (i.e. the output data) to evaluate the quality of the distribution plan—the higher the sum of the predicted sale prices, the better the distribution plan. Hence, there was a strong relationship between the prediction and optimization models, as is the case within most Decision Optimization Systems.

The optimization model was comprised of several different AI-based algorithms that used different solution representations. For instance, evolutionary algorithms (see Chapter 6.7) used solutions based on indirect representation, where all available auction sites were sorted by *distance* from a particular car. In other words, auction 1 was the closest (distance-wise), auction 2 was the second closest, and so forth. Hence, each solution was represented by a vector of auction site indices (relative to a particular car), and the length of the vector was equal to the number of cars being distributed:

3	4	4	...	1	1

The vector above represents a solution where the first car is shipped to the third closest auction (for this particular car), the second car is shipped to the fourth closest auction (for this particular car), the third car is shipped also to the fourth closest auction (note, however, that the second and third car are most likely shipped to different auction sites, as the fourth closest auction for the second and third car need not be the same), and so on, with the last two cars being shipped to the closest auction sites. In this particular implementation of evolutionary algorithms, the optimization model applied the elitist strategy, which forced the best solution from one generation to the next, as well as various mutation and crossover operators that were discovered through experimentation. For additional information on the optimization model used within the GMAC Decision Optimization System, please watch the supplementary video at: www.Complexica.com/RiseofAI/Chapter6.

To enable learning within the Decision Optimization System, both the prediction and optimization models updated themselves with the arrival of

---

13 Data about sold cars was also used to tune the prediction model (explained later in this section).

new data. The prediction model contained numerous parameters (different values for various adjustments) that were automatically updated to capture changing trends in the used car marketplace at regular intervals (as discussed in Chapter 7.2), and in terms of optimization, each day brought a different "instance" of the same problem, as changes occurred in the number and type of cars to be distributed. For this reason, the optimization model was based on several optimization algorithms where each algorithm contained a few parameters that were adapted (as discussed in Chapter 7.3), and the usage of several optimization algorithms together generated a result that was better than the result of any single algorithm. For additional information on the learning components of this case study, please watch the supplementary video at: www. Complexica.com/RiseofAI/Chapter7.

The graphical user interface of the Decision Optimization System allowed GMAC to add, modify, or delete various constraints and business rules (as discussed earlier), as well as "visualize" the distribution plan. In the screen below, there are icons for each distribution center and each auction site, and four performance graphs. The white "horseshoe" icons represent distribution centers where off-lease cars are collected, cleaned, and conditioned for eventual sale at an auction site.[14] The red "hammer" icons represent auction sites, and the lines between the distribution centers and auction sites represent the volume of cars transported between these points (the thicker the line, the more cars are transported):

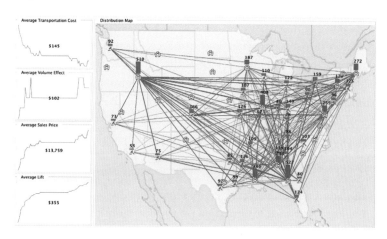

---

14 Only the largest leasing companies—such as GMAC—have such distribution centers. For leasing companies that do have them, an off-lease car is dropped off at a dealership, then shipped to the nearest distribution center for cleaning and conditioning, and then the Decision Optimization System ships the car to the best auction site. For leasing companies that don't have distribution centers, the car would be cleaned and conditioned at the dealership, and the Decision Optimization System would ship the car to the best auction site directly from the dealership.

The four graphs on the left-hand side display the optimization objectives:

- *Average Transportation Cost.* The Decision Optimization System calculates the total transportation cost and then displays the average cost per car.
- *Average Volume Effect.* The Decision Optimization System calculates the total lost revenue due to sending too many similar cars to the same auction sites and then displays the average value lost per car.
- *Average Sale Price.* The Decision Optimization System calculates the expected sale price for all the cars and then displays the average value per car.
- *Average (Net Sale Price) Lift.* This corresponds to the average "profit improvement" per car. The Decision Optimization System calculates this as the difference between the predicted average net sale price for the optimized solution (i.e. the sale price after subtracting all auction fees, transportation costs, etc.), and the predicted net sale price for the standard solution (which was based on expert rules that were developed by GMAC over the years).

In the screen above, the average transportation cost per car (first graph) has steadily decreased during the optimization run, while the average volume effect per car (second graph) has increased. The Decision Optimization System has chosen a distribution plan with a higher average volume effect, because it was more than offset by a lower average transportation cost and higher average sale price per car. This in turn resulted in a higher average (net sale price) lift per car (fourth graph).

Once the optimization process is complete, the Decision Optimization System generated an output file with the recommended distribution of cars, specifying the distribution center, recommended auction site, predicted sale price, transportation cost, and other data:

No.	Make	Model	Trim	Year	Distribution Location	Auction	Sales Price	Volume Effect	Distance	Transportation Cost	Net Price	Lift
1	Ford	F150	Base	2001	Augusta, ME	ADESA Buffalo	$9,452	$0	445	$180	$9,272	$113
2	Jeep	Grand Cherokee	Limited	2003	Albany, NY	ADESA Buffalo	$17,786	$0	241	$123	$17,663	$225
3	Toyota	Land Cruiser	VX	2002	Annapolis, MD	ADESA Buffalo	$31,662	$0	300	$153	$31,509	$318
	Total	3					$19,633	$0	328	$152	$19,481	$218
4	Dodge	Durango	Base	2002	Boise, ID	ADESA Seattle	$15,548	$94	385	$68	$15,480	$74
5	Dodge	Grand Caravan	SE	2002	Boise, ID	ADESA Seattle	$10,025	$61	385	$68	$9,957	$48
6	Ford	Expedition	Eddie Bauer	2003	Boise, ID	ADESA Seattle	$25,502	$154	385	$68	$25,434	$122
7	Ford	Expedition	Eddie Bauer	2003	Boise, ID	ADESA Seattle	$24,858	$150	385	$68	$24,790	$119
8	Ford	Mustang	GT	2002	Boise, ID	ADESA Seattle	$16,361	$99	385	$68	$16,293	$78
9	Ford	Mustang	Base	2001	Boise, ID	ADESA Seattle	$11,083	$67	385	$68	$11,015	$53
10	Honda	Accord	EX	1997	Boise, ID	ADESA Seattle	$5,334	$32	385	$68	$5,266	$26
11	Honda	Accord	LX	2003	Boise, ID	ADESA Seattle	$12,083	$73	385	$68	$12,015	$58
12	Honda	Accord	EX	2002	Boise, ID	ADESA Seattle	$12,054	$73	385	$68	$11,986	$58
13	Jeep	Grand Cherokee	Limited	2001	Boise, ID	ADESA Seattle	$12,373	$75	385	$68	$12,305	$59
	Total	10					$14,522	$87	385	$68	$14,454	$69
14	Dodge	Durango	Base	2003	Saint Paul, MN	ADESA St. Louis	$18,574	$0	478	$205	$18,369	$199
15	Dodge	Durango	Base	2003	Springfield, IL	ADESA St. Louis	$18,969	$76	109	$55	$18,914	$87
16	Dodge	Neon	HIGHLINE	2002	Springfield, IL	ADESA St. Louis	$7,497	$30	109	$55	$7,442	$22

An auction inventory report (below) was used to show inventory at each auction site, the number of cars being sent to each auction, the projected number of cars at each auction, and whether or not any inventory constraints are violated:

No.	Auction	Inventory	Distributed	Projected	Inventory Min.	Inventory Max.	Over(+)/ Under(-)
1	ADESA Atlanta	261	33	294	200	300	0
2	ADESA Birmingham	254	7	261	150	300	0
3	ADESA Boston	390	7	397	300	400	0
4	ADESA Buffalo	99	3	102	100	150	0
5	ADESA Charlotte	120	19	139	100	200	0
6	ADESA Cincinnati Dayton	123	8	131	100	200	0
7	ADESA Colorado Springs	297	0	297	150	300	0
8	ADESA Dallas	289	3	292	200	300	0
9	ADESA Des Moines	103	1	104	100	150	0
10	ADESA Golden Gate	141	10	151	100	150	+1
11	ADESA Houston	213	0	213	150	300	0
12	ADESA Indianapolis	135	2	137	100	150	0
13	ADESA Jacksonville	190	9	199	200	300	-1
14	ADESA Kansas City	185	16	201	150	300	0
15	ADESA Knoxville	204	2	206	150	300	0
16	ADESA Lansing	258	2	260	150	300	0
17	ADESA Lexington	103	1	104	100	200	0
18	ADESA Little Rock	207	3	210	150	300	0
19	ADESA Los Angeles	257	0	257	150	300	0
20	ADESA New Jersey	154	6	160	150	300	0
21	ADESA Phoenix	154	10	164	150	300	0
22	ADESA Pittsburgh	156	0	156	150	300	0
23	ADESA San Antonio	286	1	287	150	300	0
24	ADESA Seattle	224	10	234	150	300	0
25	ADESA Shreveport	162	6	168	150	300	0
26	ADESA St. Louis	182	6	188	100	200	0
27	ADESA Tampa	214	5	219	150	300	0
28	ADESA Wisconsin	173	2	175	100	200	0
	Total	**5,534**	**172**	**5,706**			

When used in a high-volume setting—where thousands of cars are returned off-lease each day—the Decision Optimization System generated a net profit lift in the hundreds of millions of dollars per year (by predicting the auction site at which GMAC could maximize the resale value of each car, and then optimizing the logistics). There were a few ways to validate this financial result:

- One way was by dividing the daily intake of returned off-lease cars into two equal groups with an almost identical division of makes/models. One group would be distributed using the manual method, whereas the Decision Optimization System would distribute the other group, and then the results would be compared.
- Another way was by using the manual method on selected days of the week (e.g. Mondays, Wednesdays, and Fridays) and the Decision Optimization System for the remaining days (e.g. Tuesdays and Thursdays). Again, the results could be compared when all cars were sold and the aggregate prices known.

- And the third way was by using the Decision Optimization System for one year and then comparing the average sale price with that of the previous year (before the system was implemented).

Using this last method (year-by-year comparison), the benchmark would need to be a trusted pricing source, like the Black Book price guide. GMAC applied this particular method by selecting a subset of cars with the same makes/models, year, trim, etc. and compared the average sale price of these cars with the average Black Book sale price for 2003 (before the Decision Optimization System was implemented). A chart depicting this comparison is presented below:

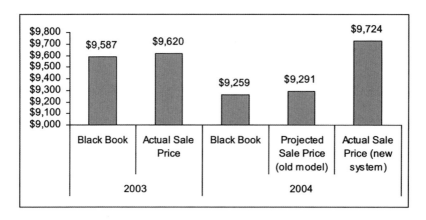

In this example, the average Black Book sale price for a particular mix of makes/models, year, trim, etc. in 2003 was $9,587 per car, and GMAC sold these cars for an average of $9,620 per car, or 0.344% higher than the Black Book sale price. The next step would be to compare the sale prices in 2004 (when the Decision Optimization System replaced the manual method of distributing cars) against the Black Book sale prices for that year. In this example, the average Black Book sale price was $9,259 per car in 2004, and the average actual sale price obtained by the system was $9,724. If the cars had been distributed using the manual method in 2004 (termed "old model" in the chart above), then GMAC would have attained similar results to those of the previous year (i.e. a 0.344% improvement over the Black Book benchmark, or an average of $9,291 per car). Using this approach, GMAC could credit the Decision Optimization System with the increased average sales price of $9,724 minus $9,291, or $433 per car. With one million cars being distributed on an annual basis, this result represented $433,000,000 in additional revenue, not to mention automation of a business process that required a substantial amount of time and human effort.

And lastly, as discussed in Chapter 6.10, the concept of global optimization was highly relevant to this distribution problem given the scale of the problem (e.g. million cars per year) and complexity (e.g. number of possible solutions, volume effect, price changes, transportation costs, etc.). A tempting approach for dealing with this scale and complexity would have been to break the overall problem into smaller "pieces"; for example, breaking up the United States into six regions that each dealt with a subset of the overall problem. This approach could have been taken further by breaking the problem apart into individual states or distribution centers, solving these individual pieces, and then "assembling" the pieces into an overall distribution plan. Although this would have made the problem easier to solve, the result would have been substantially inferior in comparison to solving the problem in its entirety (because the Decision Optimization System could consider and balance the volume effect, price changes, depreciation, and more across the entire United States, rather than being constrained to a particular state, site, or region). As this case study illustrates, the financial result of global optimization can be substantial, but the amount of complexity that needs to be addressed is equally substantial.

PART IV

Implementing AI in Your Organization

Overview

In Part I of this book we demystified the topic of Artificial Intelligence, its history, and primary areas of research focus, paying particular attention to the process of solving complex business problems and making optimized decisions. By examining the many characteristics of complex business problems, we gained a better appreciation of what makes them so difficult to solve and the limitation of information and knowledge for bridging the gap between past data and future (optimized) decisions:

In Part II we presented a technical exploration of Decision Optimization Systems, focusing on how they use data to predict, optimize, and learn; while in Part III we brought to life the application of Artificial Intelligence within sales, marketing, and supply chain business functions, for the purpose of generating revenue and margin growth.

Having covered so much ground, the natural question that now remains is how to introduce Artificial Intelligence into an organization, and how to test the technology and develop a business case. We'll attempt to answer these questions in Part IV, along with our thoughts on the commercial and practical aspects of deploying Decision Optimization Systems.

CHAPTER 11

The Business Case for AI

"The biggest issue I see with so-called AI experts is that
they think they know more than they do, and they think
they are smarter than they actually are. This tends to plague
smart people. They define themselves by their intelligence
and they don't like the idea that a machine could be way
smarter than them, so they discount the idea—which is
fundamentally flawed."

Elon Musk, *visionary and entrepreneur*

When it comes to assessing the application of Artificial Intelligence for any given organization—as well as determining which business problems and approaches are most suitable—we should take several considerations into account from both a commercial and practical nature, including:

- Selecting the right problem
- Starting large or small
- Executive sponsorship
- Return on investment and payback
- Technology partner alignment

Presented in sequential order, this chapter provides practical guidance on these topics for assessing the extent to which the application of Artificial Intelligence (and the associated business case) stacks up.

11.1 Selecting the Right Problem

When thinking about Artificial Intelligence and what benefit it might bring, the first step is defining a business problem worth solving or an opportunity worth capturing, and then working backward into the required technology and data. The cardinal sin that many organizations make is beginning with the data and then trying to *move forward* into the required technology and business problem. Most often, managers will say, "We have lots of data; let's see what we can do with it!" Such an approach is likely to result in disappointing results for a number of reasons, including:

- The organization's starting viewpoint is constrained by what data exists, its quality, and what can be done with it, meaning that any undertaking is unnecessarily restricted and limited in scope from the very onset.
- Within the problem-to-decision pyramid, "good data" refers to any data that can help us diagnose, explain, and assess a specific problem (as explained in Chapter 2.2). Hence, data that's disconnected from any given problem is data without context, without meaning—data that's difficult to work with because the goal is unclear.
- For organizations wanting to explore what AI can offer, data exploration is typically a poor "showcase" of Artificial Intelligence methods (at that end of the data exploration exercise, some manager will inevitably ask: "Where's the AI?").

By starting with the business problem first, our thinking isn't limited by data or technology, and allows us to focus on various business processes that, if improved, could add substantial value. Hence, the first step in finding a suitable application area for AI is being clear that our focus should be on a specific business problem or opportunity, rather than what data exists and its quality (which comes later).

The second step is identifying a *high-value* problem or opportunity, so that any improvement is meaningful for the organization and impactful on relevant KPI metrics. High-value problems usually attract executive sponsorship and considerable attention, making them ideal candidates for showcasing the capabilities of Artificial Intelligence. This is a critical consideration, because solving obscure problems that no one cares about is a pointless exercise (even if the solution is grand, the results are likely to be ignored with a shrug—"so what?" will be the overall assessment).

The opposite is also true. Going after "moonshots" to test the value of Artificial Intelligence is likely to backfire, bringing disappointment to those that sponsored the moonshot and potentially making the organization gun-shy of any future AI projects. Hence, the business problem must be high value on the one hand, but also realistic and attainable on the other. With this in mind, selecting the right problem should involve the following considerations:

- *Complexity*: The greater the complexity of any business process or workflow, the greater the potential applicability and benefit of Artificial Intelligence. As an example, making pricing decisions for 100,000 products involves far greater complexity than making pricing decisions for 100 products. As discussed in Chapter 2.1, complex problems have an astronomical number of possible solutions, making it impossible for any team of human experts to make optimized decisions (especially in fast-moving business environments). Hence, when looking for candidate problems to address with Artificial Intelligence, the complexity of the problem should be a primary consideration, as simple problems rarely require a sophisticated solution.

- *Scale*: The greater the scale of any given problem or opportunity, the greater the value that can be realized through improved decision making. Continuing with the example from above, making pricing decisions for 100,000 products sold to 100,000 customers carries greater scale than making pricing decisions for the same number of products sold to only 100 customers. Because problems with scale and complexity have the potential to generate the greatest return on investment, they should be considered as prime candidates for potential applications of Artificial Intelligence.
- *Frequency or impact of decisions*: Business processes or workflows where decisions are either made frequently (such as pricing decisions), or where the impact of decisions is significant and consequential (such as the implementation of structural changes within an organization) represent higher-value problems that are better suited for the application of Artificial Intelligence over business processes or workflows where decisions are infrequent or of low importance.
- *Feasibility and effort*: And lastly, consideration should be given to the feasibility and effort of addressing the problem. For example, a high-value problem that sits within a well-defined workflow and requires minimal change management represents a better candidate problem than one spread over multiple business functions and beset by significant change management challenges.

The above should rule out many problems or opportunities from consideration, while enabling a comparison of the remaining problems, the level of effort required to address them, and their potential value and benefit.

After selecting a business problem—which might represent "low hanging fruit" from both a feasibility and effort point of view, as well as potential value—we can then work backward into the required technology and data. The easiest way to do this is by describing the problem in terms of "inputs" and "outputs," with the step in the middle being the Decision Optimization System comprised of prediction, optimization, and learning. We first define the "outputs" (i.e. What will the system do? Will it generate a prediction? A recommended decision? What supporting features and functionality are needed?), and then determine what "inputs" are required to achieve the desired output. This is similar to what we described in the opening of Chapter 4:

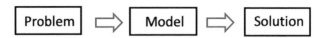

except we can view it in a slightly different way:

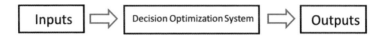

Once the outputs have been defined in the context of a specific business problem, then we can assess the available data in terms of its quality and quantity. Most likely, the internal datasets will exhibit many shortcomings—potentially even serious ones—requiring an assessment on whether this data can be cleaned or augmented to provide a minimal level of input into the Decision Optimization System (as discussed in Chapters 4 and 5). If workarounds can be devised for the imperfect data—which is often the case—then the project can proceed. For this reason, it's important to start with the business problem and move backward into an investigation of the data, because starting with the data and seeing its shortcomings would unnecessarily preclude many (or even most) business problems from consideration. Also, any initiative focused on a high-value problem can create the necessary impetus for an organization to resolve its data challenges, which, in the absence of such an initiative, often go unaddressed.

Furthermore, two additional points are worth mentioning: First, business problems should be addressed with the simplest, most cost-effective solution. This means that if a cheaper and simpler non-AI solution can solve the problem in full (rather than solving it "halfway," as many simpler solutions are apt to do), then that solution should be favored over a more sophisticated approach. However, if a non-AI solution can solve the problem, then it means the problem wasn't overly complex to begin with in terms of the criteria presented in Chapter 2.1:

- The number of possible solutions is so large that it precludes a complete search for the best answer
- The problem is set in a dynamic environment
- There are many (possibly conflicting) objectives
- The problem is heavily constrained

The second point is that whatever problem is addressed with whatever technology, the outcome should be measured against a pre-defined benchmark (as discussed further in the next section). Without such a measure in place, we'll find it difficult to prove value and develop trust with end users.

11.2 Starting Large or Small

The previous section can be summarized as "start with a high-value problem and then work backward into the required technology and data." This provides an understanding of feasibility and effort, which leads to yet another important question: "How large or how small?" There are many possible approaches for deploying a Decision Optimization System, with further consideration required to identify the best option for any particular problem or situation. Each approach presents its own set of benefits and limitations, but ultimately, the main consideration is whether the high-value problem warrants a software project from the start, or if it requires some initial proof of concept for business case purposes. Since the former is self-explanatory, we'll focus on the many different approaches that are possible for the latter, which fall into two broad categories:

Analytics proof of concept

The focus of an analytics proof of concept is on establishing the extent to which AI-based methods can solve the selected problem, as well as quantifying the potential benefits. If we take customer churn as an example, we might hypothesize that early signs of customer deflection can be identified within historical sales data,[1] and we might test this hypothesis through an analytics proof of concept (often referred to as a *data science engagement*). Doing so requires access to different datasets (e.g. transactional sales data, customer complaint data, demographics data, customer attribute data) and a series of analytical and modeling steps as presented in Part II of this book. However, it's important to note that such projects are not aimless data exploration engagements to see "what can be done with the data," but rather, business problem investigations to determine whether the data and algorithmic methods can adequately address the problem.

This type of project can be undertaken within short timeframes and limited budgets (as there's no software deployed) to conduct modeling or simulation on historical data. Such analytics proof-of-concept projects are usually conducted in parallel with software specification engagements in order to create a high-level blueprint of how the output of the algorithmic models can be operationalized and integrated into the target business process or workflow. These two engagements are synergistic and often lead to a better understanding of:

- The required datasets, as well as how the Decision Optimization System will integrate into the data and IT landscape

---

1 By "early," we mean before the customer reaches the point of actually leaving and can still be influenced to stay.

- The business and technical requirements to operationalize the AI-based recommendations (or whatever the desired output from the Decision Optimization System might be) within a specific business process or workflow—in other words, defining how the predictions or optimized recommendations will be "actioned" by end users in their everyday roles
- The business case, with well-defined return-on-investment and payback periods

Although this type of proof of concept is nimble from a cost and duration perspective, it comes with its own limitations, which might not make it suitable for more complex problems that require some sort of software to "prove" the concept. As a case in point, the customer churn example above relies on a single prediction model for predicting churn, but other business problems may require several algorithmic models to be combined and trained on many different datasets. Promotional planning is one such example, requiring multiple prediction and optimization models working in parallel to prove the "concept"—something that would be difficult to capture within a standalone analytics proof of concept.

Software proof of concept

Although many complex business problems can be addressed through analytical proof of concepts to quantify a business case and prove the adequacy of the technology and data, other problems might require a more sophisticated approach, such as when the business benefit needs to be tested and proven during a live test (meaning that the trained AI models are required to run on new, live data and generate real outputs, such as a predicted value or optimized recommendations that can be assessed). This necessitates the configuration and deployment of a Decision Optimization System, which can be used for some length of time to test and prove the concept.

This approach is more involved than an analytics proof of concept, because it requires the specification, configuration, and deployment of a Decision Optimization System that can address some aspect of the problem. To keep costs down, we can:

- Limit the software functionality to a minimum set of features necessary for conducting the live test
- Use manual data loading mechanisms instead of live integrations with the IT environment
- Release the software to a select group of users (e.g. one department, one state, or one team) to test the functionality and measure the results against some pre-defined benchmark

Although the above will reduce the cost of a software proof of concept, we must be mindful that a "small-scale" software deployment doesn't necessarily mean "low cost" or "low effort." The problem is that certain proof-of-concept projects may require the training and implementation of many models, the definition of all relevant business rules and constraints, and so on, turning the proof of concept into something almost as large as the full project (with the only difference being integration and additional functionality). Hence, a software proof of concept necessitates the curtailing of scope to the smallest possible subset of the problem, which when tested, provides a result that's representative of the overall problem.

Selecting the right approach

Although we've discussed two separate and distinct approaches for proof of concepts and established that certain business problems may warrant a software project from the start (which in itself may come in different "sizes"), in reality a continuum exists, with different levels of effort, investment, and benefits:

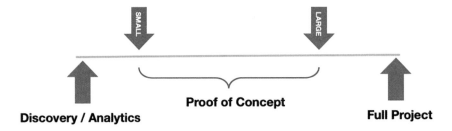

On one end of the spectrum, we can "dip our toes" into Artificial Intelligence through low-cost, consulting-based *discovery* or *analytics* engagements, which can assist us in evaluating and selecting a high-value problem, as well as determining the suitability of the technology and data. On the other end of the spectrum, we can deploy a full Decision Optimization System into the IT environment and all relevant business processes, along with trained prediction models, tuned optimization models, and automated feedback and learning mechanisms.

Within these two extremes, there's a proof-of-concept spectrum in the middle that represents a broad range of possible approaches within a more limited scope than a full project. However, if the proof of concept requires testing on live data, then it may represent 50% or more of the cost of a fully deployed and integrated Decision Optimization System—a fact that often comes as a surprise to many organizations, as the term "proof of concept" is usually associated with "low cost"! The easiest way to understand this misconception is by imaging an "ERP proof of concept," where we would like to "test" the system before buying it. Such a proposition would be ridiculous, of course, as the entire ERP system would need to be implemented for the "proof of concept" to be satisfied. To some extent, the same holds true for many proof of concepts for Decision Optimization Systems, where we need to prepare and analyze the data, train and tune all the models, define the objectives, constraints, and business rules—all of which might require the same level of effort as a full project.

Having said that, selecting the right approach usually comes down to answering a few key questions:

- Is the high-value problem well defined?
- Does the return on investment calculation or validation require a live system to be used on live data (to demonstrate a real KPI improvement)? Or can the return on investment be calculated through modeling/simulation on historical data?
- If a live software deployment is necessary, can the scope be limited to a small subset of the problem, which when tested, will provide a result that's representative of the overall problem?
- If the scope can't be reduced, then can the proof of concept be limited to a small geography or sub-set of users (as well as manual loading of data)?
- Are there any time pressures or constraints for implementing the software system? In other words, does the problem or opportunity require urgent action, thereby limiting the available time to conduct any detailed proof of concept?

Armed with these answers, it's usually straightforward to identify the most suitable approach; for example, if the problem isn't well defined, then the organization needs to undertake a high-level discovery or analytics engagement to better shape the eventual software project. On the other hand, if an organization already has a documented high-value problem and internally validated business case, then it should select a technology partner and begin the full project (especially if there are time pressures to act). Hence, when it comes to starting "large" or "small," some organizations are earlier in the discovery and evaluation process, while others have already made the decision and are simply searching for the right technology partner (a topic we'll return to in Section 11.5). Whichever approach is used, it's important to remember the fundamentals: Start with a high-value problem (but not moonshot), measure the outcomes, prove value, and develop trust with end users.

11.3 Executive Sponsorship

In Section 11.1 we discussed the importance of selecting a high-value problem (but avoiding moonshots), because such problems are likely to align to an organization's drivers of opportunity and risk, as well as feature in the executive agenda or overall strategy. Selecting a high-value problem for the application of Artificial Intelligence is also more likely to generate executive sponsorship, with an effective sponsor ensuring that:

- The high-value problem/opportunity is aligned to the longer-term vision of the organization
- Change management issues are identified and addressed throughout the project and upon deployment
- The approach for addressing the problem aligns with internal processes and standards at the corporate level
- The right internal political support is harnessed, especially if the high-value problem or opportunity spans multiple operating silos and/or geographies
- The metrics used for return on investment calculations align with the executive agenda (as funding is typically dependent upon a business case based upon various return on investment calculations)
- The business case considers less tangible, yet still important business benefits, especially when the return on investment calculations fall short or when payback periods are too long
- Any conflicts or risks are promptly addressed during the project, which might relate to change management, conflicting KPIs or internal agendas, or disagreement over the scope or project direction

For these reasons and others, it's not only important to secure executive sponsorship, but to secure the *right* executive sponsor; ideally, someone who has:

- *Problem-specific knowledge*: which provides them with a clear understanding of the problem, as well as any issues or circumstances that may impact upon the project and what's needed to resolve them.
- *Track-record with similar initiatives* (e.g. digital transformation, supply chain optimization, sales enablement, and so on): especially on projects where they played the role of advocate and influencer among senior executives, overcame resistance where required, and maintained ongoing support.
- *Strategic communication skills*: to drive clarity on the importance of the initiative and harness organizational support. This includes winning the hearts and the minds of end users of the Decision Optimization

System, as well as those affected by the project or impacted by any changes in business processes.

- *Project vision*: to deliver ongoing support of the initiative throughout the project life cycle and communicate the big picture to the organization and project team.

Depending on the size of the organization, the executive sponsor may also play a role in:

- *Steering Committees*: with other stakeholders involved in the project, whether directly or indirectly.
- *Alignment to broader strategic initiatives*: if the project is part of a larger, corporate-level initiative, such as "Project 2030" or "Customer Experience Transformation." In such circumstances, the Artificial Intelligence component might be the core enabling technology of the overall initiative or just an adjunct technology to another system that's being deployed.
- *Departmental budget and capacity planning:* to manage constrained resources—especially during peak periods—to ensure that neither the project nor business-as-usual activities are negatively impacted.

And lastly, as with any technology project, the inevitable changes and surprises must be dealt with from both a cost and time perspective. Such events—whether major or minor—require direction from the executive sponsor to ensure the project focus remains on track. In our experience, the right executive sponsor (who is strong, decisive, and has a clear vision) can deliver transformational projects and outcomes, versus the executive sponsor that is weak, hesitant, and unable to articulate what the organization is trying to achieve. In other words, the right executive sponsor can make all the difference.

11.4 Return on Investment and Payback

Once we've selected the high-value problem and secured an executive sponsor, the next task is usually to estimate the return on investment. To assist with such calculations, this section proposes a straightforward framework and related set of considerations for this stage of the process. It's important to acknowledge, however, that different organizations have their own internal business case methodologies, each tied to the level of investment and/or how advanced the project is within the consideration criteria. When it comes to calculating return on investment early on, a rough figure based on many assumptions might be acceptable, but later on, this figure will usually require revision and refinement.

Just like most software projects, AI-based Decision Optimization Systems offer several categories of benefits, namely:

- *Intangible benefits*: are often difficult to quantify and may not be included in the business case. Although these benefits might be absent from return on investment calculations, they're still important and typically included elsewhere in the supporting documentation.
- *Tangible/soft benefits*: are more likely to be calculated in dollar terms based upon a broad set of assumptions, some of which may include dependencies outside the direct control of the project team or software.
- *Tangible/hard benefits*: are the easiest to assess and attribute to a Decision Optimization System, and often form the backbone of the business case.

For a better understanding of the different types of improvements (lift or reduction) and how they're classified, please see the examples below, although this isn't an exhaustive list:

Intangible benefits	Tangible/soft benefits	Tangible/hard benefits
Improved employee job satisfaction	Improved speed to competency and new staff onboarding	Improvement in sales metrics such as "average revenue per order," "share of wallet," or "rate of conversion"
Increased employee retention	Improved salesforce capability	Profitability improvement in metrics such as "gross profit," "average margin item per order," and "basket profitability"
Improved retention of corporate knowledge	Improved customer service and responsiveness	Reduced customer churn, measured as "customer retention" or "churn rates"
Streamlined and automated workflows	Improved customer experience	Reduction in "cost to serve" metrics for the channel
Reduced number of manual errors	More effective opportunity profiling	Reduction in "average cost per lead" and other marketing metrics

Intangible benefits	Tangible/soft benefits	Tangible/hard benefits
Standardization of processes and centralization of customer records	Improved salesforce utilization and increased yield on sales resources	Improved customer penetration metrics such as "wallet share" and "customer lifetime value"
Improved operational decision-making and execution	Optimized supply chain planning	Increased stock turns and reduced inventory costs, measured as "days on hand"
Improved reporting and user experience	Increased forecasting accuracy	Improved supply chain metrics such as "DIFOT" or "OEE"
Improved supply chain visibility	Time saved on labor-intensive processes	Improved trade-spend and promotional metrics, measured as "trade spend effectiveness"

Depending on the maturity of an organization's project management office (PMO), there might be a framework in place for measuring and assessing the impact of various projects, as well as tools and methodologies for benefits realization. Based on this and other factors, the variables listed above will differ from organization to organization, as will the focus of any Decision Optimization System and its return on investment calculations. For instance, if we think of deploying an AI-based recommendation engine for cross-selling and up-selling within digital, phone-based, and in-field sales channels, then "improved average revenue per order" is an example of a *hard benefit*. On the other hand, "improved customer experience" may be classified as a *soft benefit* within the same project, given its secondary priority and dependencies outside the scope of the Decision Optimization System (as the customer experience is also dependent on the human telephone operator or in-field sales rep).

It's worth remembering that not every quantifiable metric will be of direct relevance to the business case. A common misconception is to think of "time-savings" as critical, but if we're are unable to redeploy the additional "time saved" into other workflows or tasks, then we should consider a reduction in headcount instead. In such cases, executive sponsors can provide guidance on metrics that align with the executive agenda.

Although there's no standardized approach for calculating return on investment within a business case, there are some generic steps we can follow:

1. Estimating tangible/hard benefits
2. Estimating tangible/soft benefits
3. Estimating intangible benefits
4. Calculating full deployment and usage costs
5. Deciding on the timeframe for calculations
6. Bringing it all together

Because the classification of any given metric (e.g. tangible/hard vs. tangible/soft) for return on investment calculations is problem and system-specific, we'll use a fictitious example to demonstrate these calculations. More specifically, we'll discuss the hypothetical deployment of an AI-based recommendation engine within an eCommerce portal for an organization that also has other sales channels, such as face to face and phone-based sales.

For this example, the following variables would be the most appropriate inputs for the return on investment calculation:

Intangible benefits	Tangible/soft benefits	Tangible/hard benefits
Improved customer experience and convenience	Customer migration to lower-cost channels, leading to improved "cost to serve" metrics	Profitability improvement measured as "gross profit uplift" for the digital sales channel

Step 1: Estimating tangible/hard benefits

The first step is to quantify the annualized benefit, which we've defined above as:

• Gross profit uplift for the digital sales channel

The exact calculation method and input variables will vary from one organization to the next and needs to reflect the specific workflows and key dependencies of the project, but for illustration purposes, we can use the table below to determine the potential gross profit uplift for the digital channel. For each line of the table below we can use historical data or averages (e.g. number of orders per year per channel, average order value per channel, and so on), supplemented with assumptions for key variables where historical data isn't available (such as the *cross-sell acceptance rate*, which isn't known until the AI-based recommendation is deployed and which might be a focus of an analytics proof of concept):

	Sales channel
	Digital Sales
Number of orders per year	A
Average order value	B ($)
Average number of line items per order	C
Average revenue per line item per order	D = B / C ($)
Cross-sell acceptance rate (assumption based on past experience or proof of concept result)	E (%)
Number of orders where cross-selling was accepted (one additional line item sold)	F = A * E
Revenue uplift (single year)	G = D * F ($)
Average margin for the channel	H (%)
Gross profit uplift (single year)	Y = G * H ($)

The result of this calculation provides us with the tangible/hard benefits value:

Tangible/hard benefits (per year) = $\boxed{\text{Y (\$)}}$

For example, if the number of digital orders last year was 209,003, the average order value was $497, and the average number of line items per order was 12.06, then the table would look as follows:

	Sales channel
	Digital Sales
Number of orders per year	209,003
Average order value	$497
Average number of line items per order	12.06
Average revenue per line item per order	$41.21
Cross-sell acceptance rate (assumption based on past experience or proof-of-concept result)	7.50%

	Sales channel
	Digital Sales
Number of orders where cross-selling was accepted (one additional line item sold)	15,675
Revenue uplift for a single year	$645,976
Average margin for the channel	50%
Gross profit uplift for a single year	$322,988

This calculation provides us with a tangible/hard benefit of $322,988 per year.

Step 2: Estimating tangible/soft benefits

As mentioned earlier, this category of tangible benefits can be calculated in dollar terms, although a broader set of assumptions might be required, as these benefits may have dependencies outside the direct control of the deployed system. The various types of tangible/soft benefits will vary between organizations and opportunity areas, but in the example above, we can quantify the annualized tangible/soft benefit as:

- Estimated annualized savings from customer migration to lower cost channels, as measured through "cost to serve" metrics

$$X\ (\$)$$

More specifically, we can factor in the savings associated with a subset of customers changing their default ordering method from placing orders with sales reps to placing them through the digital channel (due to the improved experience and personalization of offers within the eCommerce portal). To quantify this benefit in dollar terms, we need to make an assumption about the percentage of customers that would shift between channels, so that the final number of "migrated customers" can be multiplied by each channel's "cost to serve."

For example, if the phone-based sales channel handles 157,972 orders per year at a cost of $7.46 per order, we might assume that 10% of those orders (15,797) would shift to the digital sales channel which has a cost of $1.50 per order. In this example, these 15,797 interactions would have a "cost to serve" of $117,837 through the phone-base channel and a "cost to serve" of $36,333 through the digital channel. As an outcome, a business case based on the assumption of 10% customers migrating from the phone-based channel to the digital channel would create a saving of $81,504, which completes the calculation for this step.

Step 3: Estimating intangible benefits

Quantifying intangible benefits in dollar terms isn't straightforward, so we need to decide whether these types of benefits should be used in the business case or listed in a different section of the supporting documentation. In this example, let's say we'd like to include the positive impact of customer experience as an outcome of the project, but we're unable to calculate a dollar value. In this situation, we could create a statement under *other benefits*, articulating the nature of the business benefit:

- *Improved customer experience*: The deployed AI-based recommendation engine will provide greater convenience to our customers through personalized offers, expediting their shopping experience with less chance of errors or forgotten items.

Step 4: Calculating full deployment and usage costs

Any Decision Optimization System under consideration will likely have two cost components:

- An upfront capital expenditure to configure and integrate the software, train and tune the algorithmic models, and conduct end user training
- A monthly or yearly operational cost that covers hosting, maintenance, and support for both the software and algorithmic models, as well as licensing costs

Hence, our calculation for this step is straightforward:

Deployment costs ——————————————— | S1 ($) |

Ongoing costs (per year) ——————————— | S2 ($) |

In this example, we might assume that deployment costs for an AI-based recommendation engine are $200,000 and the ongoing costs are $120,000 per year.

Step 5: Deciding on the timeframe for calculations

Realizing the full benefit of any software project usually necessitates the full adoption of the software by end users, which takes time and effort. Different software systems require different timeframes for deployment, and the corresponding change management efforts may vary greatly depending on the complexity of the system and business process, as well as the culture of the organization towards change and technology adoption. Hence, we should give consideration to the timeframe for our return on investment calculation, as we can't expect to reap full benefit from the first day of deployment.

However, an AI-based recommendation engine embedded within the eCommerce Portal is a straightforward example, allowing for a simple calculation:

- Scoping and implementation period: 4 months
- Change management effort or duration: Nil

The implementation of AI-based recommendations into the digital sales channel doesn't require any training for end users or change management, which means this channel will produce a faster return on investment than other sales channels. If the same AI-recommendation engine was rolled out across the face-to-face and phone-based sales teams, then end users would need training on how to use the software and how to reference these recommendations within sales conversations, which would extend the timeframe of when benefits are realized. Within the digital sales channel, however, the AI-based recommendations are placed within the ordering workflow from the moment the software goes live, thereby providing customers with the opportunity to accept or decline the AI-based offers from day one, meaning that full adoption is immediate.

Such considerations are important for ascertaining the extent to which our payback period aligns with the business case, especially that many organizations have investment frameworks in place that specify a required payback period for different types and magnitudes of investment. Although these payback periods vary, they generally fall into one of the following categories: *same financial year, a few years,* or *several years.* With that in mind, we need to decide on the timeframe for our calculations by defining:

Length of time to scope and deploy the software ————| Z1 (months) |

Length of time to *near full* adoption   ————| Z2 (months) |

And then adding the above results:

Timeframe to commence benefit realization (in years) = | Z1 | + | Z2 |

Back to our example, the calculations would be as follows:

- Length of time to scope and deploy the software = 4 months
- Length of time to near full adoption = 0 months

- Timeframe to commence benefit realization (in years) = 0 + 4 = 4 months or 0.33 year

We can now move to the next step and determine the appropriate timeframe for our return on investment calculations, which needs to be longer than the timeframe of when we begin realizing benefits:

Timeframe to for calculations (in years) = $\boxed{T > (Z1+Z2) \text{ or } T > 0.33 \text{ year}}$

In this example, we might determine that two years is an appropriate timeframe for our calculations, and then use that period for the next step.

Step 6: Bringing it all together

We can assess the merit of various software projects in a number of ways, ranging from return on investment and payback calculations, all the way through to commercial accounting approaches that consider the time value of money (such as net present value calculations). For the sake of simplicity, we'll only explore return on investment and payback period calculations in this section.

A basic return on investment calculation takes the total benefit, total cost, and timeframe, and applies a formula such as:

Return on investment = (benefits − costs) / costs

So, in continuing with our template from above:

$$\textit{Benefits} = \left(\boxed{\begin{array}{c}\text{Tangible}\\\text{/ hard}\\\text{benefits}\end{array}} + \boxed{\begin{array}{c}\text{Tangible}\\\text{/ soft}\\\text{benefits}\end{array}} \right) * \left(\boxed{\begin{array}{c}\text{Timeframe}\\\text{for}\\\text{calculations}\end{array}} - \boxed{\begin{array}{c}\text{Timeframe}\\\text{to benefit}\\\text{realization}\end{array}} \right)$$

$$\textit{Costs} = \boxed{\begin{array}{c}\text{Deployment}\\\text{costs}\end{array}} + \left(\boxed{\begin{array}{c}\text{Ongoing costs}\\\text{(per year)}\end{array}} * \boxed{\begin{array}{c}\text{Timeframe for}\\\text{calculations (in years)}\end{array}} \right)$$

$$\textit{ROI (\%)} = \left(\boxed{\text{Benefits}} - \boxed{\text{Costs}} \right) / \boxed{\text{Costs}}$$

The return on investment result is directly related to the timeframe used for these calculations (step 5), and may need to be revisited within the constraints of an organization's required payback period. At this point, we can also calculate the payback period with the following steps:

- *Cumulative benefit (CB)*: is calculated by adding together the benefit of each year after the commencement of benefits realization (i.e. after the software has been implemented and fully adopted by end users) for the total number of years in the timeframe used.
- *Cumulative cost (CC)*: is calculated by taking the initial project expenditure and adding the ongoing yearly cost after deployment. We can then compare the cumulative benefit against the cumulative cost, and if CB > CC, then the project has achieved payback within our timeframe for calculations.
- *Longer payback period*: if CB < CC within our timeframe for calculations, then we keep adding the yearly benefits to CB and yearly costs to CC, one year at time, until CB = CC and we've found the payback period.

Continuing with our example and sample calculations, we would have the following cumulative benefits for the first two years:

- *CB (first year)* = ($322,988 + $81,504) * (1 year − 0.33 year) = $404,492 * 0.67 = $271,009
- *CB (second year)* = $404,492
- *CB (total)* = $271,009 + $404,492 = $675,501

and the following cumulative costs for the same time period:[2]

- *CC (first year)* = $200,000 + ($80,000 * 1) = $280,400
- *CC (second year)* = $120,000
- *CC (total)* = $400,400

giving us a return on return on investment of:

- *ROI* = ($675,501− $400,400) / $400,400 = $275,101 / $400,400 = 0.687
- *ROI (%)* = 68.7%

The project would also have a payback period of just over one year from commencement, with CB of $271,009 and CC of $280,400 for the first year. This calculation should be sufficient for most business cases, but could be enhanced

---

2   Given that we're using a one-year timeframe, the ongoing costs have been pro-rated for the first year to account for the four-month scoping and implement of the software (bringing the total to $80,000 for the remaining eight months of the first year, post implementation).

by considering the cumulative benefits (and costs) obtained over a longer period of time, such as the overall returns over five years.

11.5 Technology Partner Alignment

A final and important consideration for any Artificial Intelligence project is the chosen technology partner and its alignment to an organization's vision and desired outcomes. From our many decades of experience in researching various algorithmic methods and then commercializing the results within enterprise software systems, we've found that the following three core competencies are essential for the successful execution of Artificial Intelligence projects:

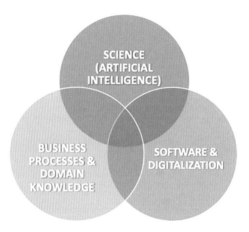

This is because scientific approaches such as Machine Learning have the *potential* to create value through better decisions, but the actual realization of this value depends on these AI-driven recommendations being embedded and actioned within operational workflows. The only way this can occur is through the deployment of enterprise software into an organization and its relevant business processes. Most technology partners excel in one or two of these areas (i.e. science, software, *or* business process understanding), but few possess all three core competencies. Given the crucial role that a technology partner plays in any Artificial Intelligence project, let's examine each of these areas in turn.

Science

"Science" refers to a technology partner's scientific capability in the area of Artificial Intelligence, which is the core ingredient for any project based on AI methods. Hence, the chosen technology partner should command a deep understanding of the various AI and non-AI methods presented in Part II of this book, and hold significant experience in applying these algorithmic methods to complex business problems in the chosen problem domain.

Technology partners that excel in this area generally employ AI scientists with Ph.D. degrees, publish scientific papers in peer-reviewed journals, and maintain research partnerships with universities and research organizations.

Most technology partners that fit this criterion develop their own proprietary algorithms and models—especially algorithms that can monitor their own performance, learn and adapt from new data and feedback, and identify any unintended consequences of algorithmic bias or effects of time-changing environments. Any technology partner that falls short in this area is likely to compromise the overall result and success of the project.

Software

A common misconception among many organizations is that "Artificial Intelligence" and "software" are somehow synonymous, when in fact they are not: Artificial Intelligence is a scientific discipline emerging from computer science departments within universities, while software is an engineering discipline concerned with the production, operation, and maintenance of software systems. However, given that prediction and optimization algorithms are embedded within a software environment (i.e. inside the Decision Optimization System), it's important for the software to fit an organization's business processes and workflows. This is often overlooked and becomes problematic in later stages of a project, when the selected software can't handle organization-specific requirements, such as non-standard business processes and workflows, complex business rules set in time-changing data environments, and variable and calendarized constraints, among many other examples.

Hence, although the software needs to be scalable and robust enough for enterprise-level use (where there might be thousands of end users, 24/7 operations, and sensitive data requiring advanced security protocols), it also needs to be flexible enough to fit the unique business requirements, processes, and workflows of the organization. For these reasons and others, the selected technology partner deploying AI-based algorithms and models within a software environment must possess a deep mastery of both.

Business process understanding & domain knowledge

Many organizations use a patchwork of spreadsheets, unintegrated tools, and manual processes to execute mission-critical activities, which creates inefficiencies and hampers agility. Because any deployed Decision Optimization System will be accessed within a specific business process, it's necessary for the technology partner to possess a business analysis team that can spend time inside an organization to understand its operation and document opportunities for business process improvement. For these business processes and workflows to be properly understood, some pre-existing, problem-specific domain knowledge is usually desirable—for example, depending on the selected problem, an in-depth understanding of pricing, customer segmentation, production scheduling, promotional planning, or demand forecasting might be needed, along with experience and expertise within a particular industry, such as food & liquor, building materials, pharmaceuticals, wholesale & distribution, financial services, consumer goods, or retailing.

Project Alignment

We mentioned earlier that most technology partners excel in one or two of these areas (i.e. science, software, *or* business process understanding). Which means that any combination of core competencies makes them better suited for specific project types. Consider a technology partner that excels in the science and business process competencies, which makes them well suited to "consulting type" analytics engagements. If we want "static" insights and recommendations in an offline, report-based approach (no software deployed), then such types of technology partners would be a good match for one-off, ad-hoc engagements. But if we'd like to access these AI-driven recommendations within an operational workflow, then a technology partner with competencies that span across the science, software *and* business process would be preferred.

Technology partners that excel in all three core competencies are unique in the sense that the enterprise software they deploy leverages a unique blend of science, software and business process understanding (a Decision Optimization System).

Additional considerations

In addition to the core competencies mentioned above, the following should also be considered when selecting a technology partner:

- *Proven track record*: working on projects that involve large data sets, challenging optimization and prediction requirements, and complex system integrations and operational workflows.
- *Measurable results*: on past projects in metrics such as revenue, volume, margin, operational costs, forecast accuracy, asset utilization, or working capital.
- *Risk mitigation*: through fixed price proposals and project management methodologies that allow for on-time and on-budget delivery.

We have now covered the fundamentals when it comes to selecting a business problem and preparing a business case for an Artificial Intelligence project, along with considerations on the right-size approach, executive sponsorship, and choosing a technology partner for all stages of the journey. And with that we move to the final set of practical considerations which warrant attention before an organization deploys a Decision Optimization System.

CHAPTER 12

Getting the Foundations Right

"Building advanced AI is like launching a rocket.
The first challenge is to maximize acceleration, but once it starts
picking up speed, you also need to focus on steering."
Jaan Tallinn, *Skype co-founder*

The previous chapter provided practical advice for organizations planning to embark upon an AI project, covering a range of topics such as selecting the right problem and technology partner, conducting proof of concepts, developing a business case, and securing an executive sponsor. We'll now build upon this advice by exploring a few additional areas which are related to the implementation of AI-based systems, such as:

- Data quality
- Digitalization
- Change management
- Requirements validation

No matter how large or small the problem—or project size for that matter—these topics have the potential to accelerate or delay a project's success and return on investment, and should be given serious consideration by the project team. We'll begin by revisiting the subject of data quality, but this time from the perspective of its potential impact on the early stages of a project.

12.1 Data Quality

In Chapters 2 and 4, we discussed the importance of collecting and using good data for solving complex business problems, and argued that what constitutes "good" is contextual to the problem we're trying to solve. That is, "good data" is a subjective concept, relative to the extent that any given dataset can be used for assessing and analyzing a specific problem. With this in mind, a common concern of many organizations is the poor quality of their data—a concern that often morphs into an outright objection for embarking upon any kind of AI project. The project team should realize, however—either through business literature or through the experience of their technology partner—that no organization has *perfect* data, and that many approaches exist for dealing with poor quality data. Hence, an organization shouldn't be hasty to conclude that a project can't proceed because of data issues (as we have already highlighted throughout Chapter 4).

In this section, we'll provide a high-level framework for assessing organizational data in the context of the selected problem. More specifically, to determine *data quality* in the context of the proposed AI project (as opposed to data quality in general). This framework is based on three distinctive steps: First, a range of datasets are catalogued and investigated in order to ascertain their *usefulness* for addressing the selected problem (which includes an assessment of their content, format, structure, and sources). Once the potentially useful datasets have been identified, the second step is to establish their *sufficiency*—in other words, their:

- *Quantity*: How much historical data is available? Is that amount of data sufficient for the algorithmic approach or method?
- *Granularity*: Is the data granular enough to provide us with the required insights and knowledge? For example, is the data available at a postcode level in daily time buckets, or only at a state level in monthly buckets?
- *Timeliness*: Does the update frequency of the input data match the required update frequency of the Decision Optimization System? For example, if a dynamic customer micro-segmentation requires weekly updates to produce new outputs, it won't work as planned if new data is only available on a monthly basis.
- *Accessibility*: Can the required data be accessed from existing IT systems? Are there any constraints that preclude using the data in certain ways? As an example, third-party data might be governed by the provider's terms and conditions, which may require analysis to be performed at an aggregated, anonymized level.

The third and final step is to determine the *cleanliness* of the data, as discussed in Chapter 4:

- *Validity*: Has the data been appropriately captured? Have the fields been consistently filled according to the predefined data structure? For example, has "stock on hand" been consistently captured as units per store, per week? Or do some records contain dollar values?
- *Completeness*: Are there any "gaps" in the data? If so, what is the size and nature of these gaps?
- *Consistency*: Is the data representative of what the organization wanted to capture? Does it contradict other data captured against the same

instance? For example, is promotional sales data reported uniformly across the organization, or differently in one system versus another?

When it comes to assessing data quality against this framework, the magnitude of the assessment can vary significantly both in complexity and effort, which is why many organizations turn to their technology partner to undertake this work (as discussed in Chapter 11.2). The outcome of these engagements is a report stating whether the data is fit for the intended project in terms of usefulness, sufficiency, and cleanliness. Or, alternatively, whether the data requires augmentation (as discussed in Chapters 4.2 and 4.3) or if the overall business problem should be reframed into something smaller that can be addressed with the available data.

Besides answering such scope-related questions, a data assessment exercise can also provide another important benefit. If we consider a situation where some datasets are inaccessible, incomplete, or lacking in granularity, then an opportunity exists to run two initiatives in parallel. That is, above and beyond the AI-project, a parallel project can be undertaken to collect the missing data. This approach enables an organization to quickly scale up the Decision Optimization System to address the larger, overall problem once the data gaps have been addressed (assuming the organization has started with a subset of the original problem due to data issues).

To better appreciate what these parallel initiatives might look like, let's return to the complex business problem of promotional planning and pricing from Chapter 3. The key to building an accurate prediction model for future promotions is understanding what drove the success of past promotions, which we can do by collating and analyzing various datasets to understand the effectiveness of past promotions (i.e. to understand not only what happened, but why it happened). After assessing the available data against the above framework, we might find that data related to the placement of products in promotional displays (usually referred to as "off-location" data) is incomplete and unreliable for several retail chains, presenting a problem given the significant contribution these displays make towards the overall success of a promotion. In this situation, we could:

- Abandon the project (which might be a rather harsh decision given that only a single dataset is missing).
- Fill the data gap by building a simulation model to approximate the impact of promotional displays (based on some partial information that could be extended with business rules—as we discussed in Chapter 4.3).

- Reframe the project by addressing a smaller subset of the original problem—for example, by focusing on retail chains for which there is sufficient data, including off-location data.
- Proceed with the AI project without using off-location data, and then collect this data over time.

Quite likely, this last option should be suitable, because even though the prediction model's accuracy would be negatively impacted by the missing off-location data, the prediction error is likely to fall within an acceptable range. Also, by implementing an appropriate data collection mechanism for future off-location promotions, the prediction accuracy of the Decision Optimization System will improve over time as this data becomes available.

As this example illustrates, most organizations are likely to find gaps when assessing their data against the above framework. This is normal—and expected—with the critical question being whether the collective quality of the data is sufficient for the AI project to proceed. If the answer is negative or uncertain, then an alternative is to revisit the high-value problem and consider it from a different angle. This may include tackling a subset of the original problem for which there is a minimum level of adequate data.

Although data gaps are common and rarely mean that an AI project can't proceed, many organizations have an "all or nothing" attitude towards data completeness, often leading to a situation where projects that could have delivered substantial business value (even with only a subset of the overall data) are postponed until the data gaps are filled. Ironically, filling those gaps becomes less of a priority once there's no live project waiting on the data, and so oftentimes, the gaps are never filled and the project never gets off the ground.

12.2 Digitalization

As mentioned in Chapter 1, improving business processes and transforming ways of working through the conversion of text, pictures, and workflows into digital formats is referred to as *digitalization*. The ongoing quest of organizations to digitalize their operation has significantly benefited the field of Artificial Intelligence, as it's difficult to apply AI to manual pen-and-paper processes and/or digitally disconnected workflows that involve multiple standalone spreadsheets.

For this reason, before an organization can reap the full benefits of an Artificial Intelligence project, it must first digitalize the relevant business processes and workflows—in other words, digitally connect the interdependent steps of a workflow to make it "AI-ready." Continuing on with the promotional planning and pricing example, let's consider a manual and disconnected workflow that a merchandising team might have to undertake to understand the effectiveness of past promotions. This workflow might involve contacting multiple departments and collating multiple datasets through a disjointed series of steps, such as:

- Accessing spreadsheets to view past promotions, with each spreadsheet containing a list of promoted products for a particular promotional period, along with all promotional prices and promotion types (see basic terminology in Chapter 3 for more details).
- Selecting a particular past promotion for investigation.
- Accessing external datasets that contain historical sales data for this past promotion.
- Contacting the marketing department to understand whether a particular product was featured on the catalogue cover during this past promotion, and if not, whether it was featured on an inside page insert (and if so, whether it was the left-hand-side page or right—all of which might have contributed to the sales result). Such information is usually held by the marketing department in their own, separate spreadsheets.
- Calling the account manager of the retail chain to understand which promoted products had off-location displays for this particular promotion (as discussed in the previous section). The account manager might have this information in yet another standalone, offline spreadsheet, which isn't readily accessible by others in the organization.
- Contacting the team that supported this promotion through in-store visits. Most likely, this team captured important information on each store's compliance with the promotion, but this information might be in people's heads, notebooks, a CRM system, or more spreadsheets!
- Collating all this disparate data into yet another spreadsheet, for analysis by the revenue management or analytics team to understand what drove the success of this particular promotion.

The inefficiencies of this workflow are obvious: The steps are time consuming, disconnected (all tasks are standalone and lack version control), with the same data being replicated across multiple spreadsheets. This leads to all sorts of errors, as these tasks involve a mix of paper-based, non-digital processes, as well as spreadsheets—the latter, despite being digital, are still disconnected and messy. The analysis of past promotions also sits alongside many other interdependent processes for planning future promotions, such as "slotting" products into different promotional periods, setting the promotional price and depth of discount, forecasting volume uplifts, exploring alternative promotional frequencies, and so on.

The difficulties this organization would encounter by embarking upon an AI project to optimize promotional planning and pricing are significant. Unless the above-mentioned processes are digitalized, a Decision Optimization System would struggle to function properly in this environment (or be used effectively by the merchandising team). With countless manual processes and offline spreadsheets scattered across a digitally disconnected workflow, this organization would have to undertake a "digitalization" project first—that is, a project to replace its spreadsheets with a centralized slotting board that enables a digitally integrated workflow. Ideally, the AI project could be split into two phases: the first phase being digitalization, followed by inclusion of AI-based capabilities (i.e. prediction, optimization, and learning) in the second phase.

Assessing whether an AI-based project requires an upfront digitalization phase is an important decision, possibly leading to a re-evaluation of the business case and payback periods (as we discussed in Chapter 11)—particularly if the benefits of the Decision Optimization System are pushed back until after the digitalization phase is complete. On a positive note, however, digitalization can drive a number of business benefits, such as reduced labor requirements for executing tasks, reduced errors and re-work, improved customer service, and so on, all of which can improve the business case.

In recent times, organizations have become increasingly cognizant of how broken processes affect their operation, making digital transformation and business process improvement a regular feature on the executive agenda. These digitalization projects are usually undertaken on their own merit, with business cases built upon benefits such as:

- Streamlined end-to-end workflows
- Transparency and visibility across interdependent tasks and processes
- Timely access to data and information by all participants in the process
- Reduction in manual handling of data, leading to less errors and re-work
- Improved customer experience and service

Moreover, these initiatives also lay the foundation for future AI projects, allowing for shorter payback periods and easier change management (as discussed in the next section).

12.3 Change Management

Let's start our discussion of change management with a humorous story:

Through the pitch-black darkness, a captain sees a light ahead that's on a collision course with his ship. He sends a signal: "*Change your course ten degrees west.*" The light signals back: "*Change yours, twenty degrees east.*" Irritated, the captain replies: "*I'm a Navy captain! Change your course, sir!*" Then: "*I'm a seaman,"* comes the reply. *"Change your course, sir."* Now the captain is really enraged. *"I'm a battleship! I'm not changing course!"* There's one last reply. *"I'm a lighthouse. Your call!"*

This simple story reinforces a basic fact about humanity: to varying degrees, we are all reluctant to change! This is an important truth for any IT project; after all, an organization can create the world's best software for its sales reps—packed with guided-selling recommendations, actionable insights, price guides, and more—but won't realize any business value if the sales reps don't use the application. Clearly, the business case for software applications rests not just on the technology working as planned, *but also on end users using the software.* And so user adoption represents a major dependency on the success of any IT project and the realization of its business-case!

Because most people are reluctant to adopt something new and change their usual way of doing things, the reality that most organizations face is that user adoption requires significant effort and a multifaceted change management approach. To give a project the best chance of success, organizations should start planning for change management at very onset—even before the project kicks off—rather than waiting until user acceptance testing. In particular, a few key points that should be considered early are:

- Executive sponsorship and communication
- End-user engagement
- Ease of use and simplicity
- KPI alignment.

Let's discuss these in turn.

Executive sponsorship and communication

In Chapter 11.3 we talked at length about the importance of securing executive sponsorship, and in particular, finding the *right* sponsor. This ensures that the organization receives guidance on how to move from a conceptual "candidate project," through to the business case and eventual project. Furthermore, depending on the magnitude of the required change, certain change management tasks (such as interdepartmental process changes) might benefit from executive engagement to ensure timely execution.

Likewise, effective top-down communication is critical to proactively removing project barriers, driving clarity on the importance of the initiative, and obtaining broad organizational support. Although the focus of this communication will vary as a project moves from requirements gathering through to configuration and then deployment, a few key messages will remain consistent on the project's intent, expected benefits, and alignment to the organization's strategy. Thus, effective change management starts early with targeted messages that are reinforced throughout the project, and which play a crucial role in engaging end users long before the go-live date.

End-user engagement

AI projects carry another peculiarity that can hinder user adoption, above and beyond the usual challenges faced by IT projects. A core feature (and change management challenge) of Decision Optimization Systems lies in their ability to make predictions and recommendations. This feature may encounter resistance, as some users will "know better" and elect to follow their gut instead, thus nullifying the system's business value.

Hence, it's essential to engage with end users early and maintain that engagement throughout the project. It's also essential to get a true understanding of the opportunities and challenges that end users face in their daily roles, which serves two important purposes: The first is to gather requirements that can be included in the Decision Optimization System (more on this topic in the next section), and the second is for end users to gain an understanding of how the system will work. Equally important is the recruitment of end users for the role of "super users," who can champion the project at the grass roots level, provide on-the-job training for other users, and play a leading role in change management initiatives.[1] The earlier an organization can engage with end users and super users, the more empowered and motivated they'll be to spread the word and get their peers onboard.

Ease of use and simplicity

The one mistake we see organizations make time and time again is "over-engineering"—meaning that a Decision Optimization System is scoped, designed, and configured with the capabilities of managers and executives in mind, rather than the capabilities of end users. The result is a system that's difficult for end users to use, and a change management process that becomes frustrating and drawn out. This generally happens if end users aren't engaged during the requirements gathering stage, or if no thought is given by management on how the system will be used by people other than themselves ("but that's the way I'd use it," managers will often say, not quite appreciating that it won't be them using the system, but someone else).

AI-based projects should be always focused on end users—their workflow, capabilities, level of sophistication, and so on—because the business case upon

---

1   For these reasons, the super user role is best suited for tech-savvy individuals who are capable of mastering a new system relatively quickly, and who are well-respected by their peers.

which these projects are based depends on end users using the system. With this in mind, every effort should be made for a Decision Optimization System to be as simple and easy to use. Although the scientific aspects of prediction, optimization, and learning can be complicated (as we discussed in Part II), it's unnecessary to surface these complexities into the user interface. In our experience, change management is easiest with systems that keep this complexity hidden in the backend, presenting a clean and uncluttered user interface that is straightforward to use. Whatever investment an organization makes into simplifying the user interface of a Decision Optimization System (as well as the end user experience of using the software), will pay dividends in the form of reduced time to full adoption and reduced change management.

Hence, instead of thinking about how to cram every feature and complex function into a Decision Optimization System, the project team should be thinking about joy of use, adoption, and other such concepts, which will ultimately deliver the business benefit when end users fully engage with the system. As a case in point, gamification can be applied to the user interface to aid with user adoption and change management. In the same way that Amazon shows the rank for each book:

Product details

ISBN-13 : 978-1925000207

ISBN-10 : 1925000206

Dimensions : 13.5 x 1.4 x 20.8 cm

Publisher : Hybrid Publishing (31 October 2018)

Language: : English

Best Sellers Rank: 67,660 in Books (See Top 100 in Books)
143 in Job Hunting (Books)

3,375 in Spiritual Self-Help (Books)

Customer Reviews: ★★★★☆ ∨   99 ratings

the same concept can be applied to Decision Optimization Systems within a number of business functions, including sales:

where each sales rep has a rank that changes hourly or daily, alongside other useful information such as their progress against financial targets or other KPIs.

KPI alignment

Change management is more controlled and measured when a project is aligned to the right mix of organizational KPIs, which in turn represent a balance between *lag* and *lead* metrics (with the former metric telling us if we've achieved the goal and the latter one telling us if we're likely to achieve the goal). To better appreciate these metrics and their impact on change management, let's assume we want to reduce customer churn (as presented in Chapter 8.1). In this particular example, the effectiveness of the Decision Optimization System can be measured by the number of churned customers each quarter, which is a lag metric telling us what happened in the past. When it comes to business improvement initiatives, however, the main limitations is that if we introduce a change to the process, then we have to wait until the process cycle runs its full course before we can assess the effectiveness of that change—that is, we have to wait until the end of the next quarter to know the actual number of churned customers (and how many sales reps followed the recommendations of the Decision Optimization System). Therefore, we should balance lag metrics with lead metrics, which can serve as a leading indicator of not only business performance, but also change management progress.

In the context of customer churn, a Decision Optimization System might determine that the most reliable indicator of early churn is when customers reduce their purchases in a specific way (the reduction being specific to the type of customer and their microsegment). The corresponding lead metric might then be the number of times the Decision Optimization System identifies these customer-specific purchasing reductions, *along with the number of times a sales rep takes action*. Hence, the lead metric also becomes a gauge of how many sales reps are actioning the recommendations, which provides insight into change management. Although this lead metric won't always point to customers at risk of leaving, it will provide a guide for how the organization is progressing during the quarter.[2]

12.4 Requirements Validation

Although AI-based systems can be applied to complex business problems of all shapes and sizes—with some organizations advancing to a software project

---

2   This is usually the case with lead metrics, which can be somewhat subjective and provide more of an approximation. On the flipside, they can be measured at any stage of the process cycle and are helpful for monitoring performance and change management.

straight away, while others progress through a discovery, analytics, or proof-of-concept engagement first (as we discussed in Chapter 11.2)—there is one fundamental recommendation we can make irrespective of the approach taken: namely, that a core group of stakeholders be involved throughout the project to provide ongoing validation of the business and technical requirements.

There are many approaches to validating these requirements, from "requirements traceability documents" that describe each requirement and how it will work in the deployed system, through to mock-ups and prototypes that demonstrate each requirement "in-action." The approach for gathering and validating requirements is essential to get right early on, as it ensures that the scope of the project is aligned to the high-value problem and organizational KPIs, as well as the capabilities and requirements of end users.

During the initial requirements gathering stage—where the overall project scope and budget is defined—the steering committee of stakeholders can provide guidance on key questions, such as: Which requirements are important and which aren't? If the project is broken into phases, do earlier phases pay for later phases through the realization of business benefits? If so, is this reflected in the requirements of these earlier phases? And so on.

The steering committee will play a balancing role, perhaps between agreeing or rejecting requirements within the context of a new business process or workflow, without losing sight of the sophistication and capabilities of end users. Once the business and technical requirements have been gathered and validated, there are many benefits of the steering committee re-validating these requirements throughout the project, including:

- Maintaining a high level of involvement with the core group of stake-holders at all stages of the project, from detailed design through to configuration and user acceptance testing.
- Validating the extent to which the Decision Optimization System solves the high-value problem using predefined business rules, constraints, and objectives.
- Prioritizing requirements as the project progresses, as well as deciding which requirements need to be added and which can be dropped.

As a final point, we'd like to emphasize that many different types of requirements need validation throughout a project, and the right approach for each requirement type depends on the nature of the requirement (e.g. "functional requirements" such as business rules, versus "non-functional requirements," such as the performance and response time of a Decision Optimization System). It's worth highlighting, however, that the earlier these requirements can move from documents (such as traceability documents) to prototypes and early-stage system configurations, the better. In doing so, an organization can begin to "see" the system, enabling stakeholders to better understand the importance of certain requirements, as well as establish a shared sense of excitement and ownership of the AI project.

12.5 Closing Thoughts

Given the "rise" of Artificial Intelligence in recent times, many business managers have been asking: *What is Artificial Intelligence truly capable of? What is it best suited for? What could that mean for my organization?* In this book, we have attempted to answer these questions.

In Chapter 2, we illustrated the role of Artificial Intelligence in the journey from problem to decision, with AI-based methods bridging the gap between past and future to form the backbone of Decision Optimization Systems:

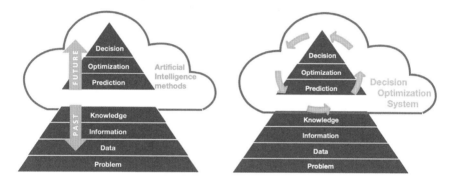

The power of Decision Optimization Systems resides in their ability to answer two fundamental questions that underpin most business decisions: *What is likely to happen in the future?* and *What is the best decision right now?* Without a doubt, organizations that can accurately answer these questions will enjoy a competitive advantage over organizations that cannot. By combining prediction *(What is likely to happen in the future?)* and optimization *(What is the best decision right now?)* into one system, business managers can reach new heights in their decision-making proficiency.

As we also emphasized in this book, another important aspect of Decision Optimization Systems is their ability to learn and adapt—otherwise the usefulness of such systems is limited. After all, most business problems are set in a dynamic environment where we must deal with unforeseen events and a multitude of external and internal forces. Thus, learning is not a "nice-to-have" feature, but a "must have" mechanism that relies on fresh data and/or direct feedback from end users:

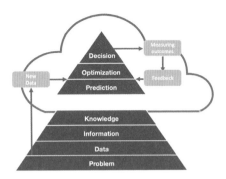

In Part II, we discussed the technical aspects of prediction, optimization, and learning, and how they can be integrated to create a Decision Optimization System; while in Part III we explored how such systems are used to generate revenue and margin growth within various industries, organizations, and business functions. And finally, in Part IV, we provided an overview on how to introduce Artificial Intelligence into an organization, including our thoughts on selecting the right problem, developing a business case, and implementing change management. However, given that most organizations are unique in some aspect of their structure, operation, or value proposition, it's likely that some of these topics would require deeper consideration and discussion. In the words of Sherlock Holmes:

> *"If you care to smoke a cigar in our rooms, Colonel, I shall be happy to give you any other details which might interest you."*

We couldn't have said it better ourselves.

Index